3D TCAD Simulation for Semiconductor Processes, Devices and Optoelectronics

Simon Li • Yue Fu

3D TCAD Simulation for Semiconductor Processes, Devices and Optoelectronics

Springer

Simon Li
Crosslight Software, Inc.
Burnaby, BC, Canada
simon@crosslight.com

Yue Fu
Crosslight Software, Inc.
Burnaby, BC, Canada
fred@crosslight.com

ISBN 978-1-4614-0480-4 e-ISBN 978-1-4614-0481-1
DOI 10.1007/978-1-4614-0481-1
Springer New York Dordrecht Heidelberg London

Library of Congress Control Number: 2011934485

Printed on acid-free paper

Springer is part of Springer Science+Business Media (www.springer.com)

Preface

3D Is for Real Life

We are living in a 3D world. Everything we see is in 3D. Seeing is believing.

3D Is Technology Trend

The movies we watch have just turned 3D. Now, even the TVs we watch, the electronic games we play are turning 3D. The same trend is happening to semiconductor technology modeling.

3D Is for Teaching and Learning

Traditionally semiconductor devices are taught with their cross sections and students may never get a chance to see the real 3D picture. It is time to explore a whole new world of semiconductor devices in their original 3D form.

A Picture Is Worth a Thousand Words

With explosive amount of information available in the information age, a new generation of students demands more efficient way of learning than reading long paragraphs of texts. Whenever possible, this book uses 3D illustrations to explain the ideas in semiconductor processing and device modeling.

What Is 3D TCAD?

The semiconductor transistor is the foundation of modern electronics technology and is undoubtedly one of the great innovations of the twentieth century. Technology Computer Aided Design (TCAD) refers to numerical modeling of semiconductor technology process and device characteristics using software tools. Some of the earlier iterations of these tools were developed at Stanford University in the 1980's and were in 1D or 2D due to limitations in computer speed and memory. This book extends these concepts to a new dimension.

3D TCAD is now made possible because of advances in computer hardware. 3D TCAD gives us an unprecedented experience in device design and new insights into their behavior. For the first time, we are able to view the device as they actually exist rather than using 2D cross sections.

Why 3D TCAD?

Semiconductor devices are all in 3D as we pointed out. Many devices have strong 3D effects which conventional 2D simulation cannot provide the accuracy required in TCAD projects.

How to Setup 3D TCAD?

This might be the very first book that deals exclusively with 3D TCAD simulation. The book will be focused on how to set up a 3D TCAD simulation, from mask layout all the way to electrical/optical device simulation. We offer plenty of 3D examples with step by step guidance, making the book a useful reference for those who wish to set up their first 3D simulation using any TCAD tools with 3D capability.

This book is arranged in ten Chapters. The first chapter provides an overview of semiconductor industry and TCAD usage. The second and third chapter are devoted to giving advanced user an overview of the physical models used in both process and device simulation. The fourth chapter prepares the reader with basic knowledge of 3D TCAD. It will bridge the gap between layout mask and simulation. Topics like how to set up 3D TCAD, the file structure and an overview of the simulation tools are included. Use of GPU acceleration in 3D TCAD is also explained.

Device examples start from Chap. 5 and are divided into several categories: P-N junction diodes, MOSFET and CMOS technology, power devices, interconnect, CMOS image sensor and Laser diodes. These examples cover most of the popular devices in use today and provide useful technology and physics insights.

Who Is This Book For?

Senior undergraduate students and graduate students, working professionals and engineers, professors who wish to find a teaching reference book and others who are interested in learning about semiconductor devices or simulations.

3D Is for Real Life

We are living in a 3D world. Everything we see is in 3D. Seeing is believing.

Acknowledgments

We are especially grateful to our colleague, Mr. Michel Lestrade, who spent countless hours reviewing and editing of this book.

Special thanks to Dr. George Xiao from Crosslight, who has made significant contributions to Chap. 9. We would also like to express our deep appreciation to Professor Maggie Xia from the University of British Columbia, Professor John Shen from University of Central Florida, Mr. Gang Xie from University of Toronto and Mr. Yuanwei Dong from the University of British Columbia for their careful reviews and suggestions.

Burnaby, BC, Canada

Simon Li
Yue Fu

Contents

Chapter 1
Semiconductor Industry and TCAD

1.1 The Semiconductor Industry

Technological advances in the past few decades have taken place at a breakneck pace. From the now-ubiquitous smartphone to new technologies like the Apple iPad™ and 3D TV, new ideas continue to change our daily lives and innovation shows no signs of slowing down. Unbeknownst to most of the public, these new products are only made possible by matching innovations in semiconductor technology.

The semiconductor industry is one of the key driving force in today's global economy. It is also one of the most dynamic economy sectors. Ever since its first appearance in the 1960s, it is now a $250 billion industry [1]. Figure 1.1 gives some key applications of semiconductor technology.

The semiconductor industry is structured much like a food chain pyramid (Fig. 1.2) where the top levels of the pyramid require the lower levels to function properly. In terms of industry size, the pyramid is somewhat inverted and most players in the industry occupy the topmost levels. The lowest level of the pyramid is the wafer suppliers and Electronic Design Automation (EDA) tools suppliers. TCAD or Technology Computer Aided Design is a special category within EDA. It is used widely by semiconductor device engineers to design process flows and various kinds of devices. The second level is device engineering, which consists of both process integration and device design. Engineers in this group will generally utilize TCAD tools to help them understand and design process recipes and structures for various devices. TCAD may even help them to predict their electrical, thermal and optical properties under different operating conditions without actual fabrication (i.e. virtual prototyping) provided that proper calibration has been done. The level above device engineering is called circuit level, which uses the process and device designed by the device/process engineers to build functional ICs. These functional ICs are the building blocks for system engineers, who eventually design

S. Li and Y. Fu, *3D TCAD Simulation for Semiconductor Processes, Devices and Optoelectronics*, DOI 10.1007/978-1-4614-0481-1_1,

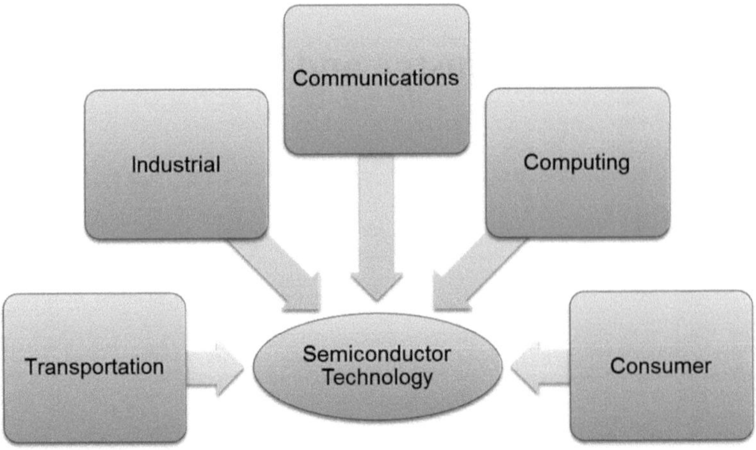

Fig. 1.1 Semiconductor technology and some of its applications

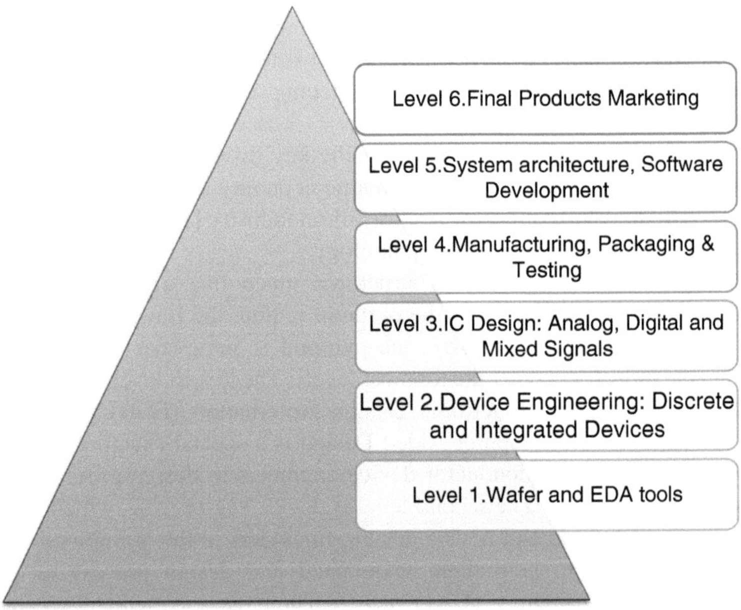

Fig. 1.2 Pyramid level illustration of semiconductor industry

hardware of cell phones, laptops, etc. for end users. This of course should be combined with software development from the upper level, level 5. The topmost level, which is often neglected by engineers but actually very important is the product marketing, without which all the effort from level 1 to level 5 are in vain.

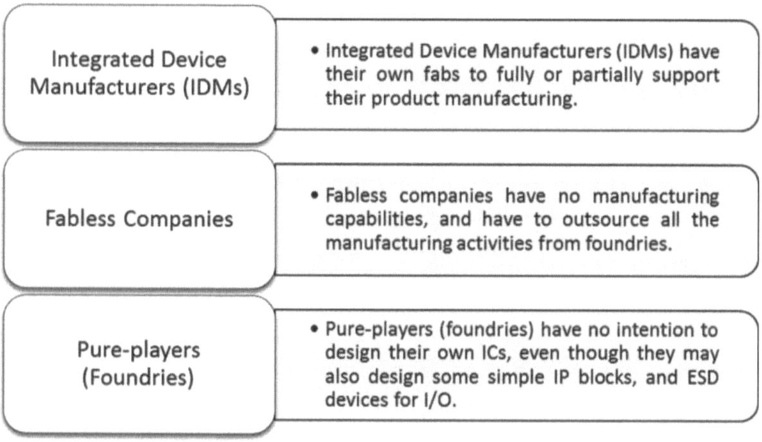

Fig. 1.3 Types of semiconductor companies

Table 1.1 Global semiconductor companies ranking as of Q2 2009 [2]

Rank	Company	Headquarters	Type
1	Intel	U.S.	IDM
2	Samsung	South Korea	IDM
3	Toshiba	Japan	IDM
4	Texas Instruments	U.S.	IDM
5	TSMC	Taiwan	Foundry
6	STMicroelectronics	Europe	IDM
7	Qualcomm	U.S.	Fabless
8	Renesas	Japan	IDM
9	Sony	Japan	IDM
10	Hynix	South Korea	IDM

1.1.1 Types of Semiconductor Companies

In order to give reader a clear picture of who in the industry will be using the TCAD tools, some basic knowledge of company types in the semiconductor industry is necessary. Generally, semiconductor companies can be divided into three categories, namely: IDM (Integrated Device Manufactures), Fabless and Pure-players (Foundries) as shown in Fig. 1.3.

Until recently, most of the largest semiconductor companies in the world were IDMs. Industry giants like Intel and Samsung are clear examples. Table 1.1 shows the top ten semiconductor companies in the second quarter of 2009, with data coming from EE times [2]. As previously discussed, the semiconductor industry is vibrant and dynamic so the rankings change every quarter.

Generally, engineers in the IDMs and foundries use TCAD tools to help them build semiconductor devices and simulate electrical, thermal and optical

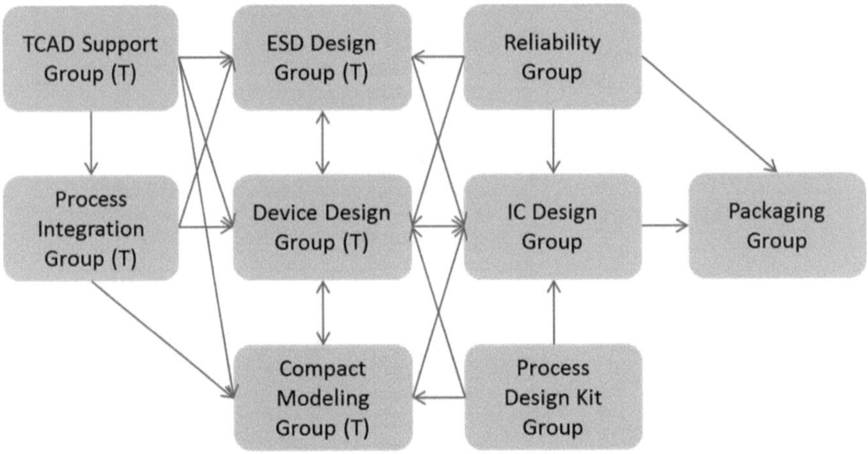

Fig. 1.4 Device engineering groups in a typical IDM

characteristics under different bias conditions. While most fabless do not have manufacturing capabilities, only a few of them need to use TCAD on a daily basis. Some fabless design houses of discrete power devices use TCAD to help them reduce the design cost and time.

1.1.2 Semiconductor Engineering Groups in a Typical IDM

TCAD users for microelectronics application can be divided into several categories:

- Advanced/nano technologies, with typical companies like IBM and Intel
- Companies developing smart power IC/BiCMOS-DMOS(BCD) technologies, like Texas Instruments and Freescale Semiconductor
- Discrete Power device providers, such as International Rectifier
- Compound semiconductor providers, like Cree
- Optics and Photonics companies

University students who specialize in semiconductors but do not have any industry experiences may wonder how the R&D groups in a typical semiconductor company might be organized. While it is true that the exact name and functions differ from company to company, the basic group functions are similar. A typical IDM has several groups of people focusing on different aspects of device and IC design.

Figure 1.4 illustrates an example for the device R&D groups of an analog/mixed-signal/smart power (BCD) company: groups with "(T)" use TCAD tools extensively. The arrows indicate the direction of support between groups. For instance, the device design group will support the IC design group, while coordinating closely with the compact modeling and ESD groups. However, the arrow direction does not necessarily mean single direction support: most of the work is done cooperatively within a company, even if those groups operate in different locations or even in other countries.

Note that different companies have very different group definitions based on their unique market situations and company strategy. For example, some smaller companies may not have individual TCAD support group. The device engineers tend to be capable of performing TCAD simulation, device design and process integration at the same time, while some larger enterprise may hire specific TCAD engineers to do the simulation work. Some largest players even develop their own TCAD software and calibrate it to fit their own process. This is especially true for advanced CMOS technology development, where a lot of work needs to be done to have the TCAD tools accurately fit the advanced process, which often involves quantum effects.

1.2 A Typical Analog/Power Technology Development Flow

As discussed previously, many semiconductor companies with different product focus run TCAD simulations. Analog/power IC/discrete power devices companies are chosen as an example here since, as a group, these companies are more homogeneous than those in the optics and photonics field.

Figure 1.5 shows a simplified development flow for analog and power technology involving nine steps. The first step is planning, which determines the fundamental questions such as the kind of wafer to be used, what technology node to apply (e.g. 22, 45 nm, etc.). This is followed by the device design stage, at which process integration steps should also be determined. TCAD virtual fabrication and testing

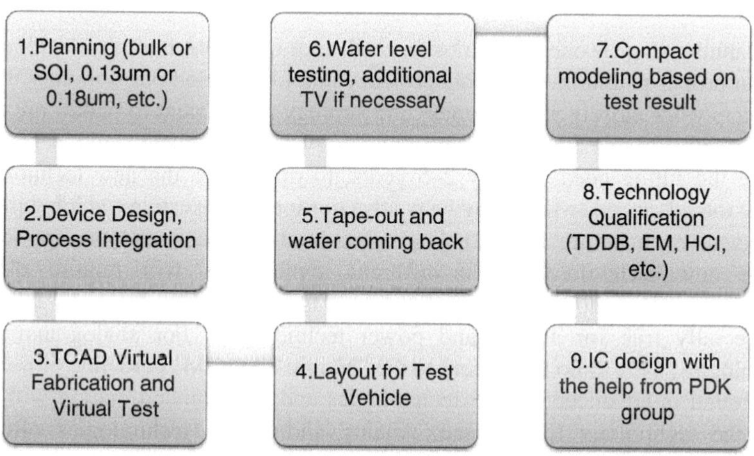

Fig. 1.5 Simplified process flow of analog and power technology development

always accompanies the device design stage: extensive modeling needs to be done well in advance of the first physical prototypes. In most cases, TCAD calibration is necessary; very often, results from previous technologies and other measurements can serve as a guideline.

After everything is ready, the fourth step is creating the layout for test vehicles (aka "testing wafers"). The devices and shapes for reliability testing should be drawn even for the first test vehicle to provide a "look ahead" practice and prevent last-minute failures. During the waiting period before a wafer comes back from the foundry, device engineers must work closely with process integration engineers and fab personnel to monitor the fabrication process and make sure everything is proceeding according to plan. This step may involve different measurements including TEM (Transmission Electron Microscopy) and SEM (Scanning Electron Microscopy). In comparison with TEM, SEM is capable of capturing comparatively large area of the specimen while TEM has better resolution to image individual atoms [3].

After fabrication, wafers are sent back for wafer-level testing and the results are used to build a compact model. After several run of test vehicles, the new technology is ready for qualification. TDDB (Time Dependent Device Breakdown), EM (Electron Migration) and HCI (Hot Carrier Injection) tests are routinely carried out. Depending on the desired application, different tiers (automotive tier, industry tier, consumer electronics tier, etc.) of reliability will be required. After all qualification tests are completed and the compact models are built, the device design is ready. The IC design engineers then may carry out for the IC design and testing work.

1.2.1 Planning Stage

The planning stage is one of the most critical steps of new technology development; without proper planning, the whole project cannot be carried out smoothly. Device engineers work closely with business and strategy department to make sure everybody is comfortable with the new technology. At this stage, engineers should foresee the future need, usually 2–5 years, to make sure the new technology is neither too advanced (which may be costly) nor too conservative (which will make it obsolete even before release). Engineers need to understand the applications of the new technology, since different applications will require different technologies. The choice of technology node is an important but risky task. This is especially true for analog and power technologies. For analog and power technologies, there is no equivalent to the "Moore's law" [4] guideline of improvement found in advanced CMOS technologies and the intended market will determine the technology to be used. Analog and power technologies often use technologies that are two or three generations older than their digital counterpart as well as smaller wafer sizes (e.g. 8 in. vs. 12 in.) (Fig. 1.6).

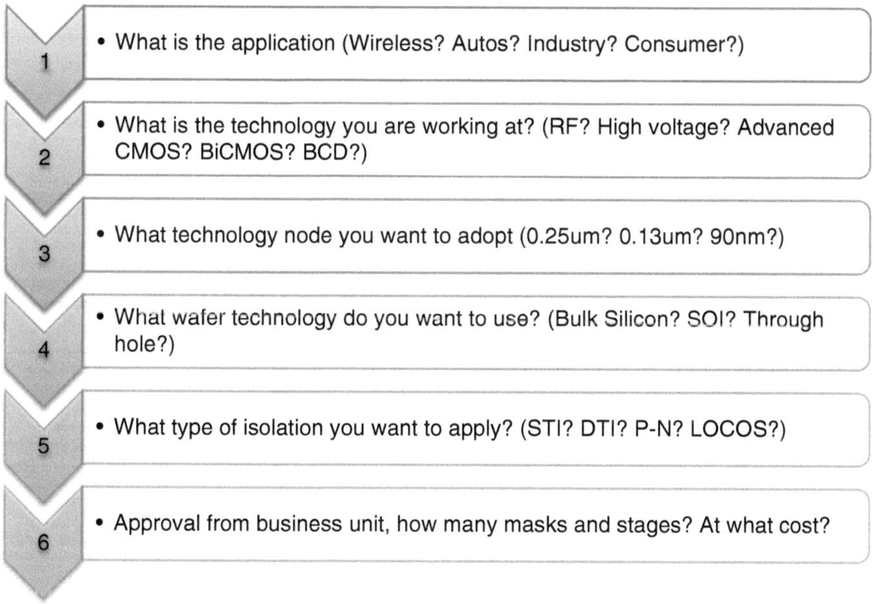

Fig. 1.6 The planning stage of a simplified analog and power technology flow

1.2.2 *Device Design Stage*

In the device design stage, device engineers work closely with process integration engineers to design novel devices for various applications. The goals of the design can be varied but often involve improving devices from earlier generations. Device design is a complicated job. It requires a lot of considerations. Even highly skilled device engineers need to turn to TCAD simulators from time to time for verification of his or her thoughts and to have a better picture of the underlying physics. Of course, like any other software the results of the simulation depend on the input parameters and the results should not be trusted blindly. Experience and instinct play a much more important role in design.

Figure 1.7 gives an outline of important aspects during the device design stage. TCAD tools are frequently used by device design engineers to save time and cost as well as understanding the physics behind.

1.2.3 *Fabrication and Test Stage*

In this stage, engineers layout the test devices, tape-out the design and perform wafer-level tests after wafers come back from the fab. Before fabrication, a Design Rule Check (DRC) is usually performed to ensure the final IC layout

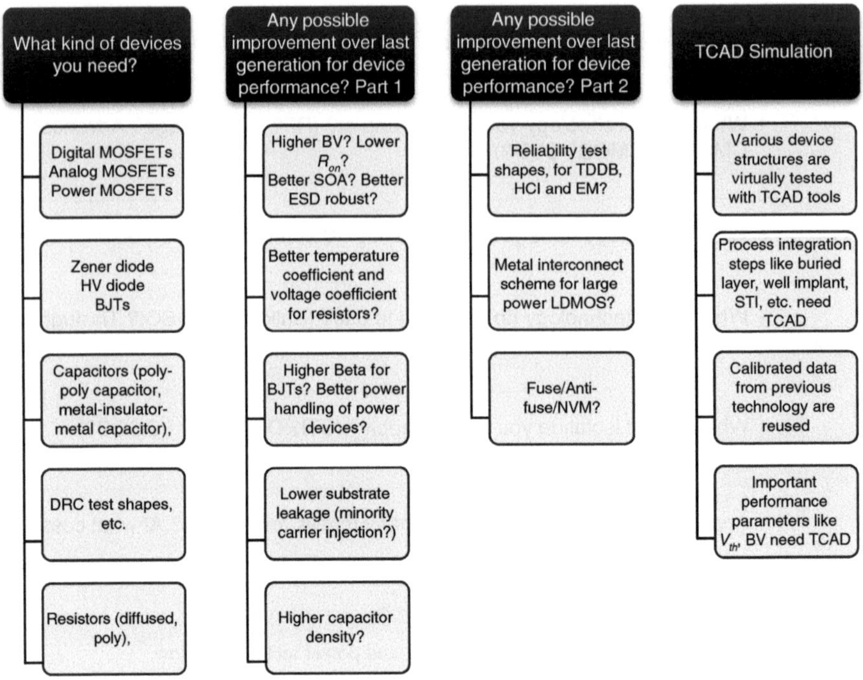

Fig. 1.7 Device design stage considerations

obeys the rules specified by PDK (process design kit) engineers at the fab. This helps increase the process yield by checking important parameters like the minimum width, spacing and coverage of all the elements of a complex design [53]. The steps involved at the fabrication and test stage of a new technology are shown in Fig. 1.8.

The term "tape-out" means to send the finished design masks to the fab. It derives from the early days of the technology when the enlarged "artwork" for the photomask was manually "taped out" using black line tape and adhesive-backed die cut elements on sheets of PET film [5, 6].

1.2.4 Reliability and Qualification Stage

After several test vehicles, the technology is ready to be tested by reliability engineers. Typical reliability tests include TDDB (Time-dependent dielectric breakdown), HCI (Hot Carrier Injection) and EM (Electron Migration). Different tiers of reliability will be required for different product applications: applications that have safety implications (e.g. automotive) have much stricter requirements than other consumer products. Other tests like IDDQ (measuring the supply current

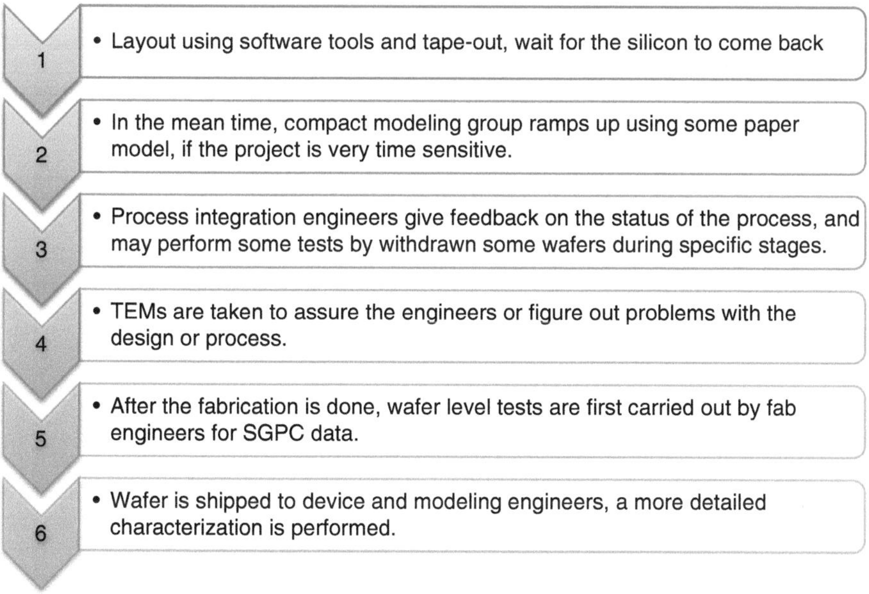

Fig. 1.8 Device fabrication and test stage

(I_{dd}) in the quiescent state) should also be performed as needed. Note that look-ahead reliability tests should often be carried out in parallel with the device design in order to avoid last-minute failures.

1.3 About Technology Computer Aided Design (TCAD)

1.3.1 Difference Between IC CAD and TCAD

Unlike the IC design CAD (IC-CAD) tools, which use compact models to simulate circuit level behavior, TCAD is purely physics–based. TCAD tools model the behavior of semiconductor devices using fundamental physical models like the current continuity (drift-diffusion) and Poisson equations. TCAD tools typically operate at the device level and usually only have very basic models for external circuit elements. Tools that go beyond simple models and integrate TCAD with SPICE [7] (Simulation Program with Integrated Circuit Emphasis) models are usually referred to as "mixed-mode" simulators.

A modern TCAD software package includes several simulation tools: the process simulator, the device simulator, the Graphical User Interface (GUI) and plotting tools. This book will give step by step instructions on how to set up a simulation. A typical TCAD software suite is shown in Fig. 1.9 and TCAD GUI tools example are shown in Fig. 1.10.

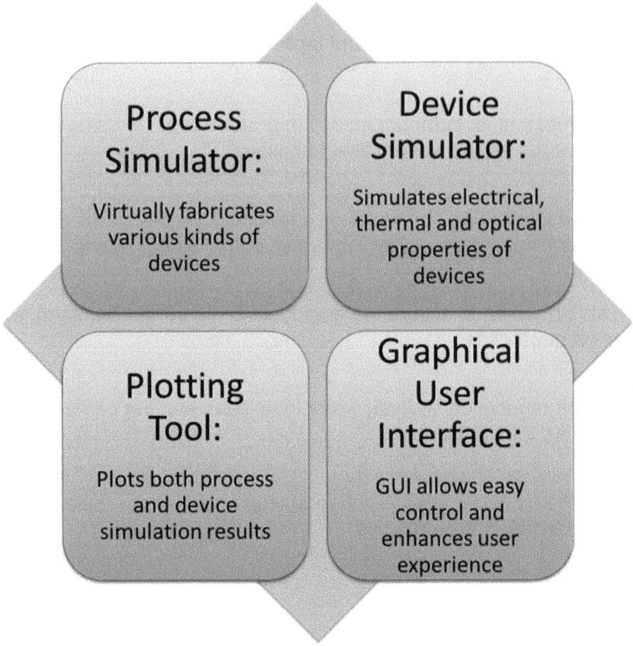

Fig. 1.9 A typical TCAD software suite

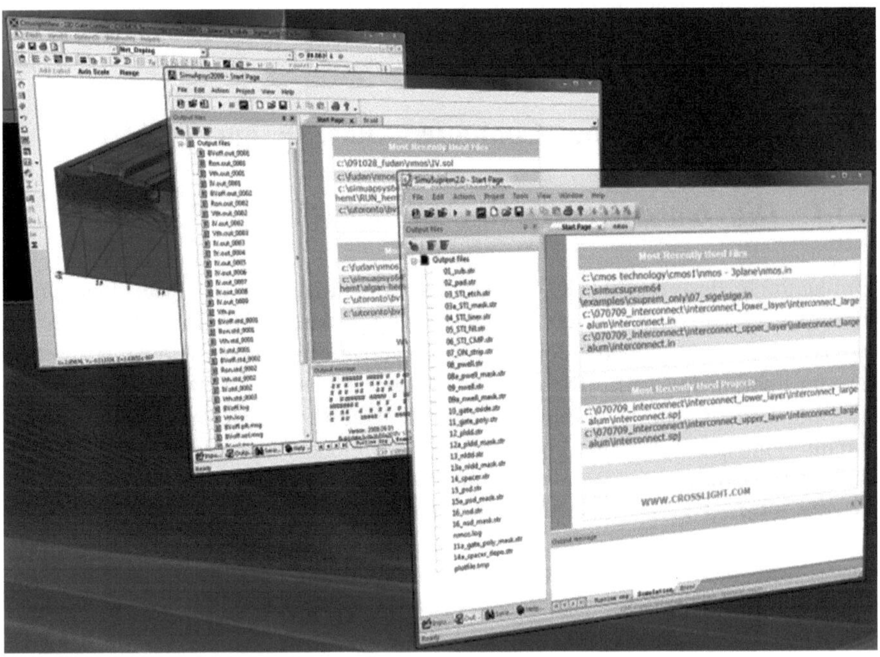

Fig. 1.10 An example of TCAD GUI tools

1.3.2 Semiconductor Process Simulator

A semiconductor process simulator models the effects of the various steps involved in semiconductor fabrication: growth, etching, depositing, etc. Most process simulators in use today are derived from the SUPREM.IV.GS code which was originally developed at the Integrated Circuit Laboratory of Stanford University [8]. SUPREM.IV.GS has long been recognized as the industry standard in process simulation. SUPREM is very easy to understand and its intuitive commands make the process simulation input files resemble the process flow sheet at a manufacturing site.

CSUPREM [9] is one of the tools which owe its existence to SUPREM.IV.GS. It not only inherits the essential physical models in the original version from Stanford but also contains substantial enhancements and extensions. Most significantly, the entire mesh system has been redesigned and extended to allow for full 2D/3D modeling at a very low computational cost. The syntax of CSUPREM is very similar to that of the original and differs only in the areas where the software has been enhanced.

In this book, most of the CSUPREM code examples are directly convertible to other TCAD tools that have evolved from SUPREM.IV.GS. In cases where the syntax differs, equivalent commands can often be found by consulting the process simulator's reference manual. The easy to read process simulation code can serve as a good tutorial to understand the more complicated process flow sheets used in the industry. The user should also keep in mind that the scope of this book is about 3D TCAD simulation, rather than a semiconductor device physics or process technology flow textbook. The process conditions used in this book are just for tutorial purposes and are oversimplified. This book is by no means a guide to perform a real device manufacturing: a device designer must rely on their experience and follow fab guidelines and recipes for most process steps. On the other hand, this book may serve as a valuable reference for TCAD simulations, no matter which software suite the reader is using.

As an example, a 3D process simulation of FINFET structure is shown in Fig. 1.11. Nine steps are used to virtually fabricate the device.

1.3.3 Semiconductor Device Simulator

Semiconductor device simulators can model the electrical, optical, thermal and sometimes even the mechanical properties of devices. They can be used in standalone mode by defining the device structure or used in conjunction with a process simulator. These tools allow the user to work on band structure engineering, quantum confinement and other optimizations that would be difficult to accomplish using experimental results alone.

Most device simulators are based on 2D/3D finite element analysis of electrical, thermal and optical properties of compound and silicon semiconductor devices.

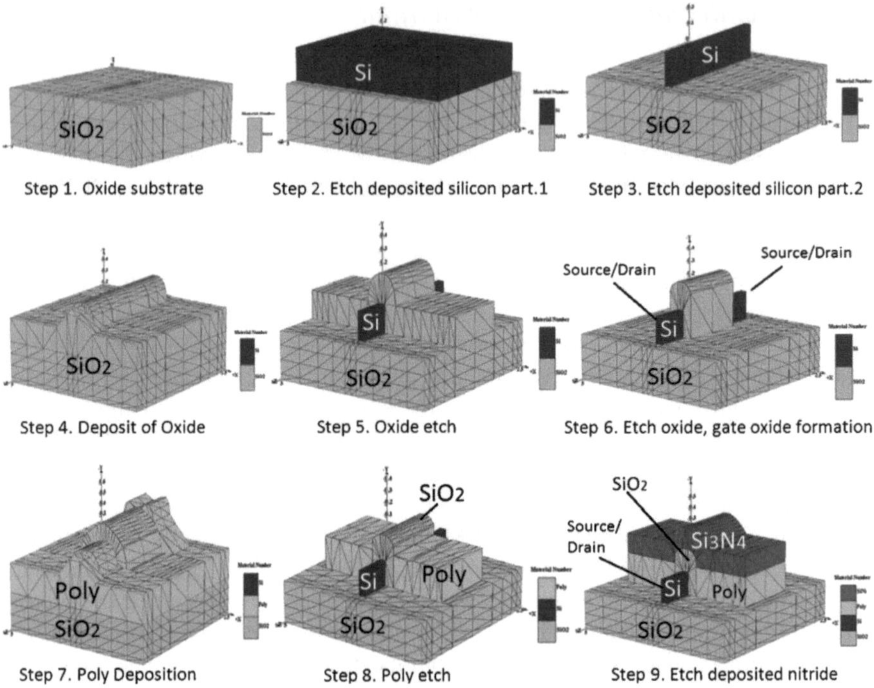

Fig. 1.11 3D process simulation of FINFET structure

The simulator solves the Poisson and current continuity equations and includes additional models like carrier energy transport (hydrodynamic model), quantum mechanical wave equations and scalar wave equations for photonic wave guiding devices.

Different TCAD software suites have different names for their individual simulators, the fundamental physics are similar. This book focuses on methods rather than software tool itself. Even though the device simulations for all the examples in the book use Crosslight's APSYS device simulator [10], the reader can apply the simulation methods used in this book to other TCAD software as well. It is true that it takes some time to familiarize oneself with a brand new software suite. But just like driving a new car requires less adaptation than learning to drive in the first place, a skilled TCAD user can easily switch to a different TCAD tool.

In this book, all APSYS device simulation codes are named as "device simulation code". This is because the skills taught from the examples can be applied to other device simulators. Figure 1.12 shows an example of device simulation result: the S11 and S22 parameters of a device plotted on the Smith Chart after small signal AC modeling.

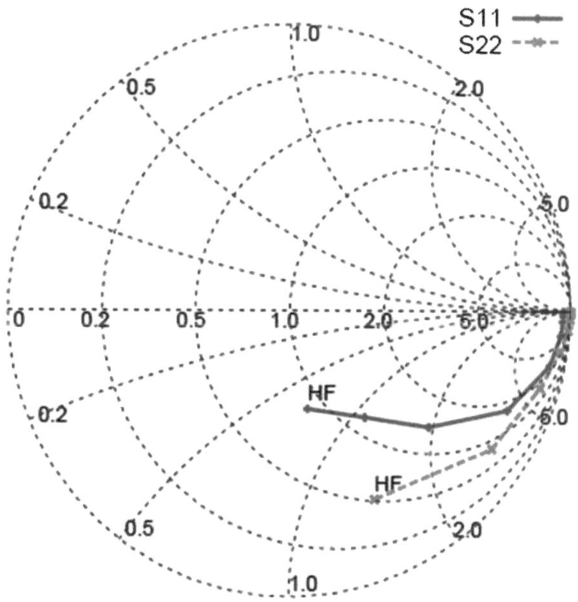

Fig. 1.12 S11 and S22 plotted on the Smith Chart from APSYS device simulation

1.3.4 Why 3D TCAD?

There are many benefits to using 3D TCAD, most notably that it shows the full device structure rather than a 2D cut. The real world is 3D and having a better representation and understanding of reality has always been the goal of CAD and TCAD tools.

Some devices do not have any natural variation along the z direction so a 3D simulation brings very few benefits. In many other devices, the structure of the device itself requires 3D simulation. For example, for the large interconnect structures coupled with underlying devices which will be presented in Chap. 8, 2D simulations are meaningless. MEMS devices never appear in 2D form and superjunction LDMOS is best understood if 3D simulation is performed to name only a few examples.

Figure 1.13 is a 3D simulation of CMOS image sensor. Do not be intimidated by this structure: this book will explain the process steps needed to build structures such as this.

1.3.5 Stacked Planes Method vs. Traditional Bulk Method for 3D TCAD

Given the usefulness of 3D simulation, one may wonder why this has not been routinely used in the past. The answer to this question is simple: there is always a tradeoff between the accuracy of the model and the speed of the calculations.

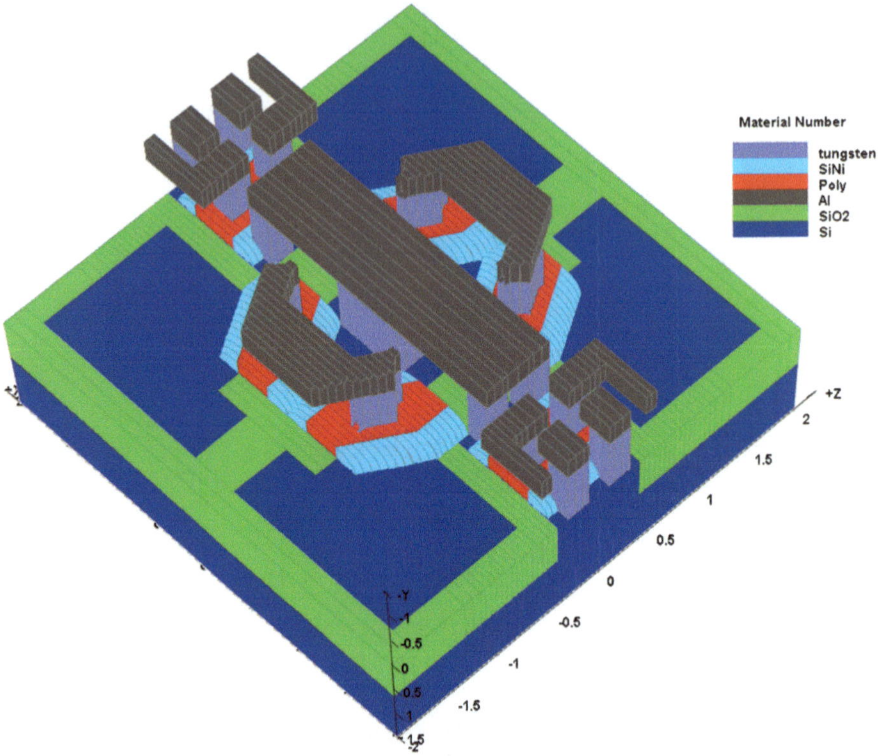

Fig. 1.13 3D TCAD simulation of CMOS image sensor

The amount of time required to perform a simulation can vary depending on the number of mesh points involved, the level of approximation in the physical models and the computing power available. A moderately complex 3D structure modeled with a modern quad-core PC can take several hours to model but this can increase substantially in certain cases. In the extreme case, the modeling time for a device can become days or weeks and it becomes comparable to the turn-around time to fabricate the actual wafers. At this point, there is almost no benefit in using TCAD tools at all, especially given the short lifecycle of today's consumer electronics.

The method used to generate the 3D mesh for the simulation is also of vital importance. Mesh generation is a complex subject and automated methods of mesh generation developed in fields like mechanical engineering do not work as well for semiconductor physics. Typical 3D mesh tools try to generate tetrahedral elements with a regular shape which, when combined with a requirement to respect internal boundary conditions, can often greatly oversample certain areas of a device. This can lead to a substantial increase in computational requirements.

An alternative method was developed to avoid these problems and generate a smaller mesh count. Rather than using 3D tetrahedral elements, the mesh is based on a set of stacked planes: 2D mesh triangles in the x-y plane are extruded

Fig. 1.14 The stacked planes method vs. traditional bulk structure

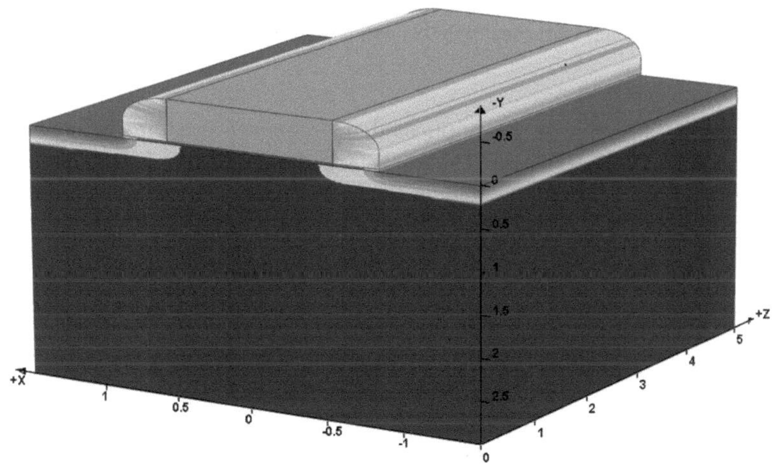

Fig. 1.15 A simple structure with two stacked mesh planes

along the z direction to form a triangular prism. This allows for the local 2D mesh to be optimized on a per-plane basis and concentrated in areas where strong lateral effects are expected. Similarly, the number of mesh planes can be increased in regions where the z direction is important. An experienced device designer quickly learns which areas are the most important to sample correctly.

Figure 1.14 shows a quick comparison between the two mesh generation methods [11].

Two examples based on the stacked planes method are given in Figs. 1.15 and 1.16. The first structure is an NMOS structure with only two stacked planes. Two planes is the absolute minimum needed to define a 3D volume. This actually brings no benefit over a 2D simulation since both planes are exactly alike. In structures where the longitudinal variation in the z-axis must be sampled, a minimum of three mesh planes should be used.

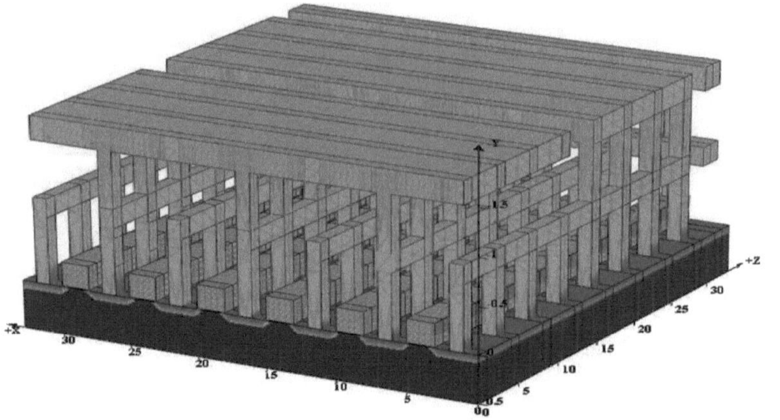

Fig. 1.16 A more complex structure with 38 mesh planes

The second structure is a complicated interconnect structure with two metal layers and one polysilicon layer. The total stacked planes count for the interconnect structure is 38 planes. Here, the longitudinal variation is well-sampled but the extra precision required increases the numerical cost of the simulation.

1.3.6 Quasi-3D vs. Full-3D in Process Simulation

Because of the unique mesh structure, there is an extra degree of freedom in the 3D modeling. In certain cases, the 3D structure itself may be important (due to different process steps acting on different regions of the device) but it may be possible to neglect interactions between mesh planes. For example, a process simulation user may not care about the amount of dopant diffusion or oxidation in the z-direction. In that case, a Quasi-3D simulation may be of use: by turning off the plane-to-plane interactions, the simulation may be sped up significantly. Note that this is not possible in cases where the plane-to-plane interactions contain essential physics and that full 3D should be used in these cases.

An example is shown with 3D segregation using both full 3D and quasi-3D. They were built to give user a clear picture of the differences between full 3D and quasi 3D. The mesh design for the x-y plane and y-z plane are deliberately made the same so that the reader can see the diffusing along both x-y plane and y-z plane. Figure 1.17 is the structural view of the segregation example, with quasi 3D and full 3D net doping chart showing below.

From figures above, one can tell the significant difference between quasi3D and full 3D. While full 3D process simulation takes substantially longer time than quasi 3D, it is often necessary when the third dimension (z direction) effect is considered.

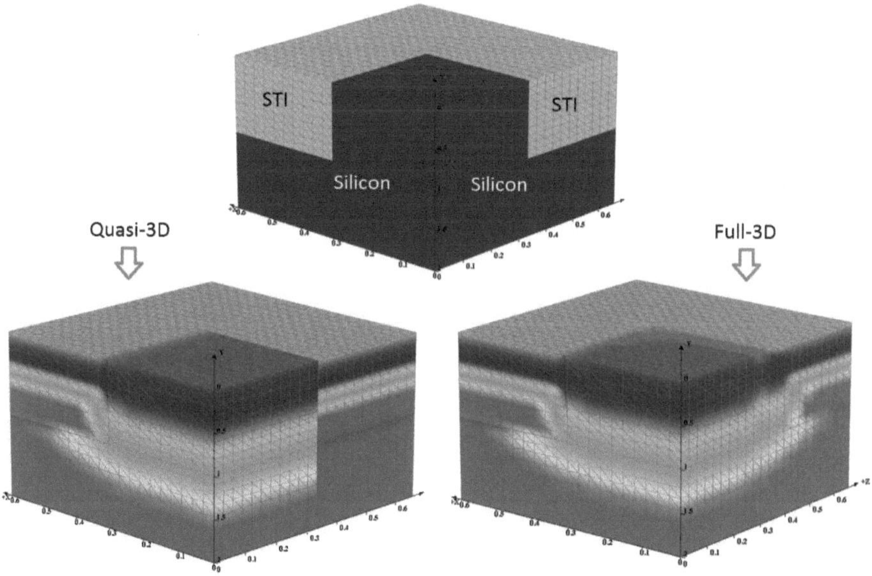

Fig. 1.17 3D segregation example

Still another method called hybrid 3D, which takes into account the 3D oxidation only. For LOCOS process growth, the oxide will grow into the z direction to form the famous "bird's beak". If the doping condition underneath the oxide layer stays the same in the z direction, we can neglect the dopant diffusion in the third dimension and consider only the oxide diffusion. By neglecting plane-to-plane dopant diffusion and only consider plane-to-plane oxide diffusion, hybrid 3D improves the simulation speed over full 3D, while trail behind that of quasi 3D.

Chapter 2
Advanced Theory of TCAD Process Simulation

2.1 Diffusion Model in TCAD

A key simulation task in TCAD process simulation is to solve the diffusion equations over the simulation mesh to predict the impurity doping profile after thermal processing. In 3D TCAD, the simulation grid size is large and diffusion is often the most time-consuming simulation procedure. Due to its importance in 3D TCAD, we describe the theoretical background of impurity diffusion based on the SUPREM-IV GS code from Stanford University [8]. SUPREM-IV GS is a widely recognized simulation software in TCAD and currently several commercial versions of it are available on the TCAD market, all of which inherit the physical models described in this chapter.

The latest models of impurity diffusion are often based on the concept of pair-diffusion. These models not only account for the effect impurity interactions via space charge, but also accurately describe the interaction between impurities and lattice point-defects such as interstitials and vacancies. The impurity atoms cannot diffuse by themselves and require neighboring point-defects as a diffusing vehicle. It is common to refer to impurities involved in a point-defect related diffusion mechanism as dopant-defect pairs. We label a dopant A paired with a vacancy V as a AV pair, a dopant A paired with interstitial I as a AI pair. So, the diffusivities of dopants are represented by the diffusivities of dopant-defect pairs. Impurity diffusion in TCAD simulation includes several types which we will introduce as follows.

2.1.1 Vacancies

In addition to dopant-defect paired diffusion we just mentioned, vacancies can diffuse by themselves. The vacancies obey a complex diffusion equation which can be written as [12]:

S. Li and Y. Fu, *3D TCAD Simulation for Semiconductor Processes, Devices and Optoelectronics*, DOI 10.1007/978-1-4614-0481-1_2,

$$\frac{\partial C_V}{\partial t} = \vec{\nabla} \cdot \left(-\overrightarrow{J_V} - \sum\nolimits_{\text{imp}} \overrightarrow{J_{AV}} \right) - R \tag{2.1}$$

where C_V is the vacancy concentration, $\overrightarrow{J_V}$ refers to the vacancy flux, $\overrightarrow{J_{AV}}$ refers to the flux of the dopant-defect pair AV, and R describes the recombination of vacancies. This model represents the diffusion of all vacancies, both paired and unpaired. It can be formally derived by assuming thermal equilibrium between the species and that the simple pairing reactions are dominant [12].

The pair fluxes are the contributions to the total vacancy flux from the impurity diffusion and will be described in a subsequent section. The unpaired vacancy flux can be written as [12]:

$$-\overrightarrow{J_V} = D_V C_V^* \vec{\nabla} \frac{C_V}{C_V^*} \tag{2.2}$$

It is important to note that the equilibrium vacancy concentration C_V^* is a function of Fermi level [13]. This flux accounts correctly for the effect of an electric field on the charged portion of the defect concentration.

The bulk recombination is simple interaction between interstitials and vacancies and can be expressed:

$$R = K_R (C_I C_V - C_I^* C_V^*) \tag{2.3}$$

where K_R is the bulk recombination coefficient, and C_I and C_V^* are the interstitial and interstitial equilibrium concentration, respectively.

The defects obey a flux balance boundary condition, as described by Hu [14]:

$$\overrightarrow{J_V} \cdot \vec{n} + K_V (C_V - C_V^*) = g \tag{2.4}$$

where $\vec{n}$ is the surface normal, K_V is the surface recombination constant, and g is the generation, if any, at the surface.

2.1.2 Interstitials

Interstitials diffuse according to an equation similar to that for vacancies. They obey a complex diffusion equation which can be written [12]:

$$\frac{\partial (C_I - C_{ET})}{\partial t} = \vec{\nabla} \cdot \left(-\overrightarrow{J_I} - \sum\nolimits_{\text{imp}} \overrightarrow{J_{AI}} \right) - R \tag{2.5}$$

where C_I is the interstitial concentration, C_{ET} is the number of empty interstitial traps, J_I refers to the interstitial flux, J_{AI} refers to the flux of impurity A diffusing with interstitials (to be detailed later in this chapter), and R is all sources of bulk recombination of interstitials. This model represents the diffusion of all interstitials, both paired and unpaired. It can be derived from assuming thermal equilibrium between the species and that the simple pairing reactions are dominant [12].

The trap reaction was described by Griffin [15]. This model explains some of the wide variety of diffusion coefficients extracted from several different experimental conditions. The trap equation is:

$$\frac{\partial C_{ET}}{\partial t} = -K_T\left[C_{ET}C_I - \frac{e^*}{1-e^*}C_I^*(C_T - C_{ET})\right] \tag{2.6}$$

where C_T is the total trap concentration, K_T is the trap reaction coefficient, e^* is the equilibrium trap occupancy ratio. Instead of the reaction, the time derivative is used in the interstitial equation because it has better properties in the numerical calculation.

The pair fluxes are the contributions to the total interstitial flux from each of the models for the impurities (e.g. antimony, arsenic, boron, phosphorus). The unpaired interstitial flux can be written as [16]:

$$-\vec{J_I} = D_I C_I^* \vec{\nabla}\frac{C_I}{C_I^*} \tag{2.7}$$

It is important to note that the equilibrium concentration C_I^* is a function of Fermi level [13]. This flux accounts correctly for the effect of an electric field on the charged portion of the defect concentration.

The bulk recombination is the same process described for vacancies in the previous section. Like vacancies, interstitial defects obey a flux balance boundary condition, as described by Hu [14]:

$$\vec{J_I}\cdot\vec{n} + K_I(C_I - C_I^*) = g \tag{2.8}$$

Where $\vec{n}$ is the surface normal, K_I is the surface recombination constant, and g is the generation, if any, at the surface.

2.1.3 Active Impurities

Some dopants diffuse only when they are activated and the diffusion equation depends on the activated concentration of the impurities. Here activation is used to describe how well the impurity atoms are incorporated into the host lattice. In process simulation dopant activation is achieved via thermal annealing and this depends on solid solubility and other factors. For n-type dopants, this model applies to arsenic and selenium. The diffusion equation is as follows:

$$\frac{\partial C_T}{\partial t} = \vec{\nabla} \cdot \left(-\overrightarrow{J_{AI}} - \overrightarrow{J_{AV}} \right) \tag{2.9}$$

The paired diffusion fluxes are given by the following.

$$\overrightarrow{J_{AV}} = -D_V C_A \frac{C_V}{C_V^*} \vec{\nabla} ln\left(C_A \frac{C_V}{C_V^*} \frac{n}{n_i} \right) \tag{2.10}$$

$$\overrightarrow{J_{AI}} = -D_I C_A \frac{C_I}{C_I^*} \vec{\nabla} ln\left(C_A \frac{C_I}{C_I^*} \frac{n}{n_i} \right) \tag{2.11}$$

The p-type dopants boron, beryllium and magnesium follow similar equations:

$$\overrightarrow{J_{AV}} = -D_V C_A \frac{C_V}{C_V^*} \vec{\nabla} ln\left(C_A \frac{C_V}{C_V^*} \frac{p}{n_i} \right) \tag{2.12}$$

$$\overrightarrow{J_{AI}} = -D_I C_A \frac{C_I}{C_I^*} \vec{\nabla} ln\left(C_A \frac{C_I}{C_I^*} \frac{p}{n_i} \right) \tag{2.13}$$

In the above, n and p refer to the electron and hole concentrations and n_i refers to the intrinsic carrier density.

We also note the total concentration of impurities (C_T) appears on the left-hand side but only the activated concentration (C_A) is found on the right-hand side. The relationship between these two quantities is complex and beyond the scope of this book.

2.1.4 *Inactive Impurities*

Some types of impurities are able to diffuse when paired with point-defects without being activated. Examples of this behavior include n-type dopants like phosphorous and antimony. Please note that the total concentration C_T appears on both sides of the equation here.

$$\frac{\partial C_T}{\partial t} = \vec{\nabla} \cdot \left(-\overrightarrow{J_{AI}} - \overrightarrow{J_{AV}} \right) \tag{2.14}$$

$$\overrightarrow{J_{AV}} = -D_V C_T \frac{C_V}{C_V^*} \vec{\nabla} ln\left(C_T \frac{C_V}{C_V^*} \frac{n}{n_i} \right) \tag{2.15}$$

$$\overrightarrow{J_{AI}} = D_I C_T \frac{C_I}{C_I^*} \vec{\nabla} ln\left(C_T \frac{C_I}{C_I^*} \frac{n}{n_i} \right) \tag{2.16}$$

2.1.5 Neutral Impurities

In TCAD process simulation, many charge neutral impurities are involved in the diffusion process. These include oxidants such as O_2 and H_2O (during oxidation steps), gold, cesium, and germanium. Inter-diffusion of germanium in silicon as neutral impurity is of great importance since Ge has been used to produce the mechanical stress in nano-scale CMOS with stress-induced mobility enhancement. Diffusion of charge neutral impurities simply follows Fick's law:

$$\overrightarrow{J_T} = -D.\overrightarrow{\nabla} C_T \tag{2.17}$$

$$\frac{\partial C_T}{\partial t} = -\overrightarrow{\nabla} \cdot \overrightarrow{J_T} \tag{2.18}$$

2.1.6 Si-Ge Inter-diffusion

While it is easy to describe all diffusion of charge neutral impurities by Fick's law, the situation gets somewhat complicated if highly concentrated Ge and Si elements diffuse into each other while the composition and induced strain/stress keeps changing. The simple model of Fick's law is useful only if the diffusivity can be written using the Arrhenius function. This is usually the case but it may not be applicable for inter-diffusion. Due to its importance in nano-scale MOSFET, we introduce different diffusion mechanisms.

- Dopant diffusion: Dopant diffusion process is such that the host lattice does not change and the dopant concentration is dilute compared with the host. Dopant diffusion is driven by the chemical potential gradient which is the case for common dopants like boron and arsenic in silicon.
- Self-diffusion: In a self-diffusion process, a dilute concentration isotopes diffuses in a homogenous host with no change in chemical potential.
- Si–Ge inter-diffusion at Si/SiGe interfaces is driven by a chemical potential gradient. Both elements transport through the interface and the lattice composition changes by intermixing. In a Si/SiGe heterostructure, Si and Ge concentrations are both on the order of 1E+22 cm^{-3}. As a result, not only is the diffusivity a strong function of the position-dependent Ge composition, but also the simple Arrhenius function cannot be used in the Fick's law model [12].

A quick example for Si-Ge inter-diffusion is built with the process simulator (Fig. 2.1).

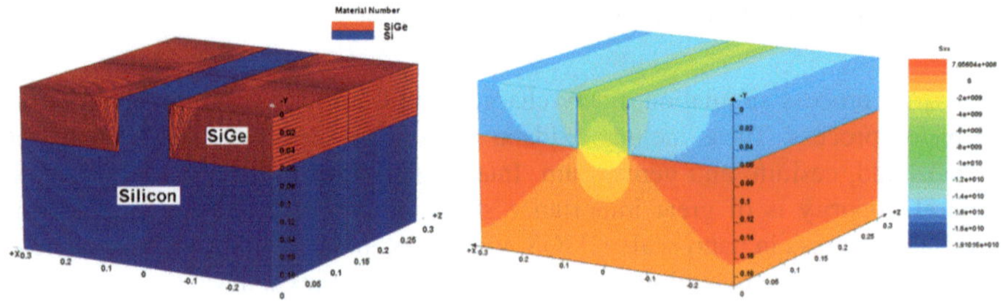

Fig. 2.1 Si-Ge inter-diffusion stress example: structure (*left*) and Sxx plot (*right*)

2.2 Stress Models

Modeling stress is one of the important tasks of TCAD process simulation especially when stress is incorporated into the device to enhance performances. The stress theories implemented into SUPREM-IV.GS are described below and can be found in most mechanical engineering books.

2.2.1 *Governing Equations*

The equation of motion for linear elasticity (i.e. Newton's second law) relates the displacement vector (u, v, w) to the applied stress and external forces:

$$\frac{\partial^2 \rho u}{\partial t^2} = \frac{\partial \sigma_{xx}}{\partial x} + \frac{\partial \sigma_{yx}}{\partial y} + \frac{\partial \sigma_{zx}}{\partial z} + F_x \tag{2.19}$$

$$\frac{\partial^2 \rho v}{\partial t^2} = \frac{\partial \sigma_{xy}}{\partial x} + \frac{\partial \sigma_{yy}}{\partial y} + \frac{\partial \sigma_{zy}}{\partial z} + F_y \tag{2.20}$$

$$\frac{\partial^2 \rho \mathrm{w}}{\partial \mathrm{t}^2} = \frac{\partial \sigma_{\mathrm{xz}}}{\partial \mathrm{x}} + \frac{\partial \sigma_{\mathrm{yz}}}{\partial \mathrm{y}} + \frac{\partial \sigma_{\mathrm{zz}}}{\partial \mathrm{z}} + \mathrm{F_z} \tag{2.21}$$

where

$$\sigma_{\mathrm{xx}} = \mathrm{C}_{11}\frac{\partial \mathrm{u}}{\partial \mathrm{x}} + \mathrm{C}_{12}\frac{\partial \mathrm{v}}{\partial \mathrm{y}} + \mathrm{C}_{12}\frac{\partial \mathrm{w}}{\partial \mathrm{z}} \tag{2.22}$$

$$\sigma_{\mathrm{yy}} = \mathrm{C}_{12}\frac{\partial \mathrm{u}}{\partial \mathrm{x}} + \mathrm{C}_{11}\frac{\partial \mathrm{v}}{\partial \mathrm{y}} + \mathrm{C}_{12}\frac{\partial \mathrm{w}}{\partial \mathrm{z}} \tag{2.23}$$

$$\sigma_{zz} = C_{12}\frac{\partial u}{\partial x} + C_{12}\frac{\partial v}{\partial y} + C_{11}\frac{\partial w}{\partial z} \tag{2.24}$$

$$\sigma_{xy} = \sigma_{yx} = C_{44}\left(\frac{\partial v}{\partial x} + \frac{\partial u}{\partial y}\right) \tag{2.25}$$

$$\sigma_{yz} = \sigma_{zy} = C_{44}\left(\frac{\partial w}{\partial y} + \frac{\partial v}{\partial z}\right) \tag{2.26}$$

$$\sigma_{zx} = \sigma_{xz} = C_{44}\left(\frac{\partial w}{\partial x} + \frac{\partial u}{\partial z}\right) \tag{2.27}$$

The C coefficients are elastic stiffness constants, the F terms are the external body forces acting on the material of interests and ρ is the density of the medium. Please note that this equation applies to any material system with cubic symmetry (such as silicon); for other crystals, a different stress tensor would apply. Given Young's modulus E and Poisson's ratio ν, the stiffness constants can be calculated as follows:

$$C_{11} = \frac{E(1-v)}{(1+v)(1-2v)} \tag{2.28}$$

$$C_{12} = \frac{Ev}{(1+v)(1-2v)} \tag{2.29}$$

$$C_{44} = \frac{Ev}{2(1+v)} \tag{2.30}$$

2.2.2 *Intrinsic Stress*

In TCAD simulation, the distribution of stress is caused by intrinsic stress which is also called initial or internal stress. It is necessary to identify the source and nature of this stress in order to apply the solver correctly.

The intrinsic stress at a location (x, y, z) is defined as the residual stress when the displacement vector (u, v, w) is zero. This happens when two materials of different lattice structures or sizes are forced to join together. At the material interface, the molecules or atoms are displaced from their stress-free state lattice positions. The interface atoms would reach a local minimum-free-energy state for which we define the displacement vector as being zero. Since the atoms at the interface at zero displacement are forced to move from their original stress-free positions, the internal force on one material from another is not zero and this is the source of intrinsic stress.

For clarity in discussion, we define lattice constant as the distance between adjacent atoms in the original stress-free crystal. We use lattice spacing to refer to the actual distance after the materials are joined together.

If we denote the intrinsic stress tensor as σ_0, the material stress tensor would be written in the following form:

$$\sigma_{xx} = C_{11}\frac{\partial u}{\partial x} + C_{12}\frac{\partial v}{\partial y} + C_{12}\frac{\partial w}{\partial z} + \sigma_{0,xx} \tag{2.31}$$

$$\sigma_{yy} = C_{12}\frac{\partial u}{\partial x} + C_{11}\frac{\partial v}{\partial y} + C_{12}\frac{\partial w}{\partial z} + \sigma_{0,yy} \tag{2.32}$$

$$\sigma_{zz} = C_{12}\frac{\partial u}{\partial x} + C_{12}\frac{\partial v}{\partial y} + C_{11}\frac{\partial w}{\partial z} + \sigma_{0,zz} \tag{2.33}$$

$$\sigma_{xy} = C_{44}\frac{\partial u}{\partial y} + C_{44}\frac{\partial v}{\partial x} + \sigma_{0,xy} \tag{2.34}$$

$$\sigma_{yz} = C_{44}\frac{\partial v}{\partial z} + C_{44}\frac{\partial w}{\partial y} + \sigma_{0,yz} \tag{2.35}$$

$$\sigma_{xz} = C_{44}\frac{\partial u}{\partial z} + C_{44}\frac{\partial w}{\partial x} + \sigma_{0,xz} \tag{2.36}$$

Please note that the intrinsic stress tensor of a uniformly flat material joining another uniformly flat material with translational or symmetry boundary means that the intrinsic stress tensor is a constant with respect to the plane of the interface. This also means that when we substitute the above stress components into Newton's second law, the derivative of the intrinsic stress would be zero and that the interface intrinsic stress does not contribute to the spatial stress profile. The intrinsic stress would cause a stress profile only if there is position dependence in the intrinsic stress tensor (i.e. $\sigma_0(x, y, z)$). This can occur when there is a change in material composition or surface flatness or due to a truncation of the film being deposited [12].

A simple treatment in computation of intrinsic stress in TCAD simulation is to assume that the device substrate is massive enough that the lattice spacing does not change. The material of smaller volume is assumed to be forced to be strained to match the substrate. Then, given the strain in the smaller material, the intrinsic stress can be calculated from the elastic constants. This is the approach taken by SUPREM-IV.GS which assumes the intrinsic stress in silicon beneath the nitride film to be zero while the top nitride film to have a constant $\sigma_{0,\,xx}$, measured from experiments.

When modeling strain/stress in multiple quantum wells (MQW) optoelectronic devices such as laser diodes (LD) and light emitting diodes (LED), the approach is similar: the whole device is assumed to have the lattice spacing of the massive substrate which we assume to be stress-free. Then, it will be relatively easy to compute the stress of each layer in the MQW system given the material elastic constants.

The above same-lattice-spacing approaching works well for calculating stress arising from interfaces between the massive substrate and the top deposits. However, for interfaces between material deposits (such as SiGe pockets) and surrounding materials in the horizontal direction, the assumption of same-lattice-spacing is no longer valid since the size of the materials involved are comparable and there is real competition between the materials for stretching the lattice spacing.

CSUPREM provides a model based on the work of van de Walle [17] which considered the case of two materials of finite size. Such a model can be used to model stress profile due to the deposit of SiGe pockets in nano-scale MOSFET.

2.3 Oxidation

2.3.1 Oxide as a Newtonian Fluid

The most sophisticated model for oxidation process is based on treating oxide and nitride as a Newtonian fluid and solving the stress equations associated with the fluid flow. This approach is accurate when oxidation temperature is sufficiently high [18].

Since the static mechanical stress equations have already been described before, we only need to explain the analogy between static stress and Newtonian fluid stress. From numerical point of view, we only need to maintain a single stress solver for both.

The following discussion of stress in Newtonian fluid follows that in ref [19]. In a linear Newtonian fluid, the stress is proportional to the rate of strain (i.e., velocity) e_{kl}:

$$\sigma_{ij} = \sigma_{ij}^{(0)} + C_{ijkl}^{(v)} e_{kl} \tag{2.37}$$

where $\sigma_{ij}^{(0)}$ is the stress distribution that can exist in a resting fluid and $C_{ijkl}^{(v)}$ is a 4th order tensor which is similar to the stiffness tensor in static mechanics. The stress for the resting fluid is just the hydrostatic pressure p:

$$\sigma_{ij}^{(0)} = -p\delta_{ij} \tag{2.38}$$

In an isotropic fluid, the 81 linear coefficients reduce to only two independent coefficients: the dynamic coefficient of viscosity (shear) μ, and the second coefficient of viscosity (dilatational) λ.

We use the variables $u_i = (i = 1, 2, 3)$ to refer to the velocities v_x, v_y, and v_z, respectively. For a Newtonian isotropic fluid, we get:

$$\sigma_{ij} = -p\delta_{ij} + \mu\left(\frac{\partial u_i}{\partial x_j} + \frac{\partial u_j}{\partial x_i}\right) + \lambda\left(\frac{\partial u_j}{\partial x_j}\right)\delta_{ij} \tag{2.39}$$

A comparison to similar equations for static mechanics shows the viscosity is equivalent to the shear modulus and λ is equivalent to Lamé's first parameter. Similarly, we can define a quantity v which is equivalent to the Poisson's ratio; in the case of a Newtonian fluid, we will refer to such a ratio as the compressibility.

Please note that a Newtonian fluid is more general case of the familiar Navier-Stokes equation. When assuming an isotropic fluid, the local mean pressure P is written as:

$$\bar{P} = -\left(\frac{1}{3}\right)(\sigma_{11} + \sigma_{22} + \sigma_{33}) = p - \left(\lambda + \frac{2\mu}{3}\right)(\nabla \cdot \vec{u}) \tag{2.40}$$

The Stokes hypothesis states that $\lambda + \frac{2\mu}{3} = 0$ and therefore, the local mean pressure is equal to the thermal dynamic pressure. Using the notation of the rate of strain:

$$e_{ij} = \frac{1}{2}\left(\frac{\partial u_i}{\partial x_j} + \frac{\partial u_j}{\partial x_i}\right) \tag{2.41}$$

The stress tensor in a Newtonian fluid can be written as:

$$\begin{aligned} \sigma_{ij} - \sigma_{ij}^{(0)} &= 2\mu e_{ij} + \lambda \delta_{ij} e_{ii} \\ &= 2\mu\left[e_{ij} - \frac{2}{3}\delta_{ij} e_{ii}\right] \end{aligned} \tag{2.42}$$

which is the familiar form of stress tensor used in Stokes-Navier equations. In SUPREM-IV.GS oxidation simulation, the oxidation process is described by the general Newtonian fluid equation rather than the Navier-Stokes approximation.

In a realistic 2/3D TCAD simulation, the viscosity is a function of the stress and the force balance equation (Newton's 2nd law) is non-linear in velocity after discretization over the simulation grid. The simulation can be made even more sophisticated if the stress balance between the solid phase and viscous fluid phase is taken into account.

In summary, the oxidation process simulator (such as SUPREM-IV.GS and CSUPREM), solves for the following variables C_{ox} (oxidant concentration), v_x, v_y, and v_z (denoted as u_i) in the following coupled equations in the fluid phase:

$$\begin{aligned} \vec{\nabla} \cdot \left(D_{ox} \vec{\nabla} C_{ox}\right) &= 0 \\ \frac{\partial \sigma_{ij}}{x_i} &= 0 \end{aligned} \tag{2.43}$$

Where the stress tensor depends on the velocity vector:

$$\sigma_{ij} - \sigma_{ij}^{(0)} = 2\mu e_{ij} + \lambda \delta_{ij} e_{ii}$$

$$e_{ij} = \left(\frac{1}{2}\right)\left(\frac{\partial u_i}{\partial x_j} + \frac{\partial u_j}{\partial x_i}\right) \tag{2.44}$$

Optionally, the static mechanical equations of the silicon substrate can be solved together with the Newtonian fluid to provide more accuracy at the expense of more computation time.

Please note that oxidation is a time-dependent process while none of the above equations explicitly depends on time. This is because time dependence comes in through the movement of the mesh grid and material boundary as determined by the fluid velocities from the above equations.

2.4 Implant Models

2.4.1 Depth Profile Model

There are two approaches to ion implantation modeling. The most accurate approach is based on atomic scale interaction using Monte-Carlo simulation techniques. The second approach is based on analytical fit to experimental SIMS data to build up a large database. Interpolation is used for ion energies not specifically included in the SIMS database. We shall detail the second approach since this is more efficient and more commonly used in practical TCAD projects. We shall refer to the 2nd approach as SIMS-interpolation approach.

The theoretical basis for SIMS-interpolation approach is to decompose the implant scattering source *S* by a product of two functions:

$$S(x_1) = P(d)R(r) \tag{2.45}$$

where x_1 is the entering position (see schematic in Fig. 2.2). *d* is the depth of implant path and *r* is the lateral scattering distance. Here we only consider vertical implant entry because any angle in implant may be eliminated by performing a rotation of the coordinate system.

The doping profile at any observation point (x,y) may be expressed as contribution from lateral scattering of all implant paths coming at the same y-value. That is:

$$D(x,y) = \int P(d(x_1,y))R(x_1 - x)dx_1 \tag{2.46}$$

Please note that the lateral function *R(r)* is normalized so that in the case of flat exposed surface and uniform material, the doping is simply given by *P(d)* as expected.

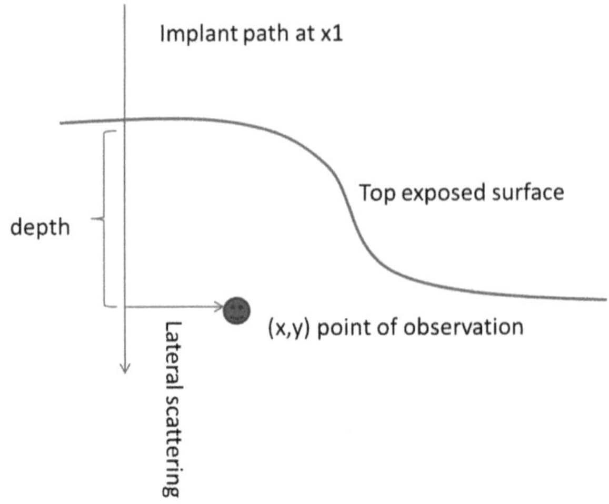

Fig. 2.2 The schematic of entering position

The lateral function *R(r)* has always been assumed to be a simple Gaussian, although some sophisticated process simulator can assume it varies with depth.

Much work in the early days of TCAD were focused on how best to use analytical function to construct the depth function P(d) to represent the real SIMS data. Gaussian functions, Pearson-V functions and dual-Pearson functions with various parameters setting and sophisticated SIMS fitting strategies have been used. The use of various analytical functions originated from limitation in speed and memory of early microcomputers: analytical functions were the only way to quickly evaluate the implant dose in the old days.

When looking back at all the fitting work of the past given today's computation power, the sophisticated fitting means very little since there is no atomic physics contained in those Pearson functions. Should we have to do it all over again, we would simply smooth out the SIMS data and directly use them in the simulation code and today's fast computer will handle the interpolation easily. We just need to keep in mind that when interpolating between different SIMS data, we take the logarithm before interpolating. This is the same as when we interpolate between two Gaussian implant functions: we interpolate the moments (projection range and standard deviation) rather than the Gaussian function itself.

The process simulator contains an option to use an implant table that is built using SIMS data from different energies and for different materials. No fitting of dual-Pearson function is needed and one need not worry whether there is any error out of fitting analytical functions.

2.4.2 Multiple Layer Model

In processing of semiconductor, it is common to implant ions through different materials. For example, a pad oxide layer is commonly formed on top of silicon before implantation. Since the implant depth profiles of the same ions in oxide and silicon are different, it is necessary to consider the situation of implant through multiple layers of materials with different implant properties. We discuss two different models commonly used in the TCAD industry: the dose matching and range matching methods.

For purpose of demonstration, we consider the most simple implant model: Gaussian function model. We normalize the doping function so that it becomes a pure math problem:

$$D(d) = \frac{1}{\sqrt{2\pi\Delta_{rp}^2}} \exp\left[-\frac{\left(\frac{1}{2}\right)\left(d - R_p\right)^2}{\Delta_{rp}^2} \right] \tag{2.47}$$

where R_p is the projected range and Δ_{rp} is the vertical projected standard deviation.

In TCAD simulation the common practice is to shift the origin of the implant function for the 2nd or more layers deeper down using an effective depth t_{eff} so that the deeper layers match better with layers on top. The meaning of t_{eff} is that the implant ions effectively go through a depth of $t_{eff,\ k}$ instead of real depth of

$$t_{tot} = \sum_{i=1}^{k-1} t_i \tag{2.48}$$

The doping function used in kth layer would become

$$P(d) = P_k(d - t_{tot} + t_{eff}) \tag{2.49}$$

In the dose matching method, $t_{eff,\ k}$ for the k-th layer under consideration would be chosen such that the dose deeper down from kth layer plus the dose already consumed in layers 1 to k-1 equals to the total dose injected. It is obvious that dose matching is a simple shift of origin and the function itself does not need renormalization if the implant function was already normalized for single layer implant.

In the range matching method, the effective depth is rescaled from the real depth according to the ranges of the different layers. The purpose is to rescale the whole layer system to match the range of the kth layer:

$$t_{eff,k} = \sum_{i=1}^{k-1} \frac{t_i R_{p,k}}{R_{p,i}} \tag{2.50}$$

We notice that in range matching, since the shift in thickness is not based on consumed dose in previous layers. This means the function of the kth layer would not integrate to the remaining dose and thus it is necessary to rescale all the doping functions.

As we pointed out previously, the implant functions are purely math and contain no atomic physics. The big question is how we should judge the merit of each method.

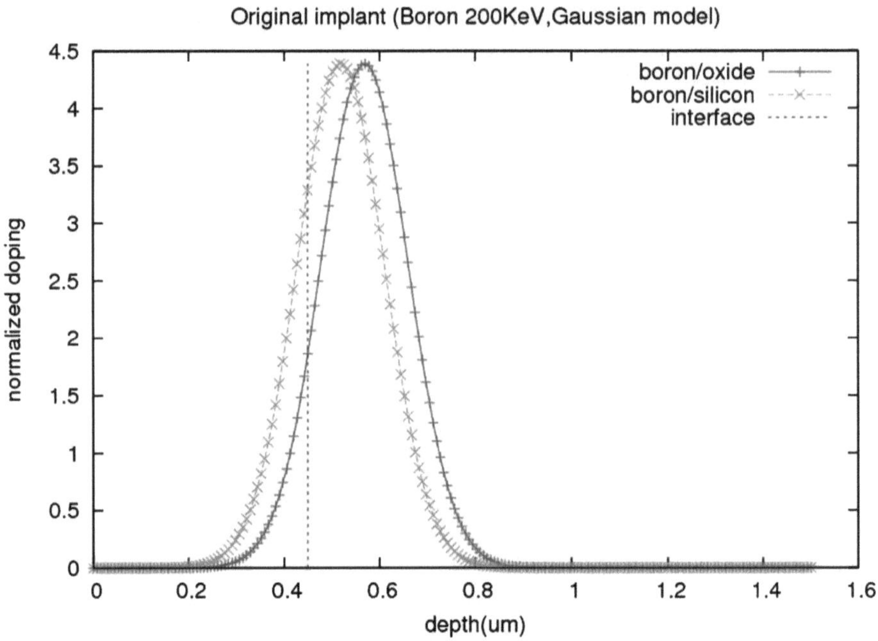

Fig. 2.3 The original implant doping functions for boron in oxide and silicon

The real test would be to experiment with SIMS on single layers and apply the methods on multiple layers, for all ion species and all energies. Another test would be to make a judgment based on smoothness of the multiple layer function. In real experiment, all doping profiles should be smooth and we can judge according to the smoothness of the profile from different methods. Let us take some examples to compare the smoothness of the multiple layer profile as follows.

To be specific, we consider implant energy of 200 keV for boron, phosphorus and arsenic. The implant is done through a thick oxide first, then into the silicon substrate. The projected range and standard deviation come from the implant table of Suprem4.gs. We notice that for boron, the standard deviation is the same for oxide and silicon while it is different for the other two ions. To make the discontinuity more obvious, we choose the oxide thickness to be smaller than but close to the projected range.

Figure 2.3 shows the original doping functions for boron in oxide and silicon. The functions are the same but there is a shift in the projected range. When we apply the two different methods, dose matching shows perfect smoothness as illustrated in Fig. 2.4. This can be explained as follows: since function shapes are identical, a shift in the silicon function according to remaining dose finds the connection point at the same value as the previous function before it is truncated.

For both cases of phosphorus and arsenic, the standard deviation in silicon is larger than in oxide as shown in Figs. 2.5 and 2.6. When we apply both methods and

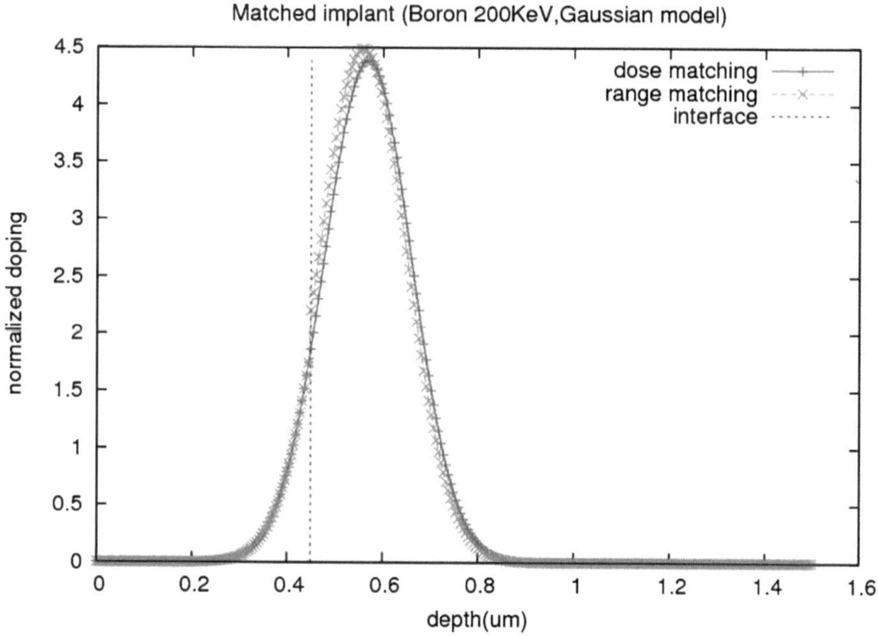

Fig. 2.4 The matched implant doping functions for boron in oxide and silicon

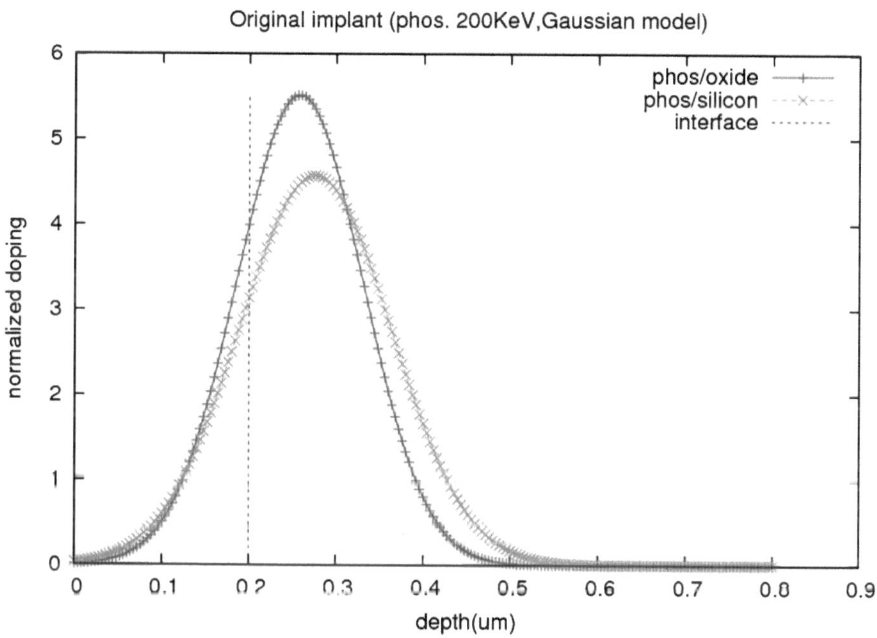

Fig. 2.5 The original implant doping functions for phosphorus in oxide and silicon

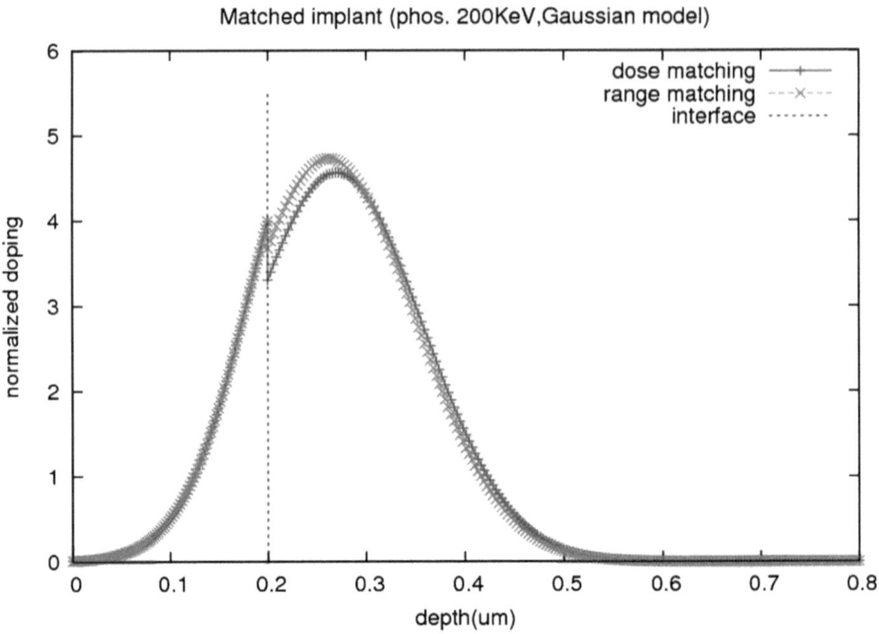

Fig. 2.6 The original implant doping functions for arsenic in oxide and silicon

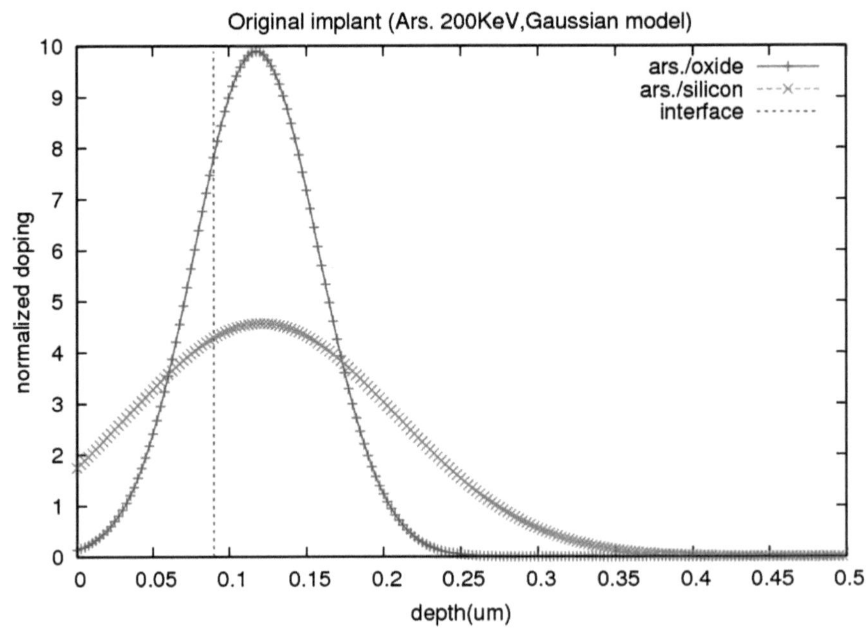

Fig. 2.7 The matched implant doping functions for phosphorus in oxide and silicon

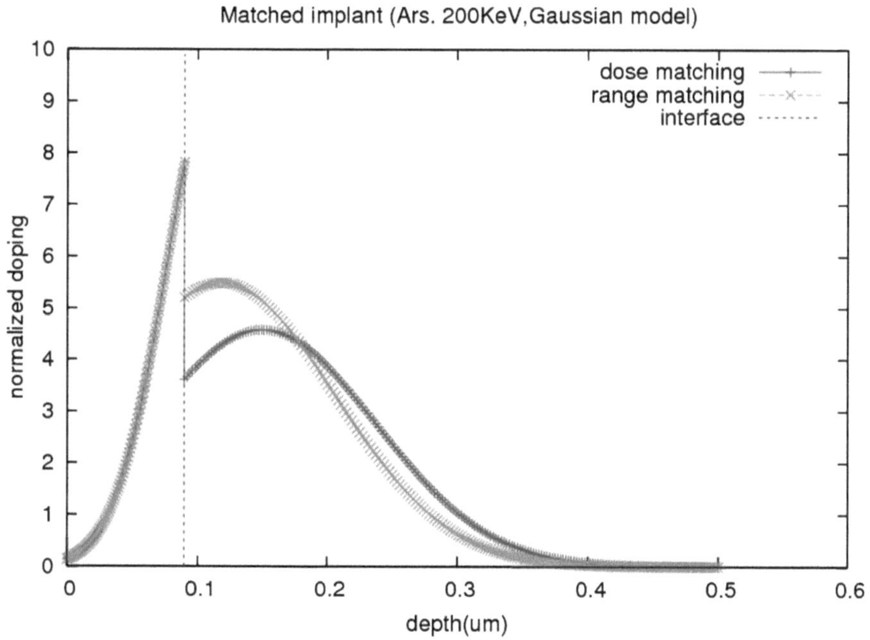

Fig. 2.8 The matched implant doping functions for arsenic in oxide and silicon

compare, we find range matching yields better smoothness as indicated in Figs. 2.7 and 2.8.

We therefore conclude that based on smoothness alone, it is better to use range matching method if standard deviation of different materials are significantly different. On the other hand, dose matching generates more smooth profile if standard deviation of different materials is similar in values.

2.5 Etch Model

The etch model in SUPREM-IV.GS was initially written as a purely geometric operation where a TCAD user define geometric shape to remove from the structure and the TCAD process simulator would cut off the mesh points accordingly.

Some efforts have been made to associate the etching operation with the actual physics of etching. Compared with other models such as diffusion and oxidation, these physically based etch models are relatively simple. We shall give an example of etch model based on parameters of reactive ion etch (RIE).

In a reactive ion etch process, the unmasked silicon surface is subjected to high energy ion bombardment which knocks out the material being exposed. The parameters we obtain from RIE process are etching rates in vertical direction and in all directions (isotropic rate). The isotropic etching originates from the

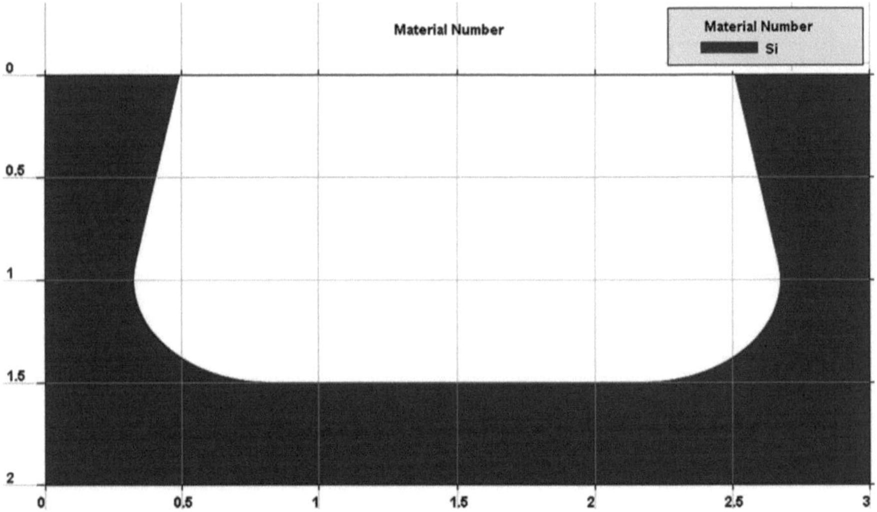

Fig. 2.9 The RIE etch simulation result

randomness of ion bombardment which causes the unmasked area to be etched sideways.

The following commands in the process simulator can be used to generate the RIE etching geometry profile:

Process Simulation Code

```
line x loc= 0.00000 spacing= 0.2 tag=lft
line x loc= 3.00000 spacing= 0.2 tag=rht
line y loc= 0.00000 spacing= 0.15 tag=top
line y loc= 2.00000 spacing= 0.15 tag=bot
region silicon xlo=lft xhi=rht ylo=top yhi=bot
bound exposed xlo=lft xhi=rht ylo=top yhi=top
bound backside xlo=lft xhi=rht ylo=bot yhi=bot
init
set_etch_geometry profile_type=1 x_length=1 vertical_rate=0.01
vertical_time=100 left_angle=etch silicon
tag_etch_geometry=label_1 shift.x=1.
struct outf=final.str
quit
```

In the above, a mock RIE condition is created to demonstrate the generation of geometry RIE etching profile. The vertical rate and vertical time would determine how deep etching would reach in the vertical direction. The parameters `isotropic_rate=0.01 isotropic_time=50` determines the etching in all directions, i.e., sideways and in any other angle. This is added on the vertical rates and the side way etching distance is used as a radius in making the round corner, as shown in Fig. 2.9. The spacing parameters are used to control the density of the mesh along the etching contour.

Since the etch feature in a process simulator is mostly geometrical, the process simulator goes one step further to make the geometry editing more convenient: to combine etch and deposit in the same command. Such an etch-then-fill-up action is called change material in the process simulator. When using change material, whatever mesh being removed would be filled up by a different material. Please refer to Chap. 4 for details about change material.

2.6 Deposit Model

The deposit model in SUPREM-IV.GS is purely geometrical. The geometric task is to find a new exposed surface above the original exposed top surface. If the original exposed surface is flat, the task is trivial and the original surface just translates upwards by the deposit thickness to form the new surface.

If the original surface contains steps, slopes or holes, it becomes a little complicated. A process simulator usually produces a new top surface that is smoother than the original. A method of generating smooth deposit surface is due to SUPREM-IV. GS and it is described as follows.

The process simulator scans the mesh points of the original surface. If it detects the adjacent mesh points form a flat line, a simple shift of original mesh point upwards is performed. If it detects a change of angle between adjacent points, it would form two vectors n1 and n2, as shown in Fig. 2.10, which are normal to the adjacent connect lines to the left and right, respectively.

The lengths of the vectors n1 and n2 are the same as deposit thickness. Then, a series of mesh points following an arc connecting the vector tips of n1 and n2 would be produced to construct a smooth top deposit surface line.

2.7 Process Simulation Examples

Here are a few simple examples to familiarize the reader with 3D structures built by the process simulator.

2.7.1 Ion Implantation Example

A concentric square shaped implantation example is shown in Fig. 2.11 with 2D center cut.

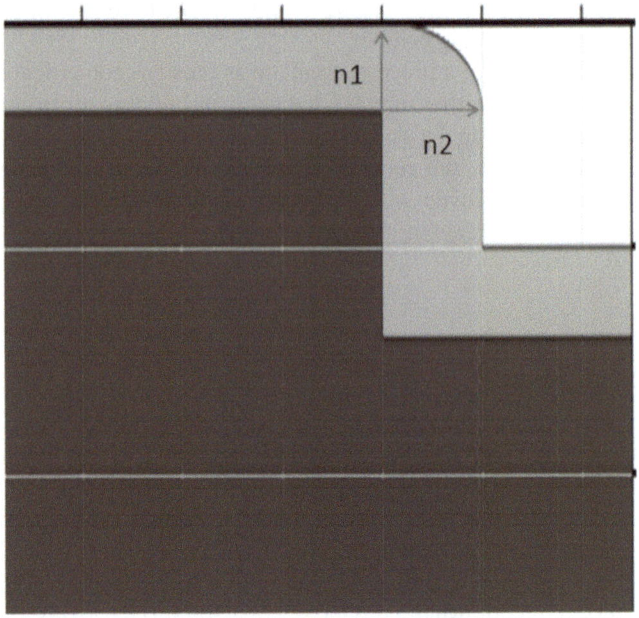

Fig. 2.10 Two vectors formed by the process simulator during deposition

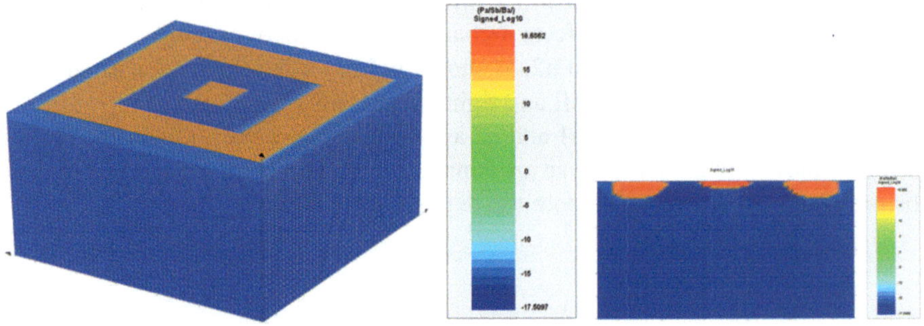

Fig. 2.11 The doping concentration showing in 3D and 2D center cut

2.7.2 Diffusion Example

The `diffuse` command can be used but not limited to:

- Activate the dopants (anneal)
- Drive in an implanted well
- Diffuse a dopant gas through the opening
- Grow dry or wet oxide on top of silicon

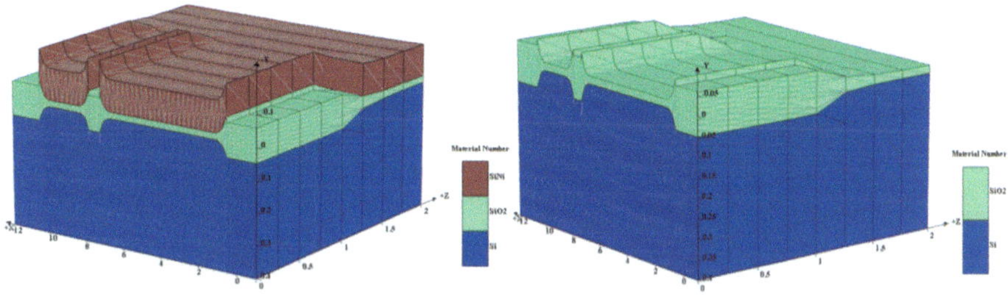

Fig. 2.12 LOCOS growth before (*left*) and after (*right*) nitride is removed

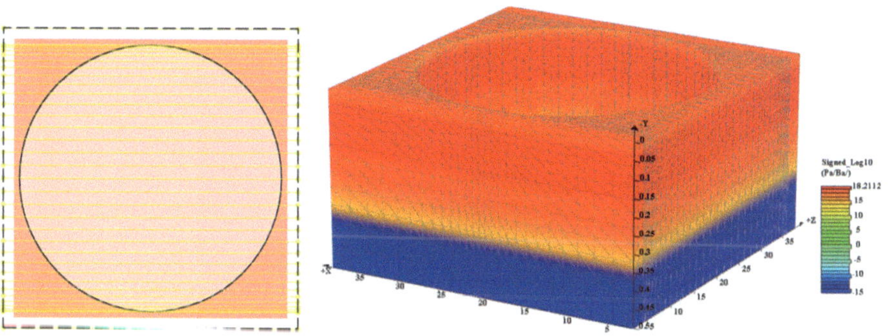

Fig. 2.13 Circular etch mask layout with cut lines (*left*) and net doping chart (*right*)

A simple 3D thermal oxide growth example is presented here. Figure 2.12 shows LOCOS growth before and after nitride is removed using full 3D simulation.

2.7.3 Etch Example

In this example, a circular etch is performed with blanket implant. The initial silicon substrate is set to have constant doping of boron, with a doping concentration of 1E + 15 cm^{-3}. Phosphorus is implanted to revert the top toping to n-type. Figure 2.13 illustrates the etch mask layout with cut lines by MaskEditor and the 3D structure of circular etch. MaskEditor is a graphic user interface tool which we use to generate the implant mask. We shall have more in-depth explanation of MaskEditor later in this book.

Chapter 3
Advanced Theory of TCAD Device Simulation

This chapter reviews the basic theories of device simulation within the framework of TCAD. Figure 3.1 shows a practical design of a device simulator of 3D TCAD capability with various modules. One may regard the drift-diffusion (DD) equation module as central building block of a 3D TCAD device simulator. Optionally, additional modules to perform quantum mechanical calculations and optical modes computation can be built around the main DD module to enhance the application of the simulation program. This is necessary for special applications of 3D TCAD such as nano-scale MOSFET, laser diodes and integrated photonic circuits.

3.1 Basic Equations

The basic equations [16] used to describe the semiconductor device behavior are the Poisson equation,

$$-\vec{\nabla} \cdot \left(\frac{\varepsilon_0 \varepsilon_{dc}}{q} \vec{\nabla} V \right) = -n + p + N_D(1 - f_D) - N_A f_A + \sum_j N_{tj}(\delta_j - f_{tj}) \quad (3.1)$$

and the current continuity equations for electrons and holes,

$$\vec{\nabla} \cdot J_n - \sum_j R_n^{tj} - R_{sp} - R_{st} - R_{au} + G_{opt}(t) + G_{imp} = \frac{\partial n}{\partial t} + N_D \frac{\partial f_D}{\partial t} \quad (3.2)$$

$$\vec{\nabla} \cdot J_p - \sum_j R_p^{tj} + R_{sp} + R_{st} + R_{au} - G_{opt}(t) - G_{imp} = -\frac{\partial p}{\partial t} + N_A \frac{\partial f_A}{\partial t} \quad (3.3)$$

S. Li and Y. Fu, *3D TCAD Simulation for Semiconductor Processes, Devices and Optoelectronics*, DOI 10.1007/978-1-4614-0481-1_3,

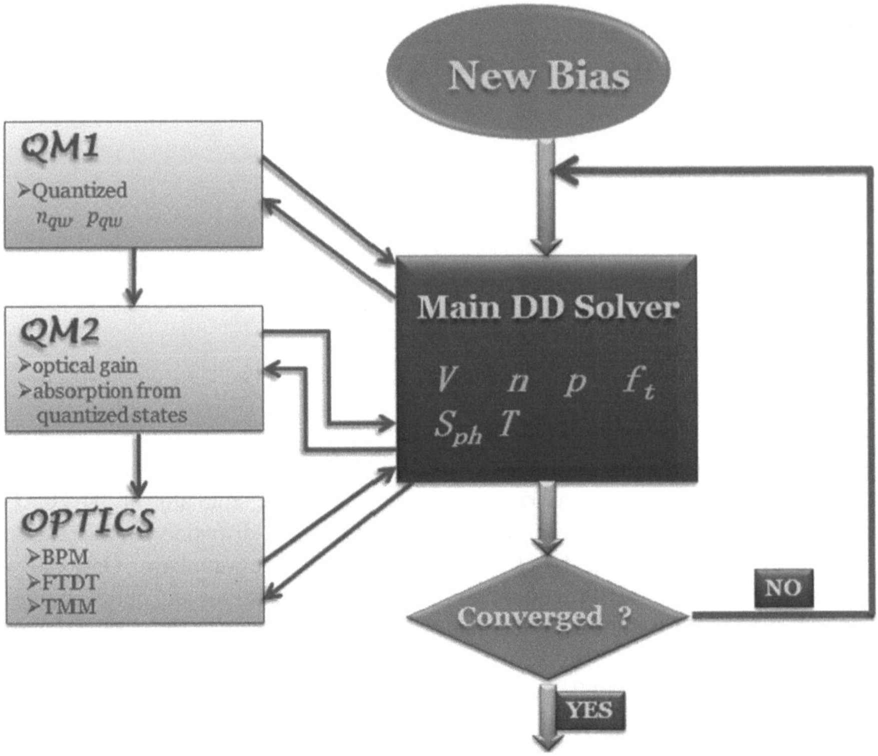

Fig. 3.1 A practical design of a device simulator of 3D TCAD capability with various modules

These equations govern the electrical behavior (e.g., I-V characteristics) of a semiconductor device. The recombination terms are as indicated by their subscripts, namely: recombination due to traps (SRH), spontaneous emission, stimulated emission, Auger recombination, optical generation and impact ionization generation.

Stimulated recombination and optical generation are similar in that they both depends on the local photon density and driven by optical field. In most applications such as pumped laser, the two processes involve different wavelengths of light of different sources. In some special application such as resonant cavity light emitting diode, they are of the same source (self-pumping) and of the same wavelength.

Optionally, two more equations are used to describe the carrier energy ω or the carrier temperature distribution. The carrier temperature is a description of how the carrier distribution deviates from the Fermi-Dirac distribution and should not be confused with the lattice temperature. Such a model is also referred to as the hydrodynamic model and the most common derivation is that of Azoff [20, 21]. For simplicity, we only present the equations for hot electrons. The corresponding equations for hot holes are completely analogous.

$$\vec{\nabla} \cdot \vec{S} + R_n\omega - \vec{\nabla}E_c \cdot \vec{J_n} + \frac{n(\omega - \omega_0)}{\tau_\omega} + \frac{\partial(n\omega)}{\partial t} = 0 \tag{3.4}$$

$$\vec{S} = -\frac{5}{3}\vec{J_n}\omega - \frac{10}{9}\mu_n n\omega\vec{\nabla}\omega \tag{3.5}$$

In the above, ω is the total energy of an electron and $\omega_0 = 3kT/2$ is the electron energy at equilibrium. $\vec{S}$ is the electron energy flux and τ_ω is the energy relaxation time.

The primary function of a TCAD device simulation program is to solve the above equations self-consistently for the electrostatic potential, V, the electron and hole concentrations, n and p, respectively, the electron and hole energy, W_n and W_p, respectively.

It is known from basic semiconductor theory [16] that the carrier flux densities $\vec{J_n}$ and $\vec{J_p}$ in (3.2) and (3.3) can be written as functions of the carrier concentration and the quasi-Fermi levels:

$$\vec{J_n} = n\mu_n\vec{\nabla}E_{fn} \tag{3.6}$$

$$\vec{J_p} = p\mu_p\vec{\nabla}E_{fp} \tag{3.7}$$

where μ_n and μ_p are mobilities of electrons and holes, respectively. For convenience, we use carrier flux density and current density interchangeably even though they differ by a factor of electron charge.

For hydrodynamic model, the expression for the electron (or hole) current is modified:

$$\vec{J_n} = \mu_n\left\{-n\vec{\nabla}[\psi + \chi + \gamma_n] + \frac{2}{3}\vec{\nabla}(n\omega) - n\omega\vec{\nabla}\ln(m_n)\right\} \tag{3.8}$$

Carrier recombination due to deep level traps are also termed the Shockley-Read-Hall (SRH) recombination and it is described by the following terms for electrons and holes, respectively,

$$R_n^{tj} = c_{nj}nN_{tj}(1 - f_{tj}) - c_{nj}n_{1j}N_{tj}f_{tj}$$

$$R_p^{tj} = c_{pj}pN_{tj}f_{tj} - c_{pj}p_{1j}N_{tj}(1 - f_{tj}) \tag{3.9}$$

where n_{1j} is the electron concentration when the electron quasi-Fermi level coincides with the energy level E_{tj} of the jth trap. A similar definition applies to p_{1j}. Under transient conditions, the following trap dynamic equation is valid [22]:

$$N_{tj}\frac{\partial f_{tj}}{\partial t} = R_n^{tj} - R_p^{tj} \tag{3.10}$$

The capture coefficients c_{nj} and c_{pj} for electrons and holes relates to the lifetime of the carrier due the jth recombination center by the following relation:

$$\frac{1}{\tau_{nj}} = c_{nj} N_{tj} \tag{3.11}$$

$$\frac{1}{\tau_{pj}} = c_{pj} N_{tj} \tag{3.12}$$

One can show that (3.9)–(3.12) reduce to the familiar SRH formula [16] under steady state condition.

The capture coefficients can be further expressed as:

$$c_{nj} = \sigma_{nj} \overline{v_n} \tag{3.13}$$

$$\overline{v_n} = \sqrt{\frac{8kT}{\pi m_n}} \tag{3.14}$$

$$c_{pj} = \sigma_{pj} \overline{v_p} \tag{3.15}$$

$$\overline{v_p} = \sqrt{\frac{8kT}{\pi m_p}} \tag{3.16}$$

A trap (or recombination center) is completely specified by its density N_{tj}, capture cross sections σ_{nj} and σ_{pj}, and energy level E_{tj}.

The Auger recombination rate is given by:

$$R_{au} = (C_n n + C_p P)(np - n_i^2) \tag{3.17}$$

where the Auger coefficients C_n and C_p depends on the type of material simulated.

Similarly, the spontaneous emission rate is written as:

$$R_{sp} = B(np - n_i^2) \tag{3.18}$$

where the B coefficient also depends on the material. Note that for regions where the optical gain is important (e.g. the active region of laser diodes and LEDs), the spontaneous emission spectrum should be explicitly computed and used to obtain the total spontaneous emission rate instead.

3.2 Fermi Statistics

The electron and hole concentrations in semiconductors are defined by Fermi-Dirac distributions and a parabolic density of states which, when integrated, yield [16]:

$$n = N_c F_{1/2}\left(\frac{E_{fn} - E_c}{kT}\right) \tag{3.19}$$

$$p = N_v F_{1/2}\left(\frac{E_v - E_{fp}}{kT}\right) \tag{3.20}$$

where $F_{1/2}$ is the Fermi integral of order one-half. For the convenience of numerical evaluation, the approximation proposed by Bednarczyk and Bednarczyk is used [23]:

$$F_{1/2}(x) \approx (e^{-x} + \xi(x))^{-1} \tag{3.21}$$

$$\xi(x) = \frac{3}{4}\sqrt{\pi}[v(x)]^{3/8} \tag{3.22}$$

$$v(x) = x^4 + 50 + 33.6x\{1 - 0.68\ \exp[-0.17(x+1)^2]\} \tag{3.23}$$

This expression is accurate to within 0.4% of error in all ranges. A more accurate and numerically friendly expression has also been developed by Li et. al. [24].

In the limit of low carrier concentration equations (3.19) and (3.20) reduce to the familiar Boltzmann statistics:

$$n = N_c \exp\left(\frac{E_{fn} - E_c}{kT}\right) \tag{3.24}$$

$$p = N_v \exp\left(\frac{E_{fn} - E_c}{kT}\right) \tag{3.25}$$

The program uses the more general Fermi-Dirac statistics of Eqs. 3.19–3.21 by default.

3.3 Dopant Ionization

The simulation program can accurately account for the incomplete ionization of shallow impurities in semiconductors. The occupancies f_D and f_A are used to describe the degree of ionization. It is assumed that the shallow impurities are in equilibrium with the local carriers and therefore the occupancy of the shallow impurities can be described by

$$f_D = \frac{1}{1 + g_d^{-1} \exp\left[\frac{E_D - E_{fn}}{kT}\right]} \tag{3.26}$$

$$f_A = \frac{1}{1 + g_a \exp\left[\frac{E_D - E_{fn}}{kT}\right]} \tag{3.27}$$

where the subscripts D and A are used to denote shallow donors and acceptors, respectively.

As discussed previously, the occupancy of a deep level trap can be determined through the trap dynamic equation (3.10). In general, deep traps are not in equilibrium with the carriers: the traps do not share the same quasi-Fermi level as the carriers. From (3.9), and (3.10), one obtains the following expression for the trap occupancy under steady state conditions:

$$f_{tj} = \frac{c_{nj} n + c_{pj} p_{1j}}{c_{nj}(n + n_{1j}) + c_{pj}(p + p_{1j})} \tag{3.28}$$

In the case of surface states or surface recombination centers, the software allows for the distribution of dense traps near the surface region. This provides a mechanism for the surface charge states as well as for surface recombination. Fermi level pinning effects on a semiconductor surface can be modeled using this approach. In a transient simulation the trap occupancy is a function of time, depending on the trap capture rates as well as on the local carrier concentrations. The program uses Eq. 3.10 to determine the trap states at each transient time step.

The Poole-Frenkel effect (also Frenkel-Poole effect or field induced emission) was originally used to describe field dependent thermionic emission from traps in the bulk of an insulator [16]. This mechanism is, however, equally applicable to incomplete ionization of impurities in semiconductors.

Under an electric field F, the electrostatic potential near an impurity center is modified and the effective work function or the ionization energy is reduced by:

The Poole-Frenkel model is implemented by shifting the ionization energy by $\Delta E_{PF.}$

$$\Delta E_{PF} = \sqrt{\frac{qF}{\pi \epsilon_0 \epsilon}} \tag{3.29}$$

The Poole-Frenkel model is important when the ionization energy is large and the temperature is low (e. g., 100 meV at 100 K). Without such a model, the semiconductor may have an unrealistic high resistance.

Another important ionization model is the Mott transition model for highly doped semiconductors. For most shallow dopants, high doping causes the material undergo a Mott transition so that more dopants are ionized than the incomplete ionization model described previously. The Mott transition model is described by a shift in the ionization energy to the negative direction to cause more ionization (see for example [25])

$$E_D = E_{D0}\left[1 - \left(\frac{N_D}{N_{crit}}\right)^{1/3}\right] \tag{3.30}$$

Where N_{crit} is a critical density determined by the Bohr radius as follows.

$$a_{br} = 0.529177^{-10}\left(\frac{\varepsilon_r}{m^*}\right)$$

$$N_{crit} = \left(\frac{1}{4a_{br}}\right)\left(\frac{\pi}{3}\right)^{1/3} \tag{3.31}$$

3.4 Carrier Mobility

The carrier mobilities μ_n and μ_p account for the scattering mechanism in electrical transport. In general the mobility is a function of the electrical field [16]. Common analytical formulas of field dependent mobility are described as follows.

1. The simplest mobility model uses constant mobilities μ_{0n} and μ_{0p} for electrons and holes, respectively, throughout each material region in the device.
2. Another simplified field dependent mobility model is the two-piece mobility model:

$$\mu_n = \mu_{0n},\ for\ F<F_{0n} \tag{3.32}$$

$$\mu_n = v_{sn}/F,\ for\ F \geq F_{0n} \tag{3.33}$$

$$v_{sn} = \mu_{0n}F_{0n} \tag{3.34}$$

for the electron mobility. F_{0n} is a threshold field beyond which the electron velocity saturates to a constant. Similar expressions can be defined for holes:

$$\mu_p = \mu_{0p},\ for\ F<F_{0p} \tag{3.35}$$

$$\mu_p = \mu_{0p}F_{0p}/F,\ for\ F \geq F_{0p} \tag{3.36}$$

$$v_{sp} = \mu_{0p} F_{0p} \tag{3.37}$$

3. A commonly used mobility model has the following form for electrons and holes, respectively:

$$\mu_n = \frac{\mu_{0n}}{\left(1 + \left(\frac{\mu_{0n} F}{v_{sn}}\right)^{\beta_n}\right)^{\frac{1}{\beta_n}}} \tag{3.38}$$

$$\mu_p = \frac{\mu_{0p}}{\left(1 + \left(\frac{\mu_{0p} F}{v_{sp}}\right)^{\beta_p}\right)^{\frac{1}{\beta_p}}} \tag{3.39}$$

4. Many III-V compound semiconductors (e.g., GaAs) exhibit negative differential resistance due to the transition of carriers into band valleys with lower mobility [16]. The software has implemented the following field dependent model for this case:

$$\mu_n = \frac{\mu_{0n} + \left(\frac{v_{sn}}{F_{0n}}\right)\left(\frac{F}{F_{0n}}\right)^3}{1 + \left(\frac{F}{F_{0n}}\right)^4} \tag{3.40}$$

5. Poole-Frenkel type of field enhanced mobility model. Highly localized carriers with small mobility can be excited by external field to make hopping movements. Commonly used for organic semiconductors, the following mobility model is implemented.

$$\mu = \mu_0 \exp\left[\left(\frac{F}{F_{cr}}\right)^{px}\right]$$

Besides the field dependence of the mobility, another important effect is the impurity dependence of the low field mobility [25]. The program uses the following formulas:

$$\mu_{0n} = \mu_{1n} + \frac{(\mu_{2n} - \mu_{1n})}{1 + \left(\frac{N_D + N_A + \sum_j N_{tj}}{N_{rn}}\right)^{\alpha_n}} \tag{3.41}$$

$$\mu_{0p} = \mu_{1p} + \frac{(\mu_{2p} - \mu_{1p})}{1 + \left(\frac{N_D + N_A + \sum_j N_{tj}}{N_{rp}}\right)^{\alpha_p}} \tag{3.42}$$

where the parameters μ_{1n} and μ_{2n} are fitting parameters from experimental data.

In many applications, the mobility is anisotropic and also depends on the vertical field. The effect of vertical field is usually expressed as an addition surface scattering term to the inverse of the mobility (thus causing a decrease to it). A commonly used vertical field dependent model is due to Lombardi [26] which combines various contribution to mobility as follows:

$$\frac{1}{\mu} = \frac{1}{\mu_{ac}} + \frac{1}{\mu_{srf}} + \frac{1}{\mu_0} \tag{3.43}$$

Where μ_0 is the mobility due to longitudinal field/hot-carrier effect. The other terms are as follows.

$$\mu_{ac} = \frac{B}{E_\perp} + \frac{\alpha N^\beta}{T_L E_\perp^{\frac{1}{3}}} \tag{3.44}$$

$$\mu_{srf} = \frac{\delta}{E_\perp^2} \tag{3.45}$$

3.5 Impaction Ionization

An important application of 3D TCAD is the design of smart high power devices with high breakdown voltages. Therefore, understanding the influence of impact ionization on the breakdown is necessary.

Impact ionization appears as a carrier generation term in the basic equations of (3.2) and (3.3) which can be explicitly written as [16]:

$$G = \alpha_n n v_n + \alpha_p p v_p \tag{3.46}$$

where α_n is the electron ionization rate defined as the number of electron-hole pairs generated by an electron per unit distance traveled; α_p is similarly defined for holes. Both α_n and α_p are strongly dependent on the electric field.

Equation 3.46 is somewhat difficult to implement in a drift-diffusion model because it is not directly concerned with the velocity and considers the mobility coefficient instead. Since impact ionization occurs only in high field region where the drift terms dominates, Selberherr [27] suggested the use of $\frac{J}{q}$ in place of nv for ease of implementation:

$$G = \alpha_n \frac{J_n}{q} + \alpha_p \frac{J_p}{q} \tag{3.47}$$

The issue remains how to choose values of α_n and α_p. A commonly used expression was proposed by Chynoweth [28] as follows:

$$\alpha = \alpha_n^{\infty} \exp\left[-\left(\frac{F_{cn}}{F}\right)\right] \tag{3.48}$$

This formula was later generalized to the following [29]:

$$\alpha = \alpha_n^{\infty} \exp\left[-\left(\frac{F_{cn}}{F}\right)^{k_n}\right] \tag{3.49}$$

Similar expression can be written down for the holes. Usually, the above formulas is valid for a certain field range and multiple parameter sets are needed to cover the entire useful range for a simulation.

Another theory of impact ionization is called the Baraff's three-parameter theory which takes into account the temperature dependence. A convenient formula for Baraff's theory was given by Crowell and Sze [30]:

$$\alpha\lambda = \exp[g(r,x)] \tag{3.50}$$

$$g(r,x) = (11.5r^2 - 1.17r + 3.9 \times 10^{-4})x^2$$

$$(46r^2 - 11.9r + 1.75 \times 10^{-2})x \tag{3.51}$$

$$- 757r^2 + 75.5r - 1.92 \tag{3.52}$$

where

$$r = <E_p>/E_I \tag{3.53}$$

and

$$x = E_I/qF\lambda \tag{3.54}$$

$<E_p>$ is the optical phonon energy and λ is the optical phonon scattering mean free path. E_I is the ionization energy for which the following formula was suggested [30]:

$$E_I = 3E_g/2 \tag{3.55}$$

The temperature dependence of $< E_p >$ and λ are given by

$$<E_p> = E_p \tanh\left(\frac{E_p}{2kT}\right) \tag{3.56}$$

$$\lambda = \lambda_0 \tanh\left(\frac{E_p}{2kT}\right) \tag{3.57}$$

To summarize, the Chynoweth model depends on α_n^{∞} (ionization rate at infinite field), F_{cn} (critical field) and k_n (field exponent). Baraff's model requires the zero temperature mean free path λ_0 and the optical phonon energy E_g. Similar quantities are needed for the holes.

3.6 Effect of Quantization

Modern semiconductor devices most likely have parts that are comparable in size to the quantum mechanical wavelength of the carriers and therefore quantization effects must be taken into account to achieve better accuracy. The common form of quantization is to utilize quantum wells to form a two-dimensional (2D) gas of carriers with 2D density of states (DOS) and quantized levels. We shall provide some details on how such quantization effect can be incorporated into a TCAD or 3D TCAD simulation software. For simplicity, we shall focus on quantum well (QW) and multiple quantum well (MQW) systems here. The same general principles (but different DOS) apply to quantum wires and quantum dots.

To incorporate MQW model, there are different levels of sophistication corresponding to different computation efficiencies. We shall describe a few implementations of the MQW model into TCAD software here, starting from the most simple form of isolated MQW with flat-band density of states.

3.6.1 Simple Quantum Wells

When a confining potential well gets narrow, it is necessary to treat it as a quantum well. We consider the case of a simple quantum well without an externally applied field. All the quantum levels for the band valleys (eg., sixfold conduction band valley in silicon, Γ-band in GaAs, light holes and heavy holes) are computed from well known formulas in quantum mechanics for a square quantum well [31] using a different effective mass and the appropriate confining potential for each parabolic band valley.

When an external field is applied, we allow the quasi-Fermi level to vary as a function of distance. The density of states and quantum levels are assumed to be the same as if there were no applied field. Such a flat-band, parabolic has the advantage of having to compute the quantization effect once only in a simulation.

3.6.2 Carrier Concentration in Quantum Wells

We use the following expression for the density of electrons and holes in a quantum well:

$$n = \sum_j \rho_j^0 kT\ ln\left[1 + e^{\frac{E_{fn}-E_j}{kT}}\right] + unconfined\ electrons \tag{3.58}$$

$$p = \sum_i \rho_i^0 kT\ ln\left[1 + e^{\frac{E_i-E_{fp}}{kT}}\right] + unconfined\ holes \tag{3.59}$$

where the subscript i denotes all confined states for the different hole bands, and j designates those for the Γ and L bands. The number of unconfined carriers are calculated using Fermi statistics as described earlier in this chapter.

For the simple quantum model, we assume that the electron states outside the quantum wells can be treated as unbounded states so that three dimensional Fermi statistics can be applied to these regions for the carrier density.

3.6.3 Anisotropic Parabolic Approximation

Inclusion of strain in a quantum-well has become a common practice for many heterojunction devices and nano-scale MOSFET . Strain offers an additional degree of freedom and produces some desirable effects, such as a lower threshold current for a laser diode.

Strain is known to cause the valence band of a III-V semiconductor to split into separate light hole (LH) and heavy hole (HH) bands (or CH for wurtsite structure) which are strongly non-parabolic. A rigorous treatment of strain effects on the band structure is rather complicated numerically costly so some useful simplifications are in order.

One can show that with proper choices of parameters, a good approximation to the non-parabolic band structure can be obtained using anisotropic parabolic bands. This approximation greatly simplifies the calculation of gain, spontaneous emission and carrier concentration, and it allows one to incorporate strain effects into the TCAD program. The following discussion mostly applies to compound materials while strained silicon can be treated in the same manner using effective masses with anisotropy.

For zinc-blende structure, a simplified analytical band structure of a strained quantum well has recently been developed [32] using an efficient decoupling method to transform the original 4×4 valence band Hamiltonian into two blocks of 2×2 upper and lower Hamiltonians. As a result of the decoupling, an analytical expression can be derived [32]. To be consistent with convention in 2D simulations of semiconductor devices, the y-axis (equivalent to the z-axis in [32]) is defined to be perpendicular to the quantum-well plane and the z-axis to be along the direction of light propagation.

We use the following expression for the bulk valence band energy [32]:

$$-E = \frac{\hbar^2 \gamma_1}{(2m_0)}\left(k_x^2 + k_y^2\right) \tag{3.60}$$

$$\pm\left[\left(\frac{\hbar^2\gamma_2}{2m_0}\left(k_x^2 - 2k_y^2\right) + \zeta\right)^2 + 3\left(\frac{\hbar^2\gamma_2}{2m_0}k_x^2\right) + 12\left(\frac{\hbar^2\gamma_2}{2m_0}\right)^2 k_x^2 k_y^2\right]^{1/2} \tag{3.61}$$

$$\varepsilon = \frac{\alpha_0 - \alpha(x)}{\alpha_0} \tag{3.62}$$

$$\delta E_{sh} = b\left(1 + \frac{2c_{12}}{c_{11}}\right)\varepsilon \tag{3.63}$$

$$\zeta = \frac{1}{2}\delta E_{sh} \tag{3.64}$$

The dispersion of the valence band is a function of the transverse momentum in the k_{xy} plane. The information on strain is contained in ζ; compressive strain is represented by a negative ζ.

The valence band mixing effect for bulk material has been taken into account since the coupling between heavy and light holes has been included in the above equations.

Note that the in-plane direction for (3.61) is the [100] crystal direction, and is different from the [110] direction. For many applications, it is desirable to average the energy dispersion in both the [100] and [110] directions. One can show that the average can be approximated by replacing γ_2 in the second term under the square root of (3.61) by $\gamma_2 + \gamma_3/2$. This is called the axial approximation. The simulator provides an option to turn on this approximation.

To convert the non-parabolic bulk band structure into a suitable form, we fit the valence band to the following anisotropic parabolic band expression, over a range large enough to cover all the optical transitions in k-space:

$$E = -\frac{\hbar^2}{2m_{vx}m_0}k_x^2 - \frac{\hbar^2}{2m_{vy}m_0}k_y^2 \tag{3.65}$$

The quality of the fit is found to be reasonable for practical cases.

Such an approximation is appropriate not only because of the reasonable quality of the fit, but also because the approximation has not lost the basic effects of strain, *i.e*, it causes the symmetry in the band structure to be broken. The splitting of the heavy and light hole bands is described as [32]:

$$\delta E_{hy} = 2a\left(1 - \frac{c_{12}}{c_{11}}\right)\varepsilon \tag{3.66}$$

$$E_g^{hh} = E_g^u + \delta E_{hy} - \delta E_{sh} \tag{3.67}$$

$$E_g^{lh} = E_g^u + \delta E_{hy} + \frac{1}{2}\left[\delta E_{sh} + \Delta - \sqrt{\Delta^2 + 9\delta E_{sh}^2 - 2\delta E_{sh}\Delta}\right] \tag{3.68}$$

Once the band structure is simplified to an anisotropic parabolic form, the program uses a one-band model (that is, without band-valley coupling) to solve for the subbands of the quantum well from the effective masses perpendicular to the quantum well plane.

This approach of a one-band model with parabolic subbands has been used extensively for many years in compound semiconductors optoelectronics (see for example, [33], [34]).

Once the parabolic subbands are found, one can apply conventional approaches to treat the carrier concentration and the optical transition probabilities. Specifically, the effective mass perpendicular to the plane (m_{vy}) determines the quantum subband levels (or quantum confinement effects) and the optical transition energies. The 2D density of states (DOS) and joint density of states (JDOS) of each subband depends on the effective mass in the plane (m_{vx}).

3.6.4 Carrier Density in Anisotropic Parabolic Approximation

We use the effective masses parallel to the plane of the quantum well for the evaluation of carrier concentrations.

$$n = \sum\nolimits_j \rho_j^{x0} kTln\left[1 + e^{\frac{E_{fn}-E_j}{kT}}\right] + unconfined\ electrons \tag{3.69}$$

$$p = \sum\nolimits_i \rho_i^{x0} kTln\left[1 + e^{\frac{E_i-E_{fp}}{kT}}\right] + unconfined\ holes \tag{3.70}$$

where the subscript i denotes all confined states for the heavy and light holes, and j designates those for the Γ and L bands. The unconfined carriers are calculated using Fermi statistics as described in Chap. 2. Note that the effective mass perpendicular to the plane, (m_{vy}), affects the quantum levels and therefore the distribution of carriers in the quantum well.

3.6.5 Valence Mixing and k.p Theory

More recent theories of quantum well subbands are mostly based on k.p theory with valence band mixing effects. These usually involve solving a 4 × 4 [32], 6 × 6

[35] or 8 × 8 [36, 37] Hamiltonian of the Luttinger-Kohn type, and imposing an envelope function approximation in solving the quantum well subband structures (see, for example, [38]). The valence subband structures obtained from this approach are complicated and heavily mixed in many cases. For some cases, the effective masses of some of the hole subbands are negative at the Γ point in k-space, making it impossible to use analytical approaches.

Generation of the subbands using a valence mixing model is not difficult and has been done previously by many authors [32, 35–37, 39]. In contrast, the computation of quasi-Fermi level and optical transition probabilities is not easy because the subbands are non-parabolic and the wave functions can be heavily mixed. A detailed treatment of k.p theory is beyond the scope of this book and we shall refer interested readers to various publications by S.L. Chuang [40].

3.6.6 Complex MQW Active Regions

In previous subsections, it has been implied that quantum wells in an MQW system do not couple with each other. This is a good approximation if the wells are far apart and the wave functions decay significantly in the barrier before reaching a neighboring well. There are circumstances in which we need to consider the effect of coupling between quantum wells. Also, it is may also be desirable to have quantum well designs with non-symmetric barriers.

Initially, it appeared that extension of a non-decoupled model to a coupled one should be straightforward: just use the potential for two wells instead of one. However in an actually simulation, we must solve problem of localizing the carriers in a complex coupled MQW structure.

Consider the simple case of two wells. As the two wells are brought closer together, the degenerate subbands start to split. From the view point of wave mechanics, the wavefunction of each energy level belongs to both wells. However, drift-diffusion theory is based on classical mechanics which requires that we know where the carriers are located.

Therefore, we face the task of deciding for each coupled confined state, which well does a carrier belong to. Similarly, for optical interband transition, the process takes place in the whole coupled complex structure. But again, we must decide for each transition, which well does it take place. The situation is similar to quantum tunneling where we have a contradiction between the wave nature implied by quantum theory versus the particle nature in drift-diffusion theory. We must create a reasonable semi-classical method to bridge the gap.

For carrier density calculation, we localize the coupled confined states as follows. Suppose the probability of finding an electron of a confined state in well 1 is p_1 and that of well 2 is p_2. In general, $p_1 + p_2 < 1$ (the remaining probability is spent in the barriers). We regard the electron is localized in well 1 with a probability of $p_1/p_1 + p_2$. A similar treatment is used for holes.

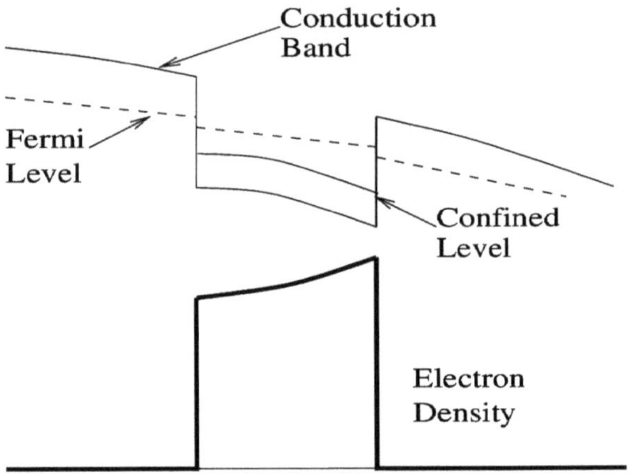

Fig. 3.2 Default model of quantum well carrier density calculation

For optical transitions from the valence band to the conduction band, we regard the optical process as taking place in well 1 with a probability of $p_1q_1/(p_1q_1 + p_2q_2)$, where q_1 and q_2 are probabilities of finding the hole in well 1 and 2, respectively.

3.6.7 Self-consistent Carrier Density Model

The quantum well levels are calculated at every bias point during an actual simulation because the bandgap of the active region is a function of carrier density. As the bias increases, the higher carrier density in the quantum well reduces the bandgap and changes the quantum well depth. Therefore the quantum levels must be re-computed at every bias point in a self-consistent simulation.

For problems of multiple quantum wells (MQW), the default approach in the device simulator is to solve the quantum confined states under flat-band conditions (assuming no electric field). When there is local variation of potential, we assume it is small enough that flat-band solution is still valid but only introduces a local correction to the confined energy level (see Fig. 3.2). To simplify numerical computation, we also assume that the carrier density is only confined within the well and is computed according to the local Fermi level and local confined energy level. The following formula for confined 2D carrier electron density is used:

$$n_{2D}(x,y) = \frac{1}{d_w}\sum_j \rho_j^0 kT \; ln\left[1+\exp\left[\frac{E_{fn}(x,y)-E_j(x,y)}{kT}\right]\right], \textit{inside well} \quad (3.71)$$

$$= 0, \; \textit{outside well} \quad (3.72)$$

where d_w is the well thickness and the subscript j denotes all confined states. Please note that both the Fermi level and confined level are assumed to be spatial dependent.

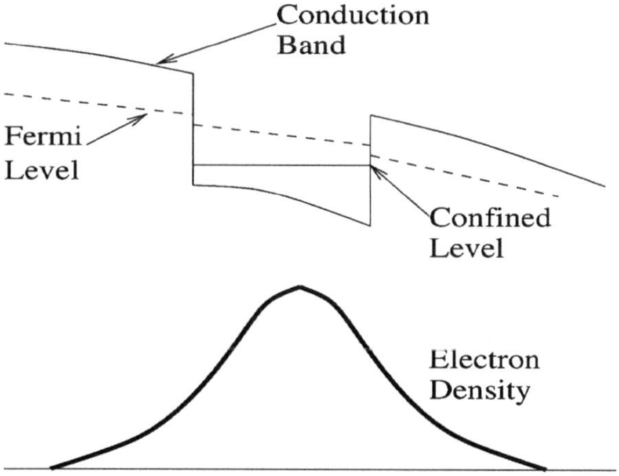

Fig. 3.3 Schematics of the self-consistent carrier density model for a single quantum well

ρ_j^0 is the 2D density of states. The variation of the confined level $E_j(x, y)$ follows that of potential, much in the same way the conduction band edge follows the potential change (see Fig. 3.2). Please note that the spatial variation in this model is uniform except for variations introduced by variations in the quasi-Fermi confined levels.

The above model is simple to implement and efficient to compute. It gives the right MQW behavior if the realistic band structure is close to flat-band condition and we can ignore the details of the carrier density on the scale of the quantum well. Such is the case of most MQW lasers under lasing condition when the band structure is close to flat-band under forward bias and high injection condition.

A more accurate model of self- consistent electron density which includes this effect is given as:

$$n_{2D}(x,y) = \sum_j g_n^j(y)\rho_j^0 kTln\left[1 + \exp\left[\frac{E_{fn}(x,y) - E_j(x,y)}{kT}\right]\right] \quad (3.73)$$

Where $g_n^j(y)$ is the electron wave function assuming the well is parallel to x-axis. The confined level E_j is no longer a function of position. Computation of the above density distribution requires more work because the density at one point depends on a global function $g_n^j(y)$. We also have to update the wave function and confined energy levels at different biases. Figure 3.3 illustrates this model. Similar expressions can be obtained for the hole subbands.

To achieve complete self-consistency, the following procedures are used:

1. At equilibrium, solve the potential using flat-band approximation. The default uniform profile model is used for the density. This gives us the initial potential distribution.
2. At equilibrium, solve the potential again with potential obtained in (1) above. The self-consistent density model is used. A new density profile is obtained in this step.

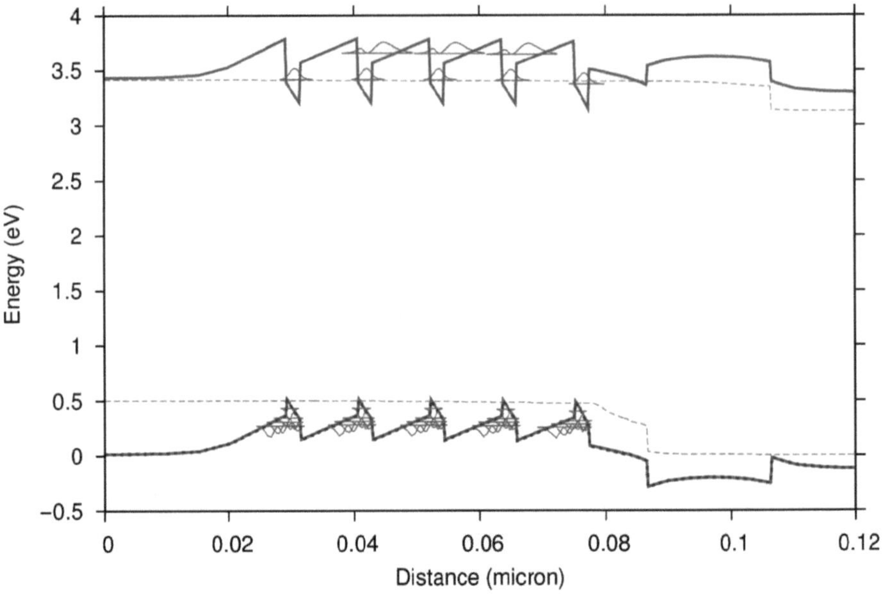

Fig. 3.4 Band diagram of a typical MQW GaN-based light emitting diode with quantum well being distorted by polarization interface charges

3. Iterate step 2 until numerical self-consistency is achieved for density profile and potential distribution.
4. Once self-consistency is achieved under equilibrium conditions, increase the bias in the main drift-diffusion solver and repeat (2) and (3) above.

3.6.8 Self-consistent Simulation of GaN-based Quantum Well LEDs

A common application of this fully self-consistent model is GaN-based LEDs. In these devices, it is well-known that the polarization interface charge due to spontaneous and strain-induced piezoelectric effects causes the quantum well to tilt to one side. This same effect also distorts the potential in other layers, thus affecting the current overflow.

As a demonstration of this effect a simple GaN-based LED is simulated in 1D. The polarization surface charge models are enabled in a self-consistent simulation causing the potential and wavefunction to distort as shown in Fig. 3.4. The LED optical power shows a totally different behavior (see Fig. 3.5) depending on whether or not the interface charges are considered. While part of this is due to the weaker overlap of the wavefunctions in the dipole moment, it is believed that a more important effect is the enhanced overflow of electrons over the MQW. As a result, the power vs. current slope decreases substantially with injection current. This effect is called LED efficiency droop which is one of the hottest research topics in the GaN world at the moment.

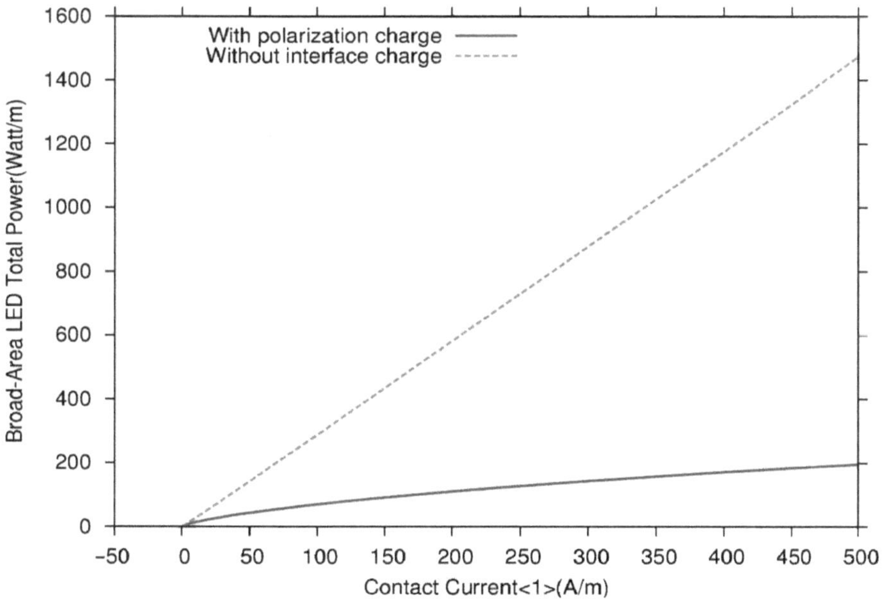

Fig. 3.5 LED light power versus injection current showing difference with and without polarization interface charges

3.6.9 Quantum Wells in Quantum-MOS and HEMT

When the gate of an MOS structure is under bias, the interface of silicon and SiO_2 forms a triangular-well for electrons or holes, depending on the polarity of the bias. Similar situation arises in high mobility electron field effect transistors (HEMT) where the internal electric field causes the heterojunction to form a triangular well to confine the 2D electron gas system.

A triangular well formed in such a way may be regarded as a special case of a quantum well discussed above. To derive such a structure from a simple symmetric flat-band quantum well, we take the following steps: (1) The left barrier is lowered to the point of zero barrier height; (2) External and internal electric field is allowed to tilt the potential in such a way that a triangular well is formed between the bottom of the well and right barrier.

Thus, a realistic quantum well model for quantum-MOS and HEMT may be achieved if we apply the complex-MQW and self-consistent models to a non-symmetric quantum well. Since the complex MQW model is well established for quantum wells with two barriers, we prefer to treat triangular wells as if they have a left-barrier with zero height.

In the case of quantum-MOS, some special treatment is required since the conduction band should be treated with anisotropic electron masses which may affect the conduction band quantum states, depending on the crystal orientation with respect to the SiO_2.

3.7 Interband Optical Transition

This section applies to compound material with direct bandgap, as mostly used in LED and LD and most solar cells except silicon based ones.

For active devices such as laser diodes and LED's, the interband optical gain and spontaneous emission spectrum are important. For photo-sensitive devices such as photo-detectors, the interband absorption (opposite of gain) is important.

Here we start with the basic theory of optical gain based on a parabolic band structure. Note that the formulas for quantum wells we consider here are rather general and are also applicable to bulk material. In the case of bulk material, only one subband with a 3D DOS needs to be considered. In the case of anisotropic parabolic approximation for strained quantum wells, the optical gain formulas can simply be extended by using the in-plane values of the carrier masses and adjusting the 2D DOS accordingly.

3.7.1 Interband Transition Model

Similar to the derivation of the absorption coefficients for a solid, the local gain due to a transition from a conduction band labeled j to a valence band labeled i can be written as [41],

$$
\begin{aligned}
g_{ij}\left(E_{ij}^0\right) &= \int P_{ij}\rho_{ij} dE_{ij} \\
&= \int \left(\frac{2\pi}{\hbar}\right)|H_{ij}|^2 (f_j - f_i)\delta(E_{ij} - \hbar\omega)\left(\frac{\epsilon_1}{\bar{n}c}\right)\rho_{ij} dE_{ij} \\
&= \left(\frac{2\pi}{\hbar}\right)|H_{ij}|^2 (f_j' - f_i')\left(\frac{\epsilon_1}{\bar{n}c}\right)\rho_{ij}
\end{aligned}
\tag{3.74}
$$

$$
\rho_{ij} = \rho_{ij}^0 h(\hbar\omega - E_{ij}^0) \tag{3.75}
$$

$$
|H_{ij}|^2 = \left(\frac{q}{m_0}\right)^2 \left(\frac{2\hbar\omega}{4\epsilon_1\epsilon_0\omega^2}\right) M_{ij}^2 \tag{3.76}
$$

Where $\bar{n}$ is the real part of the refractive index. f_i and f_j are the Fermi functions for the ith and the jth levels, respectively, and f_i' and f_j' are given by:

$$
f_i' = \left\{1 + \exp\left[\frac{E_i^0 - \frac{m_{ij}}{m_i}\left(E - E_{ij}^0\right) - E_{fp}}{kT}\right]\right\}^{-1} \tag{3.77}
$$

$$f'_j = \left\{ 1 + \exp\left[\frac{E_j^0 - \frac{m_{ij}}{m_j}\left(E - E_{ij}^0\right) - E_{fn}}{kT} \right] \right\}^{-1} \quad (3.78)$$

The gain function is the sum of g_{ij} over all the subbands for the allowed transitions. The simulator uses TE (transverse electromagnetic) mode as the default: the user should set the polarization mode to TM (transverse magnetic) mode if a large tensile strain is present in the material. We use the following anisotropic dipole moment for the heavy and light hole transitions in a quantum well [42]:

$$A_{hh} = \frac{3 + 3\cos^2(\theta_e)}{4}(TE) \quad (3.79)$$

$$A_{lh} = \frac{5 - 3\cos^2(\theta_e)}{4}(TE) \quad (3.80)$$

$$A_{hh} = \frac{3 - 3\cos^2(\theta_e)}{2}(TM) \quad (3.81)$$

$$A_{lh} = \frac{1 + 3\cos^2(\theta_e)}{2}(TM) \quad (3.82)$$

$$M_{hh} = A_{hh} O_{ij} M_0 \quad (3.83)$$

$$M_{lh} = A_{lh} O_{ij} M_0 \quad (3.84)$$

Where A_{hh} and A_{lh} are the quantum well dipole moment enhancement factors. M_0 is the dipole moment of the bulk material given by the following expression:

$$M_0 = \frac{1}{6}\frac{m_0}{m_e}\frac{E_{g0}(E_{g0} + \Delta)}{E_{g0} + \frac{2\Delta}{3}} \quad (3.85)$$

$cos(\theta_e)$ is defined as

$$\cos(\theta_e) = \frac{E_{ej}}{E_{ej} + \frac{m_r}{m_e(E - E_{ij})}} \quad for\ E > E_{ij}^0 \quad (3.86)$$

$$\cos(\theta_e) = 1 \quad for\ E \leq E_{ij}^0 \quad (3.87)$$

Where E_{ej} is the electron confined subband energy of level j and m_r is the reduced mass.

A few interesting points about dipole enhancement are discussed here. The above equations show that the heavy hole transition is favored under TE polarization while the light hole is favored for TM polarization. With a particular polarization (TE or TM) the bulk moment is split between the heavy and light holes and they average out to the bulk moment. It is interesting to note that the light hole under TM has a larger dipole moment than the heavy hole under TE. Therefore, from the point of view of the larger dipole moment, the light hole transition is very attractive. A major drawback for light hole transition is that, due to light mass, the quantum mechanic wave function tends to spread out more than for the heavy hole. As a result, the overlap integral (O_{ij} above) is reduced.

Broadening due to intra-band scattering significantly reduces the local gain function and must be considered. To describe the broadening of the quantum levels, the software uses a line shape function L in a convolution integral with the optical gain. The final expression for the gain function in the quantum well then becomes

$$g_{qw} = \sum_{i=j} \int \frac{g_{ij}(E_x)\tau}{\hbar L \left[\frac{E_x - E_{ij}^0}{\frac{\hbar}{\tau}} \right] dE_x} \tag{3.88}$$

The simplest line shape function is the Lorentzian shape function:

$$L\left(E_x - E_{ij}^0\right) = \frac{1}{\pi} \frac{\Gamma_0}{\left(E_x - E_{ij}^0\right)^2 + \Gamma_0^2} \tag{3.89}$$

where Γ_0 is the constant half width of the shape function. (3.88) is used to evaluate the stimulated emission and dielectric constants in the basic equations at all the points of the 2D mesh. The choice of τ is important since it reduces the peak gain and directly determines the threshold current.

3.7.2 Bandgap Shrinkage: Blue Shift vs. Red Shift

The effective bandgap for optical gain calculations is often reduced because of exchange effects [33]. Such a bandgap shrinkage is given by:

$$\Delta E_g = A_x \left(\frac{n+p}{2}\right)^{\frac{1}{3}} \tag{3.90}$$

Note that this term is an empirical formula which approximates more complicated models involving many-body and excitonic effects. These models are included in most advanced device modeling programs but require careful calibration of many parameters so the empirical model is still often used in industry.

Since bandgap shrinkage increases with carrier concentration, it causes a "red shift" tendency in the optical gain spectrum with increasing current. The peak optical gain also has a "blue shift" tendency due to the band filling effect (*i.e.*, the Fermi level separation becomes larger as more subbands are filled). The blue shift effect is usually stronger than the red shift effect due to bandgap shrinkage, and one often observes a blue shift in experiment [43, 44].

3.7.3 Material Gain vs. Modal Gain vs. Net Gain

A thorough treatment of laser theory is beyond the scope of this book but at its core, it can be stated simply as $RTG = 1$. This implies both net gain = mirror loss and a phase matching condition. In general, the round trip gain (RTG) is a complex value and phase matching is achieved when the real part is positive and the imaginary part is zero. The lasing wavelength is then determined by the gain spectrum and the feedback provided by the optical cavity.

Even if there is so wavelength selection by the optical cavity, it is important to realize that it is the maximum of the modal gain g_m which appears in the photon rate equation, not the local gain g. The former value is an average of the local gain weighed by the optical mode profile so a careful design of the optical waveguide is required to make the most of the available material gain. For the sake of simplicity, it is often assumed that the lasing wavelength is the peak modal gain but this is not true of all lasers.

In addition to the material gain shown above, various loss mechanisms must also be considered in lasers. This can be written as a local loss term:

$$\alpha_{qw} = -\alpha_{fn}n - \alpha_{fp}p - \alpha_{act} \tag{3.91}$$

where the first two terms are due to free carrier absorption in the quantum well. α_{act} is an adjustable background loss term to account for losses in the active quantum well for mechanisms other than interband transitions and free carrier absorption.

We finally arrive at our definition of net gain which refers to the quantity $g_{\mathrm{m}} - \alpha_m$ where α_m is the modal average of α_{qw}. This is quantity that must be positive and large enough to compensate the mirror loss to ensure lasing.

Note that the net gain can never be larger than the mirror loss under steady state conditions: increasing the gain merely increases the power output. The additional photons in the cavity ensure (via stimulated recombination) that the carrier density remains clamped above threshold.

3.7.4 Spontaneous Emission

When a laser is biased near threshold, the current is mainly determined by the spontaneous emission rate and/or Auger recombination. Therefore the evaluation of the spontaneous emission rate is important. The following expression for the spontaneous emission rate is used [45]:

$$r_{sp}^{qw}(E) = \sum_{i=j} \left(\frac{2\pi}{\hbar}\right) |H_{ij}|^2 f'_j (1 - f'_i)\, D(E) \rho_{ij} \tag{3.92}$$

where D(E) is the optical mode density in the material which has a refractive index of $\bar{n}$, given by [45]:

$$D(E) = \frac{\bar{n}^3 E^2}{\pi^2 \hbar^3 c^3} \tag{3.93}$$

One can use the same broadening line shape functions for the spontaneous emission spectrum in (3.92) as for the gain function. This allows the user to compare his/her experimental data with the broadened spontaneous spectrum. The broadened spontaneous emission spectrum is, however, not used in the main simulation. The reason is that since carrier recombination involves emission at all frequencies, it is only necessary to integrate the unbroadened spectrum in (3.92) over all possible frequencies:

$$R_{sp}^{qw} = \int_0^\infty r_{sp}^{qw}(E) D(E) \tag{3.94}$$

Note that Eqs. 3.88 and 3.94 are given in terms of the quasi-Fermi levels which can be treated directly as variables for the Newton's method used by the TCAD software.

Similar to the treatment of the gain function, the spontaneous emission rate can be written as

$$r_{sp}^{qw}(E) = \sum_{i,j} \left(\frac{2\pi}{\hbar}\right) |H_{ij}|^2 f'_j (1 - f'_i)\, D(E) \rho_{ij} \tag{3.95}$$

where $D(E)$ is the optical mode density.

3.7.5 Gain Integral with Valence Mixing

In the case of valence mixing, the valence bands are not parabolic and we must use a different formula for the optical gain spectrum. Based on theories developed in Ref. [40], the gain spectrum can be expressed in numerical integration over k_t.

$$g(E) = \frac{g_0}{2\pi t E} = \sum_{i,j} \int_0^{\infty} \frac{\left(\frac{\pi}{\Gamma}\right) f_{dip}(k_t) M_b (f_j - f_i) dk_t^2}{1 + \frac{(E_{cj}(kt) - E_{kpi}(kt) - E)^2}{\Gamma^2}} \tag{3.96}$$

where t is the quantum well thickness; $\Gamma = \frac{\hbar}{\tau_{scat}}$ is broadening due to intraband scattering relaxation time τ_{scat}; E_{cj} is the jth conduction subband; E_{kpi} is the ith valence subband from the k.p calculation; the sum is over all possible valence and conduction subbands; g_0 is a constant defined as

$$g_0 = \frac{\pi q^2 \hbar}{\varepsilon_0 c m_0^2 \bar{n}} \tag{3.97}$$

Where q is free electron charge; $\bar{n}$ is the real part of the refractive index; all other symbols have their usual meanings.

M_b is the bulk dipole momentum given by:

$$M_b = \frac{1}{6} \frac{m_0 q}{m_c} \frac{E_{g0}(E_{g0} + \Delta_{so})}{E_{g0} + \frac{2\Delta_{so}}{3}} \tag{3.98}$$

Where E_{g0} is the unstrained bandgap; m_c is the effective mass of the conduction band; Δ_{so} is spin-orbit coupling energy.

The dipole factor due to non-parabolic subbands is given by the following overlap integral for the case of symmetric well:

$$f_{dip} = \left(\frac{3}{2}\right) \left[<c|g_1(k_t, z)>^2 + \left(\frac{1}{3}\right) <c|g_2(k_t, z)>^2 \right] \tag{3.99}$$

$$f_{dip} = 2 <c|g_2(k_t, z)>^2 \quad (TM) \tag{3.100}$$

Where $|c>$ is the conduction band wave function; $|g_1>$ and $|g_2>$ are the valence band envelop function.

f_j and f_i are the Fermi functions expressed as follows:

$$f_j^{-1} = 1 + \exp\left[\frac{1}{kT}\left(E_{cj0} + \frac{\hbar^2 k_t^2}{2 m_0 m_c} - E_{fn}\right)\right] \tag{3.101}$$

$$f_i^{\ 1} = 1 + \exp\left[\frac{E_{kpi}(k_t)}{kT}\right] \tag{3.102}$$

where E_{cj0} is the bottom of jth conduction subband.

3.8 Wurtzite Strained Quantum Wells

3.8.1 Introduction

Recently, wurtzite strained quantum wells have been studied intensively because of applications in GaN blue green lasers, LED and high voltage HEMT. Previous band structures models are applicable for zincblende compounds. In this section, we will describe MQW band structure and optical transition models for wurtzite crystals here. We follow the work of S.L. Chuang [46] closely here. In the following subsections, we will describe the bulk wurtzite band structure, the MQW subband model and the optical transitions.

We must point out a couple of major differences between zincblende and wurtzite crystals here. The first issue is the base lattice. For zinc-blende material system, it is usually well known what the base lattice constant is (for example, GaAs or InP) and the strain is well defined in each layer once we know the substrate lattice. In a wurtzite system, a bulk GaN-based buffer layer is usually grown on a sapphire substrate. This results in a significant amount of lattice mismatch between the substrate and the GaN-based layer and a large dislocation and defect density. When an MQW system is grown on top of the bulk layer, it is not always clear what the base lattice constant of the MQW should be as it depends on how much the buffer layer has relaxed.

In most TCAD software, the default lattice base constant is that of GaN which may not be true for all systems. For example, if a buffer layer of AlGaN grown on a sapphire substrate before the MQW is placed on top, then the base lattice constant should be that of bulk AlGaN instead of GaN. For this reason, the material parameters of a wurtzite MQW depend on the base lattice and should be adjusted to fit the design.

Another important issue is the number of material parameters. As we will see below, there are many times more band structure and strain parameters for wurtzite than zincblende material system. Since wurtzite semiconductor band structures and optical properties are less understood, many band structure parameters are not available and our expectation for the accuracy of the simulation should be lower than that for a zincblende material system.

3.8.2 Bulk Band Structure

Consider a strained wurtzite crystal pseudomorphically grown along the c-axis (z axis) on another thick wurtzite layer. The base lattice constant is a_0 and the original lattice constant of the layer under consideration is a. The strain tensor in the well region has the following elements:

$$\varepsilon_{xx} = \varepsilon_{yy} = \frac{a_0 - a}{a} \tag{3.103}$$

$$\varepsilon_{zz} = -\frac{2C_{13}}{C_{33}}\varepsilon_{xx} \tag{3.104}$$

$$\varepsilon_{xy} = \varepsilon_{yz} = \varepsilon_{zx} \tag{3.105}$$

The conduction bands can be characterized by a parabolic band model with different electron effective masses m_e^t and m_e^z perpendicular and parallel to the c-growth direction, respectively. The hydrostatic energy shift in the conduction band can be written as

$$P_{ce} = a_{cz}\varepsilon_{zz} + a_{ct}\varepsilon_{xx} + \varepsilon_{yy} \tag{3.106}$$

We take $a_{cz} = a_{ct} = a_c$ for simplicity.

The band structure of the valence band is more complicated. Using the k.p method Chuang and Chang [47] have derived a 6 by 6 Hamiltonian. This has been further simplified using a block-diagonalization to the following upper and lower Hamiltonians.

$$H^v_{6\times 6}(k) = \begin{bmatrix} H^U_{3\times 3}(k) & 0 \\ 0 & H^L_{3\times 3}(k) \end{bmatrix} \tag{3.107}$$

where

$$H^U = \begin{bmatrix} F & K_t & -iH_t \\ K_t & G & \Delta - iH_t \\ iH_t & \Delta + iH_t & \lambda \end{bmatrix} \tag{3.108}$$

$$H^L = \begin{bmatrix} F & K_t & iH_t \\ K_t & G & \Delta + iH_t \\ -iH_t & \Delta - iH_t & \lambda \end{bmatrix} \tag{3.109}$$

When there is strain, the energy bands of shift in a complicated manner. We need to use a common reference energy level when writing down the Hamiltonian. Following the convention for the wurtzite structure, if E_c^0 is the unstrained conduction band edge, the reference valence band level is as follows:

$$E_v^0 = E_c^0 - (E_g + \Delta_1 + \Delta_2) \tag{3.110}$$

When there is strain, the conduction band is shifted by an amount P_{ce} and the new reference level is:

$$E_v^0 = E_c^0 - (E_g + \Delta_1 + \Delta_2 + P_{ce}) \tag{3.111}$$

Using the above reference energy, the matrix elements are defined as follows:

$$F = \Delta_1 + \Delta_2 + \lambda + \theta$$

$$G = \Delta_1 - \Delta_2 + \lambda + \theta$$

$$\lambda = \lambda_k + \lambda_\varepsilon$$

$$\lambda_k = \frac{\hbar^2}{2m_0}\left(A_1 k_z^2 + A_2 k_t^2\right)$$

$$\lambda_\varepsilon = D_1 \varepsilon_{zz} + D_2(\varepsilon_{xx} + \varepsilon_{yy})$$

$$\theta = \theta_k + \theta_\varepsilon$$

$$\theta_k = \frac{\hbar^2}{2m_0}\left(A_3 k_z^2 + A_2 k_t^2\right)$$

$$\theta_\varepsilon = D_3 \varepsilon_{zz} + D_4(\varepsilon_{xx} + \varepsilon_{yy})$$

$$K_t = \frac{\hbar^2}{2m_0} A_5 k_t^2$$

$$H_t = \frac{\hbar^2}{2m_0} A_6 k_t k_z$$

$$\Delta = \sqrt{2}\Delta_3 \tag{3.112}$$

Please note that the base vectors $|1>$ to $|6>$ of the above can be expressed by spherical harmonics $Y_{lm}(l = 1)$ [46]. For the upper Hamiltonian, $|1>$, $|2>$, $|3>$ corresponds to heavy hole (HH), light hole (LH) and crystal field split hole (CH), respectively. At the zone center ($k = 0$), the HH subbands are decoupled from the LH and CH subbands, while there is always a coupling between the LH and CH bands.

In the device simulation software, the above Hamiltonian is solved to generate the bulk band structure from which the density of states is evaluated for bulk materials. To achieve maximum efficiency, the bulk valence band structure is fitted to a parabolic band for each of HH, LH and CH bands under any strain condition. Such fitted parabolic bands are used for modeling the non-active bulk regions. For MQW active region, a more rigorous approach is used which we shall describe below.

3.8.3 MQW Model – Effective Mass Approximation

In general, the dispersion of bulk valence bands is non-parabolic and anisotropic with strong mixing or anti-crossing behavior in the direction perpendicular (or transverse) to the c-axis. We need an effective mass model for the computation of density of states and for a simplified MQW model.

An efficient approximation is to take the effective masses fitted from the bulk band structure. This approach takes into account the anti-crossing behavior but the quality of the fit can be poor near some range for the transverse direction (see Figs. 3.6 and 3.7).

Based on Ref. [46], we can also use the following analytical effective mass model for the valence band:

1. Within a range of small k (a situation when the valence band is lightly populated by holes), the following effective masses hold:

$$\frac{m0}{m_{hh}^{z}} = -(A_1 + A_3)$$

$$\frac{m0}{m_{lh}^{z}} = -\left[A_1 + \left(\frac{E_2^0 - \lambda_e}{E_2^0 - E_3^0}\right)A_3\right]$$

$$\frac{m0}{m_{ch}^{z}} = -\left[A_1 + \left(\frac{E_3^0 - \lambda_e}{E_3^0 - E_2^0}\right)A_3\right] \tag{3.113}$$

$$\frac{m0}{m_{hh}^{t}} = -(A_2 + A_4)$$

$$\frac{m0}{m_{lh}^{t}} = -\left[A_2 + \left(\frac{E_2^0 - \lambda_e}{E_2^0 - E_3^0}\right)A_4\right]$$

$$\frac{m0}{m_{ch}^{t}} = -\left[A_2 + \left(\frac{E_3^0 - \lambda_e}{E_3^0 - E_2^0}\right)A_4\right] \tag{3.114}$$

Where $E_i^0 (i = 1, 2, 3)$ are the valence band edges at k $= 0$. The parabolic bands of this model are shown in Figs. 3.8 and 3.9.

2. For a large range of k (a situation when the valence band is heavily populated by holes), the following effective masses formulas are valid:

$$\frac{m0}{m_{lh}^{z}} = -(A_1 + A_3)$$

$$\frac{m0}{m_{ch}^{z}} = -A_1 \tag{3.115}$$

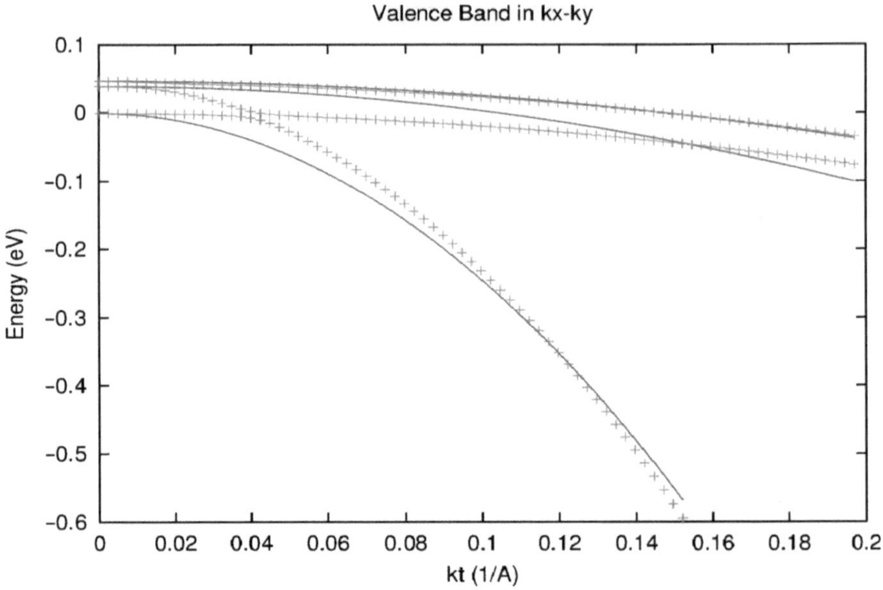

Fig. 3.6 Transverse valence bands (points) fitted to effective mass model

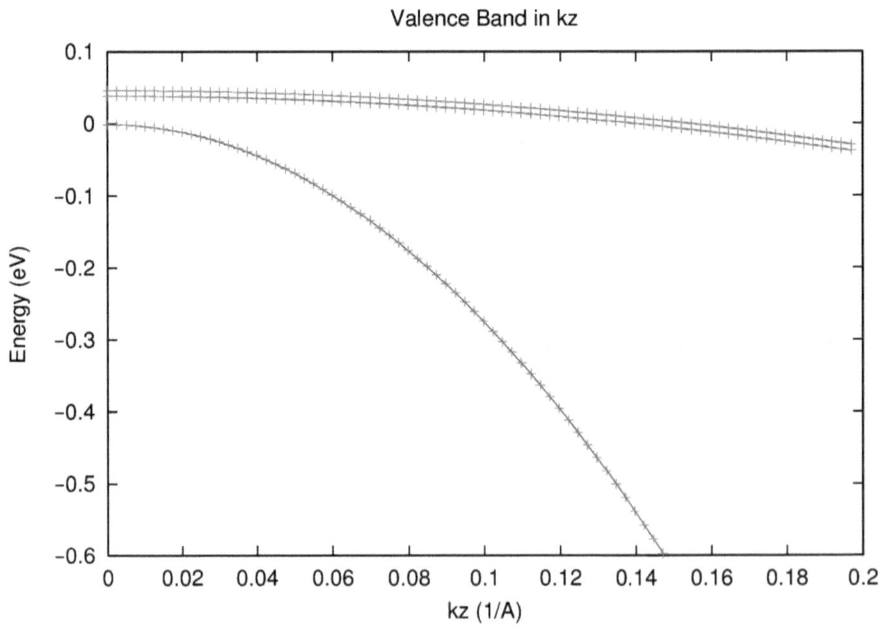

Fig. 3.7 Valence bands along c-axis (points) fitted to effective mass model

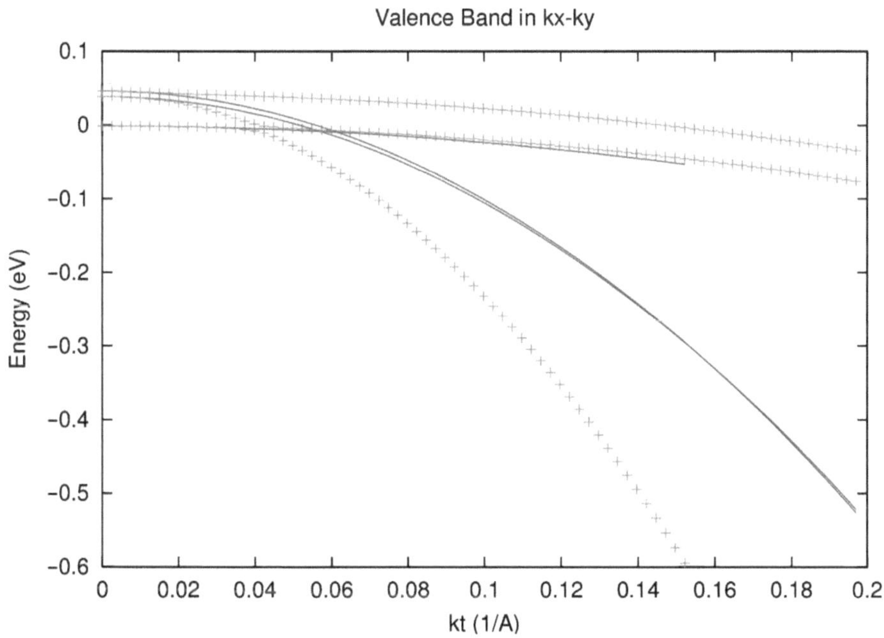

Fig. 3.8 Transverse valence bands (points) as compared with analytical effective mass model of small k-range

$$\frac{m0}{m^t_{hh}} = -(A_2 + A_4 - A_5)$$

$$\frac{m0}{m^t_{lh}} = -(A_2 + A_4 + A_5)$$

$$\frac{m0}{m^t_{ch}} = -A_2 \tag{3.116}$$

The parabolic bands of this model are shown in Figs. 3.10 and 3.11.

3. A compromise of the above two models is to average them. The parabolic bands with average masses are shown in Figs. 3.12 and 3.13.

   The default setting simulation setting is to use the compromise value but this can be adjusted in the device simulator. The choice of model should be based on the expected hole density.

   Since the crystal symmetry of wurtzite is different from that of zincblende, the dipole moment in effective mass approximation for zincblende can not be used here and we need to use more accurate models, such as k.p theory, to compute the dipole moment for optical transitions. However, study of results in k-dependent dipole moment from k.p theory for a large number of structures leads us to propose the following simplified dipole moment enhancement factors:

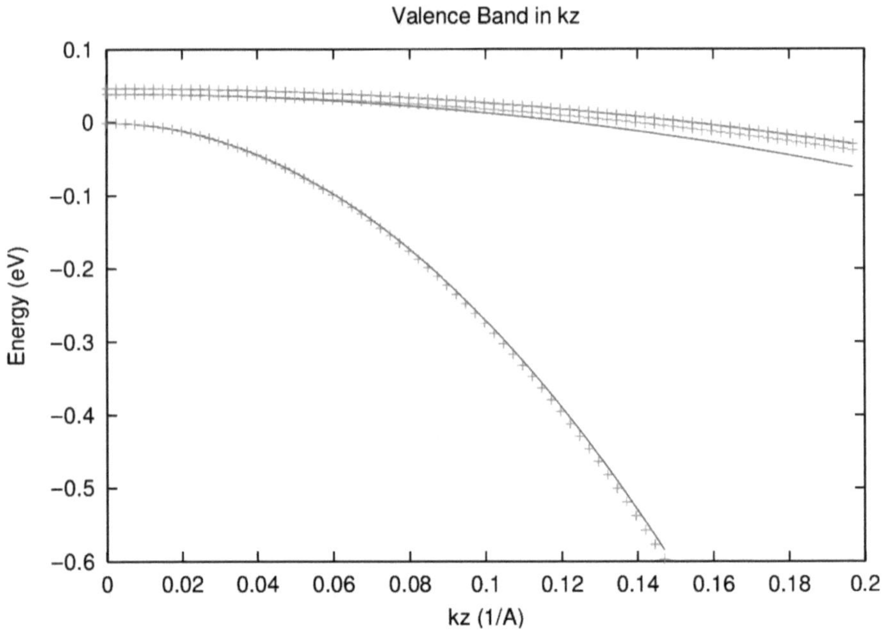

Fig. 3.9 c-Axis valence bands (points) as compared with analytical effective mass model of small k-range

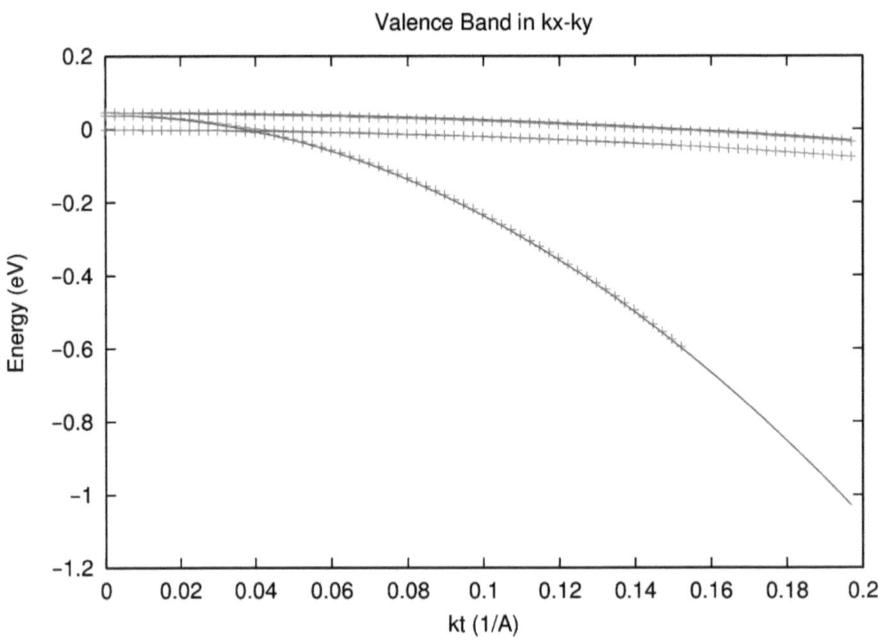

Fig. 3.10 Transverse valence bands (points) as compared with analytical effective mass model of large k-range

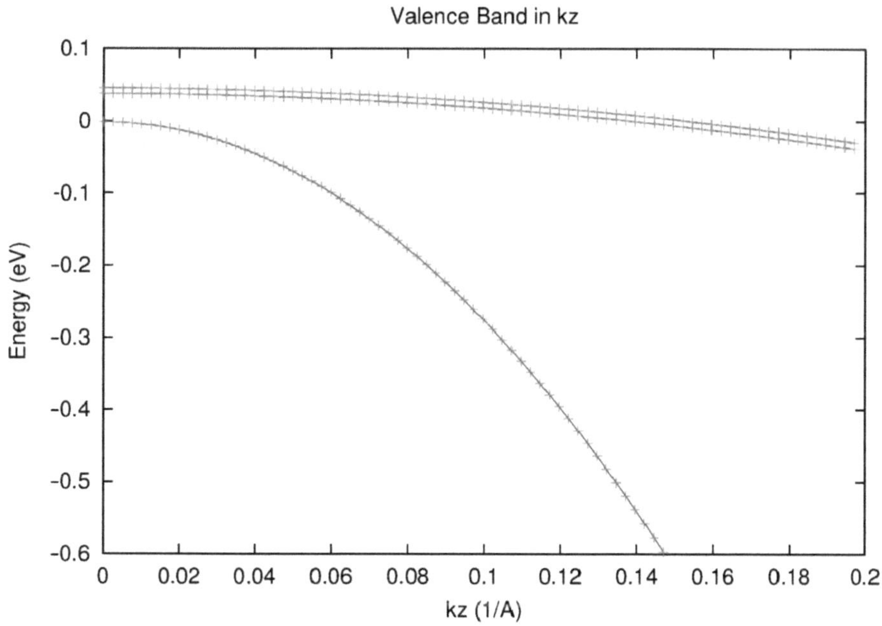

Fig. 3.11 c-Axis valence bands (points) as compared with analytical effective mass model of large k-range

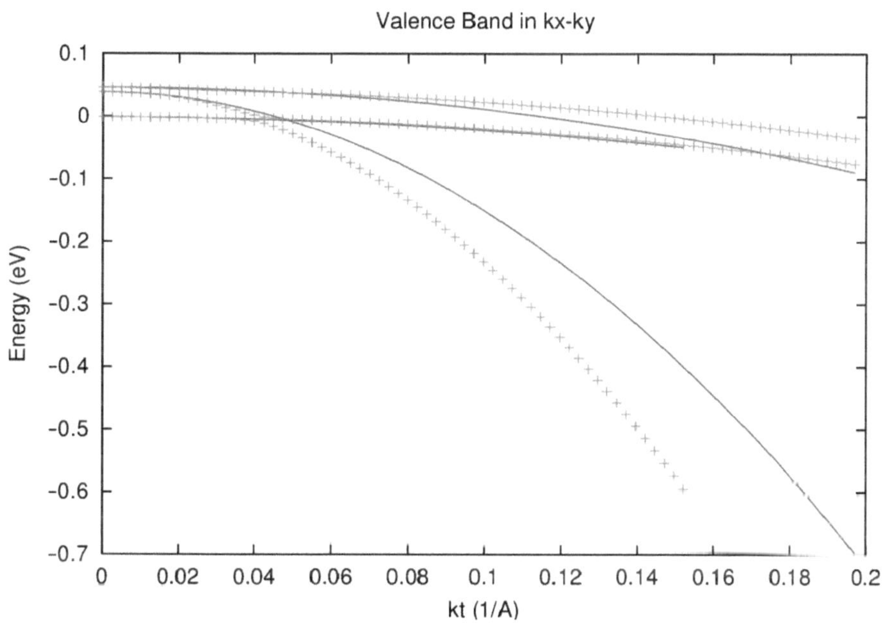

Fig. 3.12 Transverse valence bands (points) as compared with analytical effective mass model of averaged masses

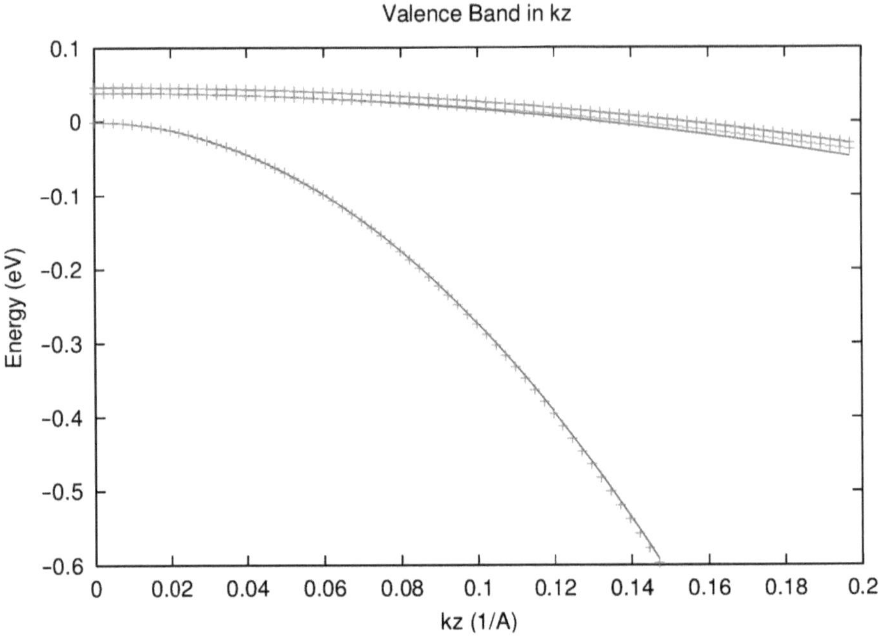

Fig. 3.13 c-axis valence bands (points) as compared with analytical effective mass model of averaged masses

$$A_{hh} = \frac{3}{2}$$

$$A_{lh} = \frac{3\cos^2(\theta_e)}{2}$$

$$A_{ch} = 0 \tag{3.117}$$

for TE and

$$A_{hh} = 0$$

$$A_{lh} = 3 - 3\cos^2(\theta_e)$$

$$A_{ch} = 3 \tag{3.118}$$

for TM. $cos^2(\theta_e)$ is defined in the same way as in zincblende. Please note that we formulate the above factors in such a way that the following conservation rule is obeyed:

$$2 \times A_h^{TE} + A_h^{TM} = 3, \ (h = hh, \ lh, \ or \ ch) \tag{3.119}$$

The dipole moment is expressed as:

$$M_{hh} = A_{hh}\, O_{ij} M_b$$

$$M_{lh} = A_{lh}\, O_{ij} M_b$$

$$M_{ch} = A_{ch}\, O_{ij} M_b \tag{3.120}$$

Where O_{ij} is wave function overlap integral and M_b is bulk dipole moment given by [46] as follows:

$$M_b^{TM} = \frac{m_0}{6} E_{pz}$$

$$E_{pz} = \left(\frac{1}{m_e^z} - 1\right) \frac{E_g(E_g + \Delta_1 + \Delta_2)(E_g + 2\Delta_2) - 2\Delta_3^2}{E_g + 2\Delta_2} \tag{3.121}$$

$$M_b^{TM} = \frac{m_0}{6} E_{px}$$

$$E_{px} = \left(\frac{1}{m_e^t} - 1\right) \frac{E_g(E_g + \Delta_1 + \Delta_2)(E_g + 2\Delta_2) - 2\Delta_3^2}{(E_g + \Delta_1 + \Delta_2)(E_g + \Delta_2) - \Delta_3^2} \tag{3.122}$$

3.8.4 MQW Model – Valence Mixing

The valence mixing treatment is given in this subsection. Within the envelope function approximation, we write the wavefunction as follows:

$$\Psi_m^U(z;k_t) = \frac{e^{ik_t \cdot r_t}}{\sqrt{A}} \left(g_m^{(1)}(z;k_t)|1\rangle + g_m^{(2)}(z;k_t)|2\rangle + g_m^{(3)}(z;k_t)|3\rangle \right) \tag{3.123}$$

$$\Psi_m^L(z;k_t) = \frac{e^{ik_t \cdot r_t}}{\sqrt{A}} \left(g_m^{(4)}(z;k_t)|4\rangle + g_m^{(5)}(z;k_t)|5\rangle + g_m^{(6)}(z;k_t)|6\rangle \right) \tag{3.124}$$

The valence subbands are determined by the following coupled differential equations:

$$\sum_{j=1}^{3}\left(H_{ij}^{U}\left(k_z=-i\frac{\partial}{\partial z}\right)+\delta_{ij}E_v^0(z)\right)g_m^{(j)}(z;k_t)=E_m^U(k_t)g_m^{(j)}(z;k_t) \tag{3.125}$$

The valence band discontinuity is represented by discontinuity in $E_v^0(z)$ (the reference energy). Similar equations can be written down for the lower Hamiltonian. It can be shown that MQW with reflection symmetry $E_v^0(z)=E_v^0(-z)$, the upper and lower Hamiltonians have the same band structures.

In our simulation software, we use a finite difference method to solve the coupled differential equations. This gives us the valence mixing subbands.

The dipole moments, taking into account the upper and lower Hamiltonian degeneracy, are given as follows:

$$M_{nm}^{TE}=M_b\frac{3}{4}\left[<\phi_n|g_m^{(1)}>^2+<\phi_n|g_m^{(2)}>^2\right] \tag{3.126}$$

$$M_{nm}^{TE}=M_b\frac{3}{2}<\phi_n|g_m^{(3)}>^2 \tag{3.127}$$

3.8.5 Nomenclature

Due to the large number of symbols used in the theory of Wurtzite material, we list some critical symbol definitions here because these are specific to wurtzite band structure.

a, c	Lattice constants of Hexagonal structure.
E_g	Bandgap.
E_g	Bandgap.
$\Delta_1,\ \Delta_2,\ \Delta_3$	Energy parameters.
Δ_{so}	Spin-orbit coupling energy.
m_e^z, m_e^t	Conduction band effective masses along c-axis (z) and transverse (x-y) direction, respective.
$A_i, (i=1,...,6)$	Valence band effective mass parameters.
a_h, a_c, a_v	Hydrostatic deformation potentials of total, conduction, and valence parts, respectively.
$D_i, (i=1,...,4)$	Valence band shear deformation potentials.
C_{13}, C_{33}	Elastic stiffness constants.

3.9 Thermal and Self-heating Effects

This section deals with thermal effects in semiconductors. As we will show that a variety of heating sources exist in semiconductors. It is critical that these sources are accurately formulated when modeling devices where self-heating is important. Examples of such devices are high power transistors and high power laser diodes. The discussion below concentrates on lasers but is equally applicable to other semiconductor devices.

3.9.1 Basic Equation of Heat Flux

It is well known that the heating effect is very important for semiconductor lasers in almost all applications. The heat generated by a semiconductor laser often forces the designer to include an additional cooling system and therefore increases the cost of the application. The heating effect is more important for high power semiconductor lasers where the device temperature often determines the achievable power output.

From the point of view of simulation and modeling, the concern is twofold. First, we must find out the temperature distribution from all possible heat sources. This involves a much larger simulation area than the small region near the p-n junction. We must consider how the heating power flows though the whole substrate as well as from any wire bonds. Secondly, we must consider how the heating affects the laser performance. This means we must accurately evaluate the degradation of power, efficiency, *etc.*, due to the non-uniform temperature distribution. This is not a trivial task because virtually all variables and material parameters are temperature dependent. Our goal is to provide a thermal modeling environment so that all possible temperature dependences can be taken into account.

We are concerned with the generation and flow of lattice heating power in a semiconductor. We first consider the heat flux and then the heat source. We introduce the thermal conductivity such that the heating power flux (in Watt/m^2) is given by

$$\vec{J_h} = -\kappa \vec{\nabla} T \tag{3.128}$$

Conservation of energy requires that the temperature distribution satisfy the following basic thermal equation:

$$C_p \rho \frac{\partial T}{\partial t} = -\vec{\nabla} \cdot \vec{J_h} + H \tag{3.129}$$

or

$$C_p \rho \frac{\partial T}{\partial t} = -\vec{\nabla} \cdot \kappa \vec{\nabla} T + H \tag{3.130}$$

Where C_p is the specific heat and ρ is the density of the material. H is the heat source.

3.9.2 *Thermoelectric Power and Thermal Current*

The temperature gradient induces a current of the form [48–50]:

$$-q\mu_n n P_n \vec{\nabla} T \tag{3.131}$$

to the electron current and

$$-q\mu_p p P_p \vec{\nabla} T \tag{3.132}$$

to the hole current.

The thermoelectric powers, P_n and P_p are given by

$$P_n = \frac{k_B}{q}\left[-\frac{5}{2} - \nu + \ln\left(\frac{n}{N_c}\right)\right] \tag{3.133}$$

$$P_p = \frac{k_B}{q}\left[-\frac{5}{2} - \nu + \ln\left(\frac{p}{N_v}\right)\right] \tag{3.134}$$

where ν is the exponent used in the field-dependent relaxation time. For phonon scattering [51],

$$\nu = -\frac{1}{2} \tag{3.135}$$

The heat source can be separated into Joule heat, generation/recombination heat, and Thomson and Peltier heating terms. Following the suggestion of Ref. [48] we discuss these terms in more detail in the following subsections.

3.9.3 *Joule Heat*

There are two sources of Joule heating. One is from the steady state or low frequency part of the electrical field:

$$H_{Joule-dc} = \frac{|\vec{J_n}|^2}{q\mu_n n} + \frac{|\vec{J_p}|^2}{q\mu_p p} \tag{3.136}$$

The other part comes from the optical frequency. When the optical wave passes a lossy semiconductor, the wave is absorbed by the lossy material. The absorbed energy can either generate electron hole pairs or be dissipated and become Joule heat. We also refer to this type of heat source as the optical part of the Joule heating. The internal loss of a semiconductor laser is the cause of the optical part of the Joule heating while the band to band absorption is the source for electron-hole pair generation in photo-detectors.

A simple derivation of this term follows. The optical power dissipated per unit volume can be expressed as:

$$power\ loss = \sigma_{op} F_{op}^2 = \epsilon_2 \epsilon_0 \omega F_{op}^2 \tag{3.137}$$

where the optical field F_{op} is the root mean square of the oscillating field. For convenience a constant γ is introduced such that the complex wave amplitude is related to the electric field by

$$|F|^2 = \gamma |W|^2 \tag{3.138}$$

To determine γ we use the following basic relation:

$$S\hbar\omega = \int \epsilon_0 \epsilon_1 |F|^2 dv = \epsilon_0 \gamma \int \epsilon_1 |W|^2 dv = \epsilon_0 \gamma < \epsilon_1 > \tag{3.139}$$

The power loss becomes

$$power\ loss = \frac{S\hbar\omega^2 \epsilon_2 |W|^2}{< \epsilon_1 >} \tag{3.140}$$

or in terms of material internal loss and local index:

$$\epsilon_2 = \frac{\bar{n}_1 \alpha_i}{k_0} \tag{3.141}$$

$$H_{Joule-op} = \frac{S\hbar\omega^2 \bar{n}_1 \alpha_i |W|^2}{k_0 < \epsilon_1 >} \tag{3.142}$$

3.9.4 Recombination Heat

When an electron-hole pair recombines, the energy either converts to a photon (radiative) or turns into heat (non-radiative). In a real device, most of the photons emitted by spontaneous radiative recombination are eventually absorbed by the semiconductor (except in the case of high efficiency LED) and are converted into heat. For simplicity we assume the spontaneous recombination is a heat source.

The stimulated recombination in a semiconductor laser should not be considered a heat source because the photons from this term are coherently amplified and are eventually emitted to the outside of the laser cavity. Some part of this energy is absorbed by the waveguide loss and that has been considered as part of the Joule heating above.

For each electron-hole pair recombined, the heat released is the difference between the quasi-Fermi levels:

$$H_{rec} = (R_{trap} + R_{Aug} + R_{spon})(E_{fn} - E_{fp}) \tag{3.143}$$

3.9.5 Thomson and Peltier Heat

The Thomson heat comes from the change in thermoelectric power when an electron-hole pair recombines:

$$H_T = qR_{total}T(P_p - P_n) \tag{3.144}$$

Where P_p and P_n are thermoelectric power for hole and electrons, respectively. The Peltier heat is related to the spatial variation in the thermoelectric power:

$$H_p = -T(\vec{J_n} \cdot \vec{\nabla} P_n + \vec{J_p} \cdot \vec{\nabla} P_p) \tag{3.145}$$

Chapter 4
Setting Up a 3D TCAD Simulation

4.1 Overview

4.1.1 Software Tools

In order to perform a 3D TCAD simulation, we first need the mask layout for all the process steps. The mask layout can be designed in a variety of software including drafting tools like Autodesk AutoCAD® and specialized EDA tools like Cadence Virtuoso® and Tanner L-Edit®. These tools are used to create a GDSII file (.gds) which is the current industry standard for IC layout work. The GDS format was originally developed by a California company called Calma for its layout design software, "Graphic Data System" ("GDS") and later, "GDS II". This format is now owned by Cadence Design Systems. Objects contained in a GDS II file are grouped by assigning them various attributes including layer number, data type or text type [52].

In this book, GDSII files are imported by a layout GUI program called MaskEditor. It can be downloaded free of charge from this website [10]. The basic purpose of this tool is to create the files that will be used in the subsequent 3D process simulation. When GDSII files are unavailable, MaskEditor can also be used to create the geometric shapes and layers used in the process simulation.

The process simulator takes the input files created by MaskEditor and performs 3D process simulation based on the user's command input. After finishing the simulation, the process simulator can export its output to the device simulator. The device simulator will then use the exported mesh and material information to perform electrical, thermal and optical simulations. The output of the device simulator can be viewed directly with a plotting GUI or saved to a graphic format such as postscript for convenient batch processing.

S. Li and Y. Fu, *3D TCAD Simulation for Semiconductor Processes, Devices and Optoelectronics*, DOI 10.1007/978-1-4614-0481-1_4,

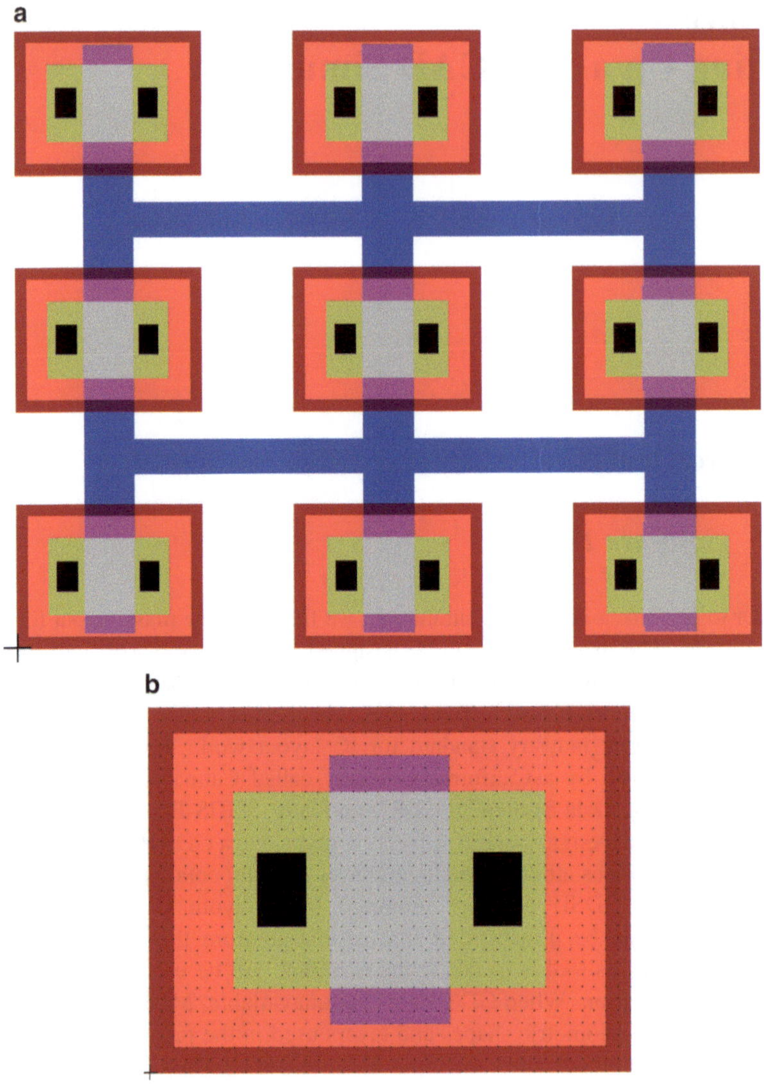

Fig. 4.1 Layout example of 3×3 matrix (**a**) and unit cell (**b**)

4.1.2 *Scope of the Simulation*

Most layout software allows the user to design complex ICs by starting with simple low-level elements which are combined to build a hierarchy of progressively higher-level elements. Figure 4.1 shows an example of this where a basic MOSFET cell is replicated to form a 3x3 matrix; this pattern can then be re-used to form even more complex elements.

As previously discussed, TCAD simulations operate at the semiconductor device level and not the IC level and will thus usually concerns itself with the basic unit cells of a design. However, a device designer should also be mindful of how the low-level components integrate into a larger design.

4.2 Semiconductor Processing Fundamentals

4.2.1 Photoresist

All large-scale semiconductor processes are based on photoresists and masks. These are used to define certain regions of a device where a process is allowed to operate while protecting all the other areas. Because much of the technology described in this section originates from the world of film photography (circa nineteenth century), it is helpful to define a few common terms for readers who are not familiar with.

Photoresists are light-sensitive materials that alter their properties when exposed to certain light sources. The term "exposure" refers to the length of time needed for the material to absorb the necessary amount of radiation to alter its properties. A chemical solution called a "developer" is used to process and dissolve the photoresist after exposure.

Most modern photoresists are viscous liquids at room temperature. A small amount of this liquid is applied to the surface of a wafer and the wafer is then spun at several thousand RPMs to spread it out. The rotation speed will determine how well the resist adheres to the wafer as well as its final thickness and uniformity [53]. The final thickness of typical photoresist film is about 1 um.

There are many kinds of photoresist materials: they are classified according to their polarity, sensitivity and resistance to etching and ion implantation. The sensitivity is critical since it determines not only the exposure time but also the technology node that can be achieved with a particular resist; small device features require small wavelengths of light because of diffraction limits. Based on sensitivity, the most commonly used photoresist include g-line, i-line and Deep UV (DUV) resists. More information on resist properties can be found in semiconductor fabrication books and photoresist manufacturer specifications.

The polarity of a photoresist refers to its solubility to the developer solution. Positive resists are normally insoluble but become soluble after exposure. A negative resists is exactly the opposite: normally soluble but becomes insoluble after exposure [54]. Figure 4.2 illustrates how positive and negative photoresist behave when exposed with the same photo mask.

For the sake of simplicity, all the examples we discuss will use positive photoresist unless otherwise noted. We also note that it is important not to confuse the polarity of the resist with that of the mask layout in MaskEditor. The convention used by MaskEditor refers to the combined effect of the resist and photo mask and will be discussed in the next section.

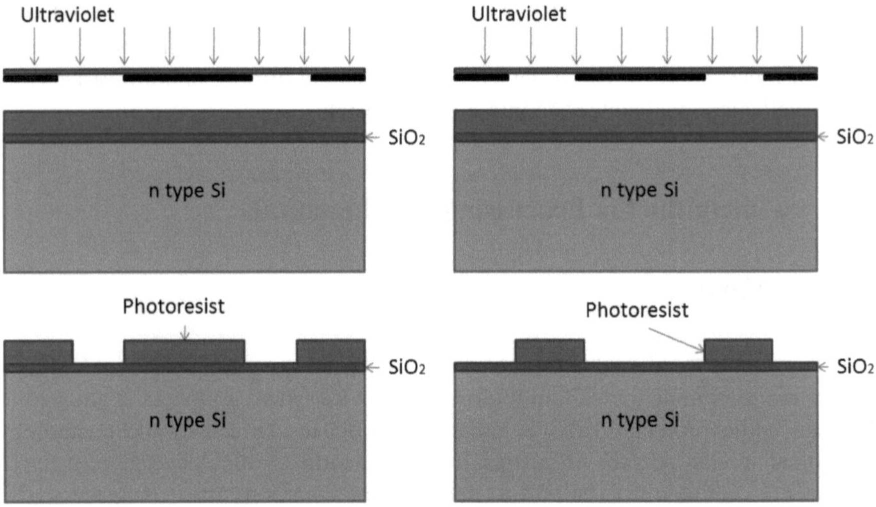

Fig. 4.2 Positive (*left*) and negative (*right*) photoresist

Fig. 4.3 Photo mask and the shape created on the wafer

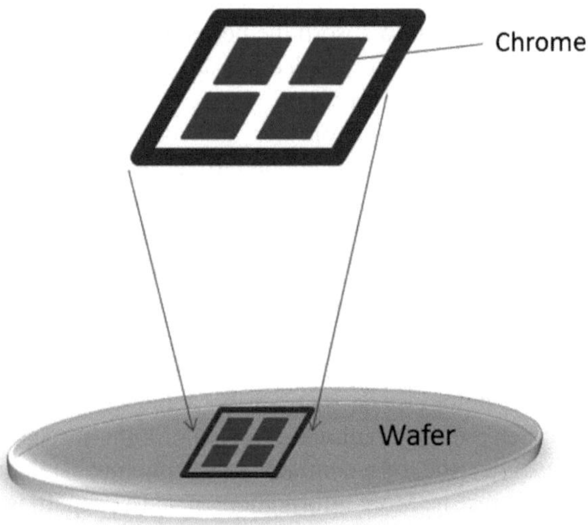

4.2.2 Masks

A photo mask is simply a glass plate on which a pattern has been drawn by depositing a layer of chrome or some other metal. Using masks is the usual method by which photoresist is exposed and processed in integrated circuit fabrication. Masks are always much larger than the final size of the pattern they define. An optical system shrinks down this pattern and transfers it to the wafer as shown in Fig. 4.3. A photolithography stepper or scanner repeats the pattern multiple times to cover the whole wafer.

A complete process for an integrated circuit will have several steps in which photoresist is applied, exposed and removed. Each process step will have its own mask design. The different masks of a process are stored as layers in the GDSII file format.

In practice, masks often come in two types: dark field or light field. They are so named because they allow user to use masks to draw the layout either as the actual desired image or its negative, depending on which is easier to draw [55].

Figure 4.4 illustrates the difference between light field and dark field. A positive photoresist is used so that the exposed area will become soluble to photoresist developer. Subtractive etching (as opposed to additive/lift off) is also used to make the effects of the mask more obvious. In this book, we will primarily focus on subtractive: lift off technology is beyond the scope of this book.

The choice between light field and dark field is a question more relevant for fab engineers and mask layout designers than for TCAD users and may cause some ambiguity for beginners. This is complicated by the fact that negative photoresist and dark field masks are roughly equivalent to positive photoresist and light field masks.

The convention chosen for the MaskEditor GUI is to assign positive and negative to the polarity of the mask/photoresist combo. Positive means the drawn area is being protected by the (positive) photoresist, just like light field in Fig. 4.4. Negative means the drawn area resist will be removed, just like dark field in Fig. 4.4. Implicitly, this means positive photoresist is always used for the examples in this book.

4.2.3 *Photoresist Processing Steps*

The steps involved in photoresist processing steps [56] are shown in Fig. 4.5.

Photoresist is first spun onto the wafer (step 1), and soft baked at 60–100°C for about 5–30 min to drive off solvents (step 2) [57]. It is then exposed using a photo mask stepper (step 3). The photoresist is developed and hard baked in an oven for 20–30 min at 120–180°C to solidify the remaining photoresist (step 4). Finally the oxide pattern is etched and the leftover photoresist is stripped (step 5).

4.3 Initial Setup

4.3.1 *Creating Layers Using MaskEditor*

This book uses MaskEditor as the interface between the GDSII layout files and the 3D process simulator; this software tool can be freely downloaded from this website [10]. Since we will mention MaskEditor extensively throughout

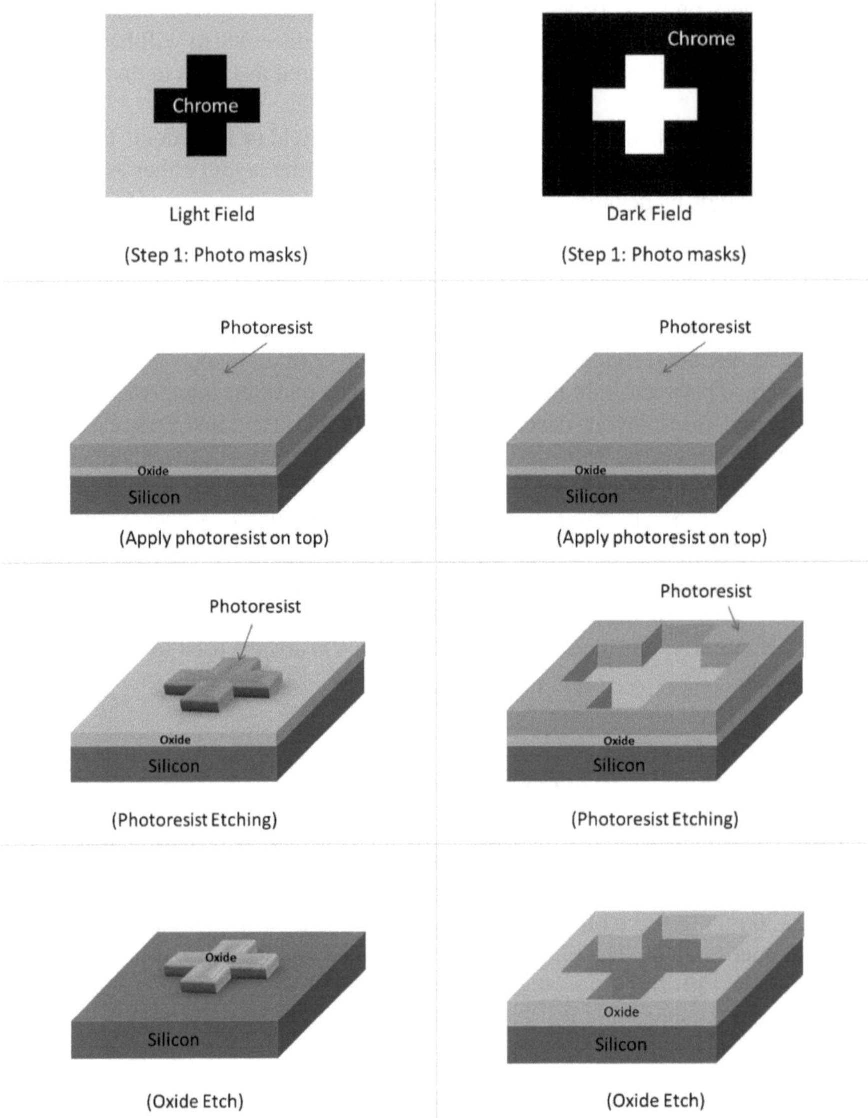

Fig. 4.4 Light field vs. dark field for subtractive etching with positive photoresist

the book, it is necessary to familiarize the reader with this tool. Again, the scope of this book is about methods to simulate 3D semiconductor devices and provide useful information for device design and TCAD. The methods used in this book are not limited to a particular software suite. Different tools for setting up 3D

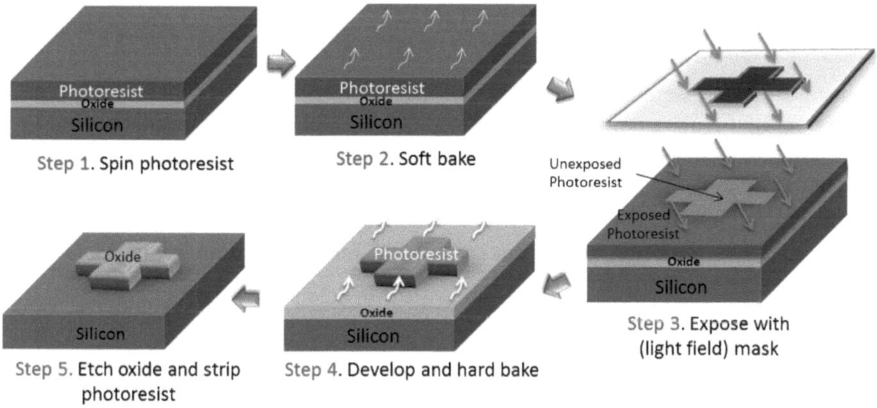

Fig. 4.5 Photoresist processing steps

TCAD may have different functions and applications, but the general ideas are the same:

- To extract individual layer mask from either GDSII or other formats
- To convert these layout masks to files containing geometric information accessible to the process simulator
- To provide further simplifications of the process simulation by letting the user define the initial mesh, substrate material, layer purpose, etc.

As an example, MaskEditor was created to work with the industry-standard GDSII format but is also capable for designing its own layout shapes and layers in stand-alone mode.

1. MaskEditor extracts and divides the layer masks from either GDSII or its own layout to individual layers. For instance, a layout mask set contains several layers. After importing the layout file, they will be displayed with different colors and patterns.
2. These layers contain basic information like layer number and geometric sizes and shapes. MaskEditor will convert the information to a format that can be read by the process simulator and save it to separate files for every layer. The process simulator will later load these files at user's request to create masks, perform etches, etc.
3. User may prefer to edit the layer properties like geometrical attribution, layer polarity (see explanations in Sect. 4.2), layer purpose (implant, etch, change material), etc. MaskEditor is designed to handle these requests.
4. To further reduce the user's workload, MaskEditor is designed to initialize the substrate with x y z mesh information and substrate material. A process simulation input file template, which contains statements like initialize 3D simulation, load the mask files, save simulation results, is created automatically after the user finishes setup.

4.3.2 Importing Layers from GDSII

MaskEditor can be used as a viewer of GDSII formatted files, besides the basic function of being an interface between layout and process simulator. Various settings are available including the choice of which layers to import into a new design. The scale of the pattern being imported is critical: the user can choose to convert the units used in the GDSII file and can also specify a scaling factor if required. Once imported, the layout can also be modified from within MaskEditor.

4.3.3 Simulation Area Define

The simulation area defines the outer boundary of the simulation. By default, MaskEditor defines this as the smallest rectangle which contains all the shapes from every layer of the layout. It is the initial area used to build the substrate mesh. Silicon is the default substrate material. Note that the simulation area is a top down view or the x-z plane. Substrate mesh refers to mesh on the x-y planes. As discussed in Chap. 1, this book adopts the 3D stacked planes method. In this method, the 3D structure is created by stacking x-y planes. The initial mesh for each plane can be defined automatically by using GUI included in MaskEditor, either through universal basic mesh define (same mesh for all planes) or "segmented" mesh define. Refer to Sect. 4.3.6 for more information.

By changing the simulation area, one can easily choose the entire device or just a small part. This is especially useful for repeating shapes where symmetry can be exploited to reduce the simulation time.

4.3.4 Layer Properties Define in MaskEditor

As can be seen in Fig. 4.5, there are many steps involved every time a photo mask and resist are used in a real process. From a software point of view though, these steps can be simplified and consolidated into a single command in order to save on simulation time. To that end, a new command named `mask` is defined in the process simulator. This command integrates the 2 process steps:

1. Photoresist is deposited with thickness defined by the parameter `thick`.
2. Photoresist is geometrically etched according to `x1.from x1.to x2.from, x2.to.` etc. (e.g. the photoresist outside the defined area of x1.from x1.to is etched away)

In order to facilitate its use, the low-level mask command is generated automatically by MaskEditor and saved to a mask file for a particular mask layer (every layer has a mask file saved to the working folder after the project is properly generated and saved).

Text Box 4.1 Content of a Typical Mask File (.msk) with General Purpose

```
mask segm=3 thick=1. x1.from=1 x1.to=2 x2.from=3 x2.to=4
```

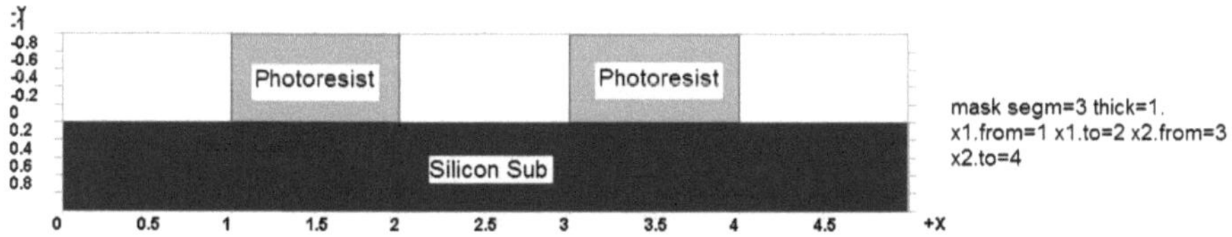

Fig. 4.6 The "general" layer purpose process simulation result

Depending on the process step that is associated with a particular mask, a "purpose" property should be defined.

The first setting of this property is "general" and is used for all process steps which merely alter the properties of the exposed material and do not add or remove material. A good example of this is ion implantation: in a real process, the left-over resist serves to block the ion implantation so that only the exposed area is doped but no material is added or removed. Text Box 4.1 is a typical statement found in mask files created by MaskEditor for the general purpose. For simplification purpose, only 1 segment is shown here. Please keep in mind that the mask file contains the commands for each plane/segment. In this book, the phases "plane" and "segment" have the same meaning, the stacking x-y plane.

This statement will create a simulation result like the one shown in Fig. 4.6. Note that after the desired process steps are performed (e.g. implantation), photoresist needs to be removed manually using the statement `etch photoresist all`.

The second setting is "etch" and is used when material must be removed from the exposed area: a typical example of this is plasma etching. This setting simply removes all the selected material in the area defined by the mask. Etch depth and etch angle can be defined in the GUI. MaskEditor will then create a mask file with extension (.msk) which contains commands that perform the following process steps on every plane/segment:

1. `mask` command. As discussed before, this command integrates two process steps: (1) to deposit and (2) selectively etch photoresist for subsequent process.
2. `etch` command. This command instructs the simulator to etch selected material in the area defined by the mask.
3. `etch photoresist all`. Remove all remaining photoresist. The etch command assumes user's only purpose of using this mask is to etch, so the photoresist can be removed for the user to simplify the simulation input file.

Text Box 4.2 Content of a Typical Mask File (.msk) with "Etch" Purpose

```
mask segm=3 thick=1. x1.from=1 x1.to=2 x2.from=3 x2.to=4
etch segm=3 silicon avoidmask depth=0.5
etch segm=3 photoresist all
```

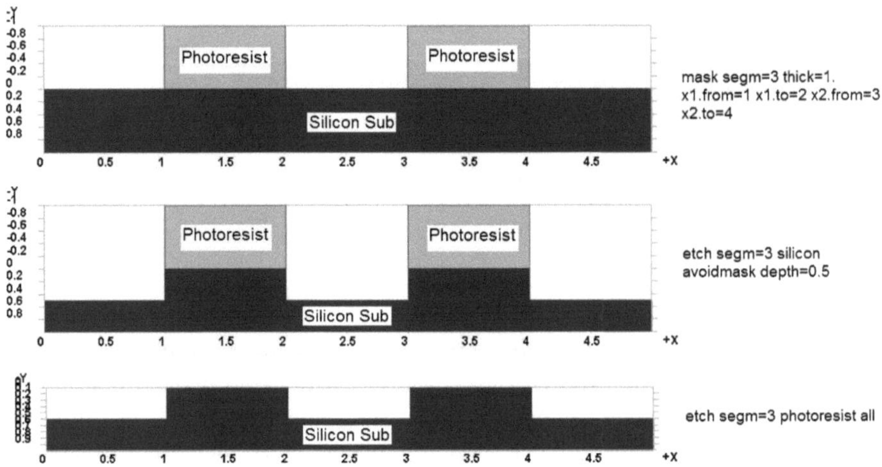

Fig. 4.7 The "etch" layer purpose process steps

As an example, the content of a typical mask file (.msk) generated by MaskEditor with "etch" as layer purpose is shown in Text Box 4.2. A photoresist layer 1 um thick is deposited and etched according to the first statement line. Then the `etch` statement instructs the simulator to etch silicon with a depth of 0.5 um. Finally the leftover photoresist is removed. Figure 4.7 visualized the process with each statement.

The last setting that can be used is "change material". This is used in cases like shallow trench isolation formation (STI) where one needs to first etch material A and fill with material B. Since one cannot precisely fill up the well, most often a Chemical Mechanical Polish (CMP) is necessary to remove excessive B material from the top of material A, which makes this a 3-step process. To simplify matters, MaskEditor supports a new method named "change material" in the process simulator. Instead of doing all the process steps explicitly, just one step is necessary: change the original material A to the desired material B at designated location with user specified depth.

The "change material" layer purpose contains the following steps:

1. Etch material A in the defined area with user specified etch depth.
2. Fill with the second material, B
3. Chemical mechanical polishing to remove excess B leftover.

Text Box 4.3 Content of a Typical Mask File (.msk) with "Change Material" Purpose

```
mask segm=3 thick=1. x1.from=1 x1.to=2 x2.from=3 x2.to=4
change_material dry segm=3 silicon /oxide thick=0.5
etch segm=3 photoresist all
```

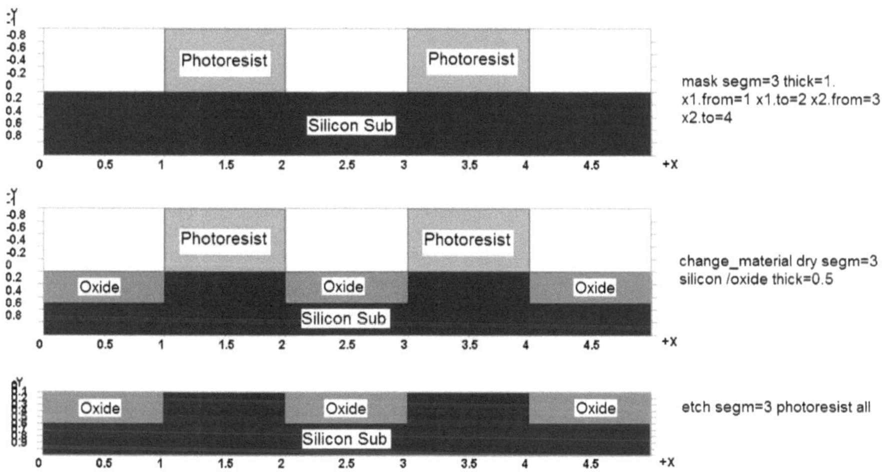

Fig. 4.8 The "change material" layer purpose process steps

As an example, a portion of a typical mask file (.msk) created by MaskEditor for "change material" purpose is illustrated in Text Box 4.3. Again, only one plane for one layer is shown here.

Figure 4.8 is the illustration of process steps for "change material" layer purpose. Readers are encouraged to compare and contrast "change material" with the process steps of Fig. 4.7.

Please note that while the use of the mask command through MaskEditor simplifies the process, it should not be used if one wants to take a look at the more detailed process simulation results. For example, in a real STI process, there is often an extra step after etching called STI oxide liner growth. This is typically a thermal oxide instead of a deposited oxide and is used to improve the silicon/SiO_2 interface. Since "change material" simplifies matters, it should not be used if one is interested in this distinction. The user must decide when it is appropriate to use masks to simplify the process simulation.

A review of the different mask settings in MaskEditor is shown in Fig. 4.9. Since "general purpose" is mostly for implantation, so ion implantation is illustrated.

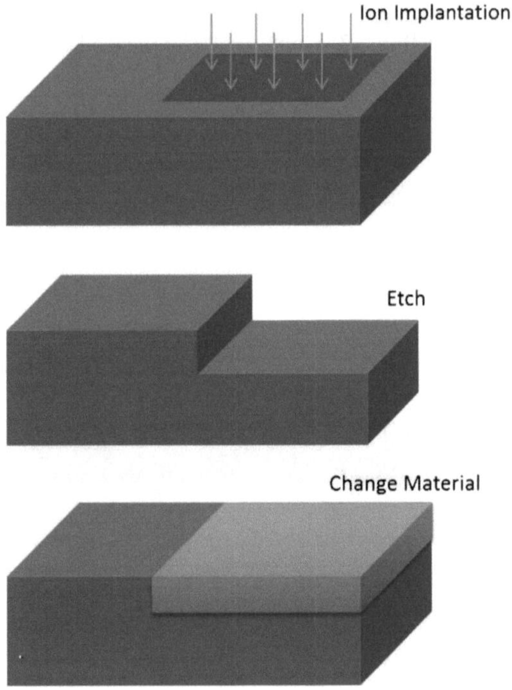

Fig. 4.9 Three mask layer purposes in MaskEditor

4.3.5 *Cut Lines*

After defining the simulation area and specified layer property, the user should define cut lines in the mask layout to start the definition of the 3D mesh. As we have discussed previously, this book uses a stacked mesh plane approach to 3D mesh generation. By convention, mask layers are in the x-z plane so cut lines at specific z values define the required x-y mesh planes. However, we also remind the reader that the z-direction is assumed to be sparsely meshed because it is where there is the least variation in material properties. The user thus has the choice of setting vertical or horizontal cut lines in the plane of the mask layer: the z-direction of the simulation is defined as being perpendicular to these cut lines.

The cut location is automatically determined by MaskEditor by detecting variations in the z direction; two cuts are applied at closely-spaced material boundaries to help with accuracy. Figure 4.10 shows an example of the automatic cut lines detection in MaskEditor.

Note that MaskEditor only adds the absolute minimum number of mesh planes required for the simulation to function correctly. Users should not attempt to remove the cut lines acting as material boundaries as it will adversely affect the simulation accuracy. However, the simulation accuracy can be improved by adding cuts to locations of interest or using the automatic "Add cut" feature. In future, it is possible to work with bended planes, see Sect. 4.12 for more information.

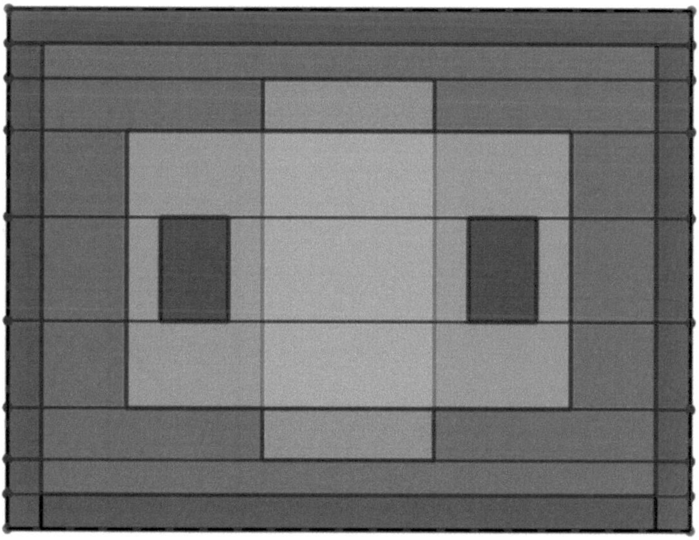

Fig. 4.10 Automatic cut line detection by MaskEditor

4.3.6 3D Mesh Definition

3D mesh generation can be done in two steps: defining the longitudinal mesh through the cut lines as discussed in the previous section and defining the lateral 2D mesh for each plane. The initial mesh definition controls the x-y mesh allocation for the substrate since that is the starting point of all process simulations. The user has control over the total number of mesh points and the uniformity of the mesh.

Areas of the substrate which will be the focal points of processes or where electrical effects are expected to be important can be meshed more densely to improve the accuracy of the simulation. This includes channel regions, p-n junctions and so on.

Note that when additional material is deposited later in the process simulation, the user does have the capability to add horizontal mesh lines. This gives the user control over the vertical resolution of the new mesh that will be added as a result. However, these new layers will automatically inherit the vertical mesh lines of the substrate upon which they are deposited so it is important to generate sufficiently dense mesh at the beginning of the process simulation. Otherwise, more complicated re-grid procedures will be required to fix the mesh.

Although the lateral (x-y plane) mesh can be defined independently for each plane, the basic setting is to use the same lateral mesh everywhere ("basic mesh"). This is usually sufficient when there is only one longitudinal (z direction) segment or when there is little lateral variation between the different cut planes.

In other cases, there is a lot of lateral variation between the various cut planes and it would be beneficial to have denser lateral mesh in different x/y locations. In that case, the user should use the "segmented mesh" option.

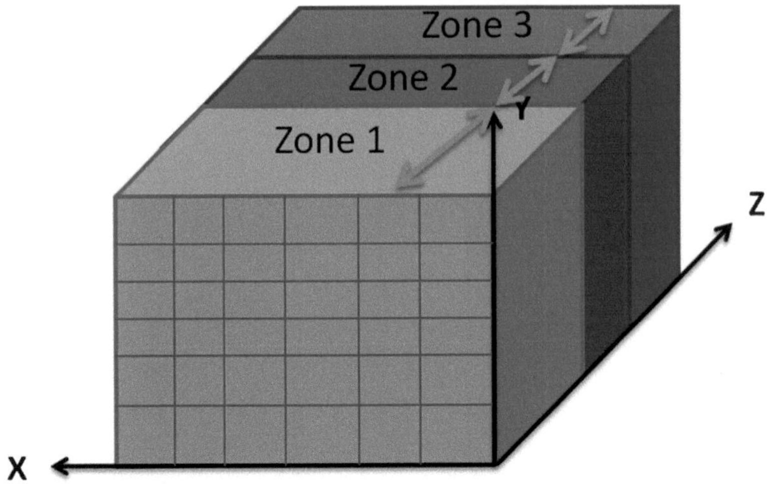

Fig. 4.11 Segmented mesh define

Figure 4.11 shows an example where three zones with different x-y mesh are being defined. These zones, labeled Zone 1, Zone 2 and Zone 3, are specified using z range (e.g. from z = 0 um to z = 10 um) where they apply. Each zone has its own lateral mesh definition. In the extreme case, every mesh plane can define its own lateral mesh.

4.4 MaskEditor Intermediate Files

MaskEditor generates a number of intermediate files which are used in the process simulation. These files are in text format and contain numerous low-level process simulation commands which would have been manually input by the user in older versions of the process simulation suite. Needless to say, MaskEditor is a vast improvement over the old way of doing things.

There are four main file types generated by MaskEditor which are used by the process simulator. Files with the pattern "geo*.in" contain geometric and lateral (x-y plane) mesh information as shown in Text Box 4.4. There is one such file for every mesh plane defined in MaskEditor; they are numbered starting from 1 for the plane at z = 0.

A separate file with the ".in" extension (main.in) is also created: this will be the main simulation file for process simulation. When originally created by MaskEditor, this file will contain only a basic template for a 3D simulation and a few basic commands like `mode` and `3d_mesh`. However, this file also tells the process simulator to read in all the other input files generated by MaskEditor through the `include` command (Text Box 4.5).

Text Box 4.4 Content of geo Files

```
line x loc= 0.0    spacing= 0.25    tag=lft
line x loc= 5.0    spacing= 0.25    tag=rht

line y loc= 0.0    spacing= 0.05    tag=top
line y loc= 2.0    spacing= 0.15    tag=bot

elimin y.dir xlo= 0.0 xhi= 5.0 ylo= 1.0 yhi= 1.0 ntimes=2

region silicon xlo=lft xhi=rht ylo=top yhi=bot
bound exposed xlo=lft xhi=rht ylo=top yhi=top
bound backside xlo=lft xhi=rht ylo=bot yhi=bot
```

Text Box 4.5 Content of main.in Template

```
mode quasi3d
3d_mesh inf=geo
#restart file=xxxx.str

init
include file=mos.gds1.msk
struct outf=01_mask.str
include file=mos.gds2.msk
struct outf=02_mask.str

export outf=npn.gds.aps xpsize=0.001
```

Here are some explanations for the code in the main.in template file.

The `struct` command is the abbreviation of structure command. The process simulator allows abbreviations provided that there is no ambivalence about their meaning. This command is used to output the physical structure. This is useful to do at various process steps since it allows the user to change a process step mid-way without redoing the entire simulation. Note that the `restart` command used here is comment out by a "#" sign. In addition to the commands defined through MaskEditor and separated in various external files specified by `include` statements, some default settings are also included through the suprem.key and modelrc files in the process simulator installation directory. The user may reissue any of these commands to change these default settings. A full discussion of the available commands can be found in the reference manual of the process simulator [9] and in later chapters of this book but let us go over a few basics:

- The `line` command in the geo file is used to specify the position and spacing of mesh lines. All line statements should come before `region` and `boundary` statements, which should in turn be followed by an `initialize` statement to actually generate the mesh. The `location` parameter specifies the location along the chosen axis (in microns) while `spacing` is the local grid spacing (also in microns). The process simulator will add mesh lines to the ones given according to the following recipe:

  – Each user-defined line has a spacing which is either specified explicitly or inferred from the nearest neighbor. These spacings are then smoothed out so no adjacent intervals will have a ratio greater than 1.5. New grid lines are then introduced so that the line spacing varies geometrically from one end of the interval to the other.
  – Each line also defines a tag which can be any words of the user's choosing; this tag is used to refer to a particular line in other commands.

- `eliminate` is used to manually control the mesh line densities in the initial stage of a simulation. Given a mesh line direction and an area, using this command once results in elimination of half of the mesh lines (spread evenly). Subsequent uses results in a further reduction of mesh lines. `x.direction` and `y.direction` indicate whether mesh lines in the x or y direction are to be eliminated. `xlo, xhi, ylo, yhi` are the low and high coordinates of an area where mesh will be eliminated. Please note that this area should be within a region defined by a previous `region` command. `ntimes` is the number of "half-mesh" action to be performed.
- The `region` command is used to specify the material of rectangles in a rectangular mesh. This statement should follow line statements. Every element must be given some material, so at least one region statement is required for each rectangular mesh. `xlo, ylo, xhi` and `yhi` are the boundaries of the rectangle being specified and should correspond to a tag previously created in a `line` statement (e.g. `top, bot`, etc.). The substrate material should be specified with this command (e.g. silicon).
- A `boundary` statement is used to specify what conditions to apply at each surface in a rectangular mesh. At present, three surface types are recognized. `exposed` surfaces correspond to the top of the wafer. Materials are only deposited on exposed surfaces: this is a common source of user error and should be investigated when `deposit` commands appear to have no effect. Impurity pre-deposition also happens at exposed surfaces, as does defect recombination and generation. Backside surfaces roughly correspond to a nitride- or oxide-capped backside. Defect recombination and generation happen here. `reflecting` surfaces correspond to the sides of the device and/or symmetric boundary conditions. This setting can also be used the backside if defects are not being simulated and is the default for surfaces.

MaskEditor also creates mask files with the extension ".msk". Text Box 4.6 shows the content of a mask file for a layer with "etch" purpose. The mask files typically contain a group of `mask, etch` or `change material` commands. The `segm` parameter controls which plane the `mask` command works on. There are four planes for this simulation; two of them have blanket etch of nitride while the other two have selective etch of nitride. Note that the ones with blanket etch do not need a photo mask and therefore there is no `etch photoresist all` statement either.

Text Box 4.6 Content of a .msk File

```
etch dry segm=1 nitride thick=0.5

etch dry segm=2 nitride thick=0.5

mask segm=3 thick=1. x1.from=1.2 x1.to=3.8
etch segm=3 nitride avoidmask depth=0.5
etch segm=3 photoresist all

mask segm=4 thick=1. x1.from=1.2 x1.to=3.8
etch segm=4 nitride avoidmask depth=0.5
etch segm=4 photoresist all
```

Text Box 4.7 Content of zmesh.zst File

```
begin_zst
3d_solution_method 3d_flow=yes
include file=suprem_taper_label.txt
z_structure uniform_zseg_from=0.0 uniform_zseg_to=0.0 zplanes=1 zseg_num=1
z_structure uniform_zseg_from=0.2 uniform_zseg_to=0.2 zplanes=1 zseg_num=2
z_structure uniform_zseg_from=0.5 uniform_zseg_to=0.5 zplanes=1 zseg_num=3
z_structure uniform_zseg_from=1.0 uniform_zseg_to=1.0 zplanes=1 zseg_num=4
z_structure uniform_zseg_from=1.5 uniform_zseg_to=1.5 zplanes=1 zseg_num=5

load_mesh mesh_inf=f1.msh zseg_num=1
load_mesh mesh_inf=f2.msh zseg_num=2
load_mesh mesh_inf=f3.msh zseg_num=3
load_mesh mesh_inf=f4.msh zseg_num=4
load_mesh mesh_inf=f5.msh zseg_num=5
output sol_outf=tmp.out
export_3dgeo file=h_cvd.3dgeo
end_zst
```

Note that when using general purpose masks, it is expected that a command such as `implant` will follow the mask step in the process simulator. In this situation, the .msk file will not include the command to remove the left-over photoresist since it is needed for the implantation; this must be done by the user in process simulation after the implantation is complete. For etch or change material masks, user simply needs to use the `include` command to import the .msk file and does not need to worry about the resist.

The last file created by MaskEditor is zmesh.zst file. This file contains the longitudinal mesh information (i.e. the location of the x-y mesh planes on the z axis). The process simulator will first load this file together with the geo files to build the 3D structure. The content of a typical zmesh.zst file is shown in Text Box 4.7. The user does not need to understand the contents of this file: it suffices to say that it defines the z locations of the mesh planes.

MaskEditor generates one additional file which is not used by the process simulator: the .cut file saves the user input from the MaskEditor GUI. It contains information such as layer settings including negative or positive polarity, purpose of layers, etc. Normally user will never have to open this file for editing but it can be reloaded by the MaskEditor GUI to continue work on a mask layout.

4.5 Running a 3D Process Simulation

At this point, all the necessary input files have been generated and one may run a process simulation. However, we first recommend renaming the main.in file generated by MaskEditor in case it is accidentally overwritten. As mentioned before, MaskEditor automatically creates a small template main.in file to help the user get started.

With this template file, we will be able to add the process simulation steps we want. Simulation always starts with the `mode` command and each mask layer is loaded to the simulator by the `include` command. It is recommended that for each step, a `structure` command be used to save the simulation results to an output file. This allows the user to restart a simulation if it fails or simply to alter the process step (e.g. the implantation dose) without having to go back to very beginning. A good practice is to add a number to each saved structure file. Since a complicated process simulation usually consists of many steps and output files, a file number will make it easy to find the corresponding step and file. Afterwards, we will need to export the process simulation result to a device simulator if electrical/thermal/optical modeling is required.

4.6 Export Process Simulation Data to Device Simulator

The details of export process which links the process to the device simulation depends on the software suite being used but the principles are usually similar. In this book, this is done using a command called `export`. This command exports all the structural and mesh data in a format readable by the device simulator. The exported mesh file has the extension ".aps". In addition to the mesh file, the `export` command of process simulator creates two other files used by device simulator.

The first file generated is a solution template named main_3d.sol. Much like the main.in template generated by MaskEditor for process simulations, this file should be renamed to avoid accidentally overwriting it. Text Box 4.8 shows a sample of solution template file for device simulator.

The solution template contains a number of simulation commands for the device simulator that can be customized by the user. For the sake convenience, the

Text Box 4.8 Solution Template Generated by Process Simulation

```
$-------Part 1: Welcome to Apsys 3D ---------------------------------
begin
convention positive_current_flow=inward

$------- Part 2: Input statement -----------------------------
$------- load mesh from csuprem output .aps file --------------
include file=zmesh.zst ignore1=load_mesh ignore2=output ignore3=export_3dgeo
load_mesh mesh_inf=sup.aps suprem_import=yes

$------- Part 3: Output statement, additional physics & alias----
output sol_outf=sup.out
$ more_output impact_ionization=yes space_charge=yes
$ define_alias alias=Vs name=voltage_1

$------- Part 4: Load material and contact information ------
$ use contact designer program to define contact
include file=contact_3d.sol

$------- Part 5: Main scan command ---------------------------
$ restart data_set=1

$ define Newton parameters
newton_par damping_step=2. max_iter=50 opt_iter=15 stop_iter=15

$ scan line #1--equilibrium sets all contacts to zero
equilibrium

$ scan line #2
scan var=voltage_1 value_to=1
end
```

template also includes some comment lines that explain the various commands and divide the file into sections:

- In part 1, `convention` determines the polarity of positive current flow. The default is that current flowing out of the device and into an electrode is positive and by Kirchhoff's current law, the sum of the current on all electrodes is zero.
- In part 2, `include` command loads the zmesh.zst file: this is the same file created by MaskEditor and used earlier by process simulation.
- In part 3, the name of the device simulation output files is chosen. The actual file names will also have a number appended to the end since output is generated at different bias steps. Also additional device simulation commands can be added to enable various physical models such as impact ionization.

  Special attention should be paid to part 4, which will load the file called contact_3d.sol. This file is not created automatically by process simulation export: instead, the user needs to set up the contacts using a small embedded program called ContactDesigner from the process simulator GUI.

We note quickly that for device simulation, contacts refer to the equipotential boundary regions used to solve the drift-diffusion and Poisson equations. Even if contact metals are defined in the process simulation, these electrical boundaries must still be defined for the device simulation.

In ContactDesigner, the user can input geometric information for all the contact boundaries (i.e. x, y, and z range) and specify the materials that are used as a boundary reference: it simplifies matter if a contact boundary "touches" only one semiconductor material. ContactDesigner then loads the material and geometric information from material_3d.sol (to be discussed later) and combines it with the user-defined contacts to generate a new file: contact_3d.sol.

A contact number is also required as the device simulator uses this to apply bias. For example, voltage_1 is the voltage on electrode #1 and current_2 is the total current on all electrodes sharing the number 2.

- Part 5 is the input area for electrical and optical bias. The device simulator also supports restarts from previously saved device simulation steps in this section. The other major command in this section is `newton_par` which controls the behavior of the sparse Newton equation solver: this setting can be very important in achieving convergence and will be discussed at greater length in subsequent chapters.

The second file created by the process simulation `export` command is material_3d.sol. This file defines the various materials found in the various segments/planes of the process simulation. For device simulation, it requires that various physical parameters like bandgap and carrier mobility be defined for these materials: this is usually done through a bundle of commands called a macro which is identified by the material name. More details on how to define material parameters and macros will be found in the reference manual of the device simulator [58].

An observant eye may notice that material_3d.sol is not explicitly included in the solution template of Text Box 4.8. This is because this file is indirectly included through contact_3d.sol when contacts are defined with the ContactDesigner application. This application reads in material_3d.sol to assign contact boundaries to the right segments as previously discussed.

4.7 Running 3D Device Simulation

The main_3d.sol template provides a good starting point for device simulation. Various types of simulation, including DC, AC, transient, etc., can be carried out with simple modification of this template file. The user may add additional physics, or make modifications to physical parameters like carrier mobility, impact ionization, material macros, etc., all within the device simulation GUI framework.

There are two types of output file associated with the device simulator. The first type has an extension of .std, which contains mesh and simulated data, such as electron concentration, current magnitude, potential, electric field, band-gap, temperature and space charge. The second file type has an extension of .out, which is used to plot curves like *I*-*V* curves, *C*-*V* curves, Smith Chart, etc.

4.8 Using Plotting GUI to View the Simulation Result

One of the most important graphic user interfaces for TCAD simulation is the plotting GUI. Without the ability to plot simulation results, there is not much point in doing the modeling in the first place. The plotting GUI allows the user to view both process and device simulation results: IV curves, band gap diagrams, doping profiles, etc.

When doing batch simulations, it is also useful to be able to do batch post-processing to extract and compare data from the various simulations. This can be a function of the plotting GUI or sometimes the device/process simulator will have built-in extract functions.

4.9 About Process and Device Simulator GUI

Process simulation can be done under DOS command or UNIX shell which has the benefit of facilitating batch processing. However, this can be quite inconvenient and imposes a learning curve on new users. Most commercial TCAD software provides easy to use GUI programs to reduce the simulation workload on the user.

Here a typical simulator GUI is shown in Fig. 4.12. There are four zones of interest. The input and output file zones show the files associated with a simulation project. The input code zone allows direct editing of the input files while the simulation output zone shows real-time output information from the simulation. Afterwards, the user can save the simulation output messages to a log file and the structural files can be viewed by opening the output files to launch the plotting GUI. Here are some typical requirements for the TCAD simulation GUI:

1. Easy to use. New users should be able to pick up the basics within minutes.
2. Real time help for the command syntax.
3. Integration with other GUIs of the software suite, like the 3D setting up tool and plotting GUI.

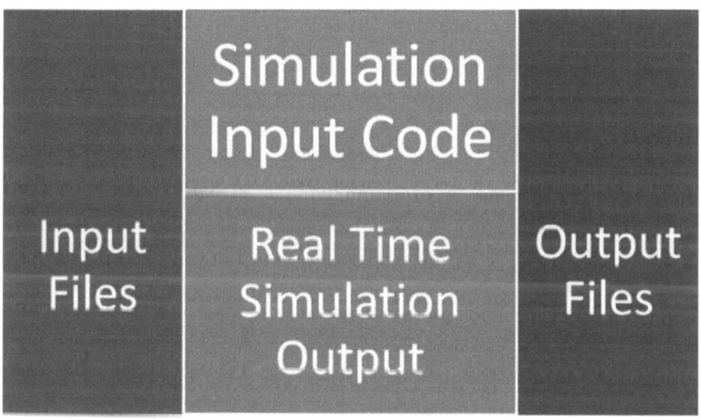

Fig. 4.12 The four zones of a TCAD simulator GUI

4. Supports restart function when the simulation was interrupted.
5. Able to perform batch simulations.

One highly demanding features of TCAD simulation GUI is the capability to perform batch simulation: this refers to a series of simulations associated with different process parameters, different device simulation parameters, etc. In the semiconductor industry this is useful because optimizing device performance often requires adjusting process parameters. There may be hundreds of possible process step combinations so it is useful to automate the process as much as possible and leave the computer to do most of the work. If multiple PCs are available, the GUI should be capable to automatically distribute the tasks to different computers that still have available computing resources.

4.10 3D TCAD Simulation Flow Chart

By now, we have presented the whole picture for 3D TCAD simulation. We can take a look at the relationship between the previously discussed tools, and see how they interact with each other to create the user anticipated results (Fig. 4.13).

4.11 About CPU and GPU Simulation

Until recently, most TCAD simulations have been done using the CPUs, either with consumer PCs or cluster-type supercomputers. Recently, the technology in graphical processing units (GPU) has begun to be applied to other fields beside 3D rendering and gaming. By exploiting the parallel architecture of the GPU and its hundreds of processing cores, GPU computing allows the creation of "personal supercomputers" [59].

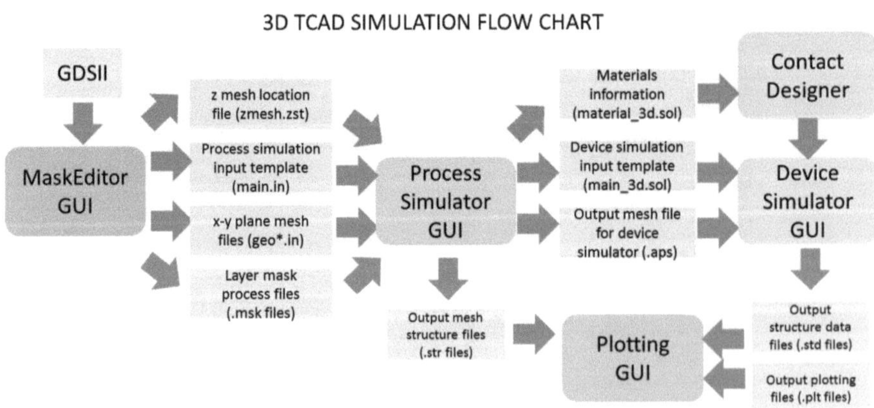

Fig. 4.13 3D TCAD simulation flow chart

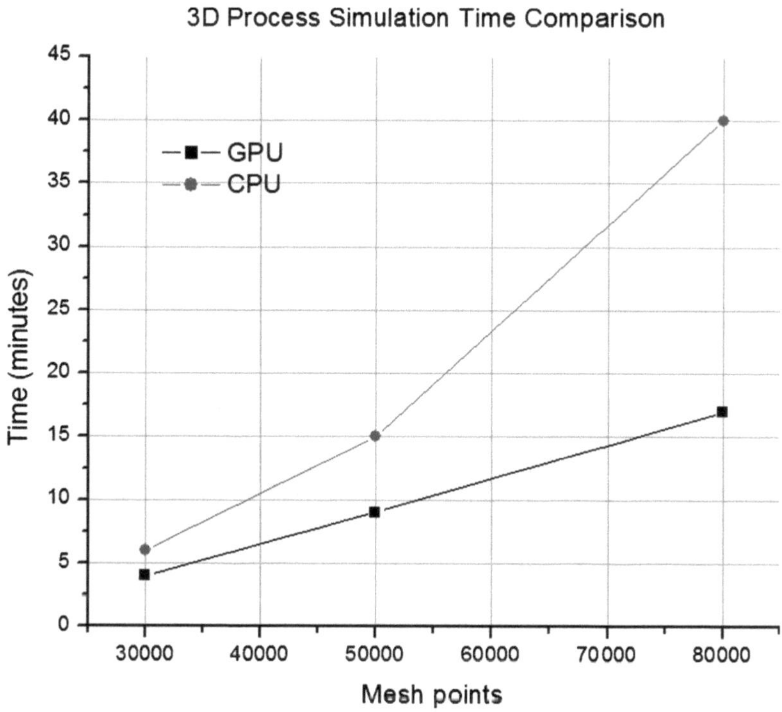

Fig. 4.14 Process simulation time comparison, CPU vs. GPU

What GPU cores lack in terms of their individual speed compared to a CPU, they more than make up in their numbers: tasks that can be parallelized benefit from GPU computing. In some areas such as FDTD speed-ups of 100× over CPU computation have been reported [60]. For TCAD simulation, an essential part of the calculation is the sparse linear solver that is at the core of the non-linear Newton method. It is expected that for devices where the mesh count is above 10^5, a factor of 2 increase in speed can be achieved using a GPU-accelerated solver.

Figure 4.14 illustrates a comparison between GPU and CPU in full 3D process simulation on the same computer. As the mesh increases, the better scaling provided by the parallel linear solver widens the gap and makes GPU-accelerated solvers a very promising approach to improving the TCAD simulation time. The computer configuration for this comparison is: Intel core i7 920@ 2.67GHz with 12G memory and 64bit Windows 7 OS. GPU: NVidia Tesla C1060.

4.12 Bended Planes

Bended planes method is a novel and futuristic meshing technology that is still under development. This method allows both bended and straight planes. This allows fewer planes to be used to represent structures with curved shapes and reduces the overall number of mesh points (Fig. 4.15).

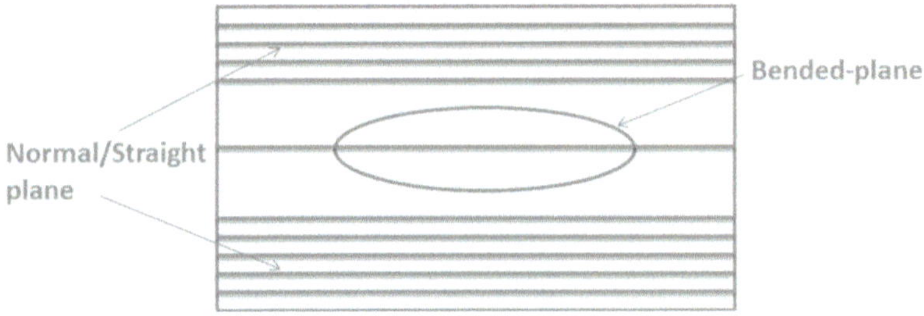

Fig. 4.15 Bended planes method

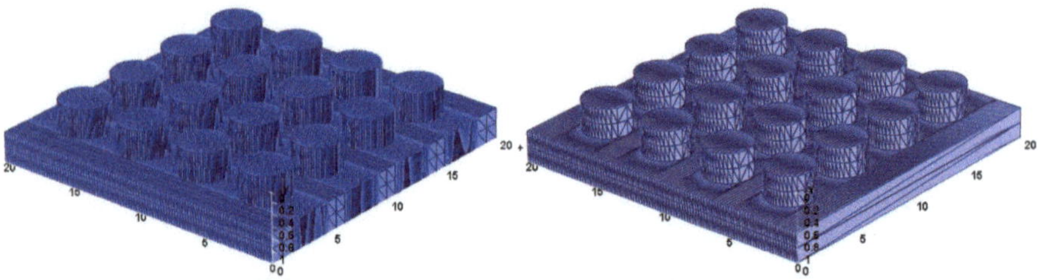

- Total Mesh Count: 52682
- Straight planes: 134
- Process Simulation time: 50 minutes

- Total Mesh Count: 4958
- Planes: 2 straight+ 16bended planes
- Process Simulation time: 2 minutes

Fig. 4.16 Comparison between straight planes only and bended planes method

As an example, a 4×4 etched pillar structure is shown with straight planes only and bended planes method. The bended planes method uses far fewer planes and thus has only 1/10th of the total mesh count. Note in Fig. 4.16, total of 134 planes are used to create the 16-pillar structure with straight planes only, while only 2 straight planes are necessary for bended planes. Extra 16 bended planes are used for the 16 pillars with the bended planes method.

Chapter 5
P-N Junction Diode

5.1 The Simplest P-N Junction Diode

To get the reader familiarize with 3D TCAD simulation, we start with the simplest p-n junction diode imaginable. This diode has just one mask layer as shown in Fig. 5.1. Strictly speaking, this simulation example is just an extension of a 2D problem since there is no variation in the z direction; it could even be modeled in 1D. Nevertheless, this is a good place to start.

From this section on, every example will include an overview section, which lists the important process and device simulation steps. Simulation code is included so that the reader can have real simulation examples. Note that except this example, most examples have the process simulation code divided into separate sections. Each section has its own purpose of simulation. At the end of each example, a simulation data is presented to illustrate the simulation time, mesh counts and number of cut planes.

5.1.1 Overview of Simulation Steps

This is the simplest example of this book so the overview of process steps in Table 5.1 is quite short. It can be broken down simply as process simulation, contact definitions and device simulation. Thumbnail figures for every process step help explain what is being done at each stage.

5.1.2 Process Simulation

An n+ substrate is first created for the n-type region of the p-n junction diode. Silicon epitaxy is then used to grow the silicon layer on top of the substrate to create the p-type region. Note that in process simulation, epitaxy is realized through deposit. The epitaxial layer can be grown at user-defined temperature, 300 K

S. Li and Y. Fu, *3D TCAD Simulation for Semiconductor Processes, Devices and Optoelectronics*, DOI 10.1007/978-1-4614-0481-1_5,

Fig. 5.1 Layout of the simplest p-n junction diode

Table 5.1 Overview of simulation steps for the simplest p-n junction diode

Simple p-n junction diode	Process simulation steps
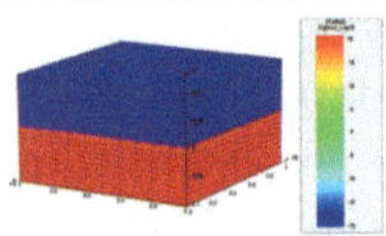	Step 1: Process simulation
Simple p-n junction diode	Contact definitions for device simulation
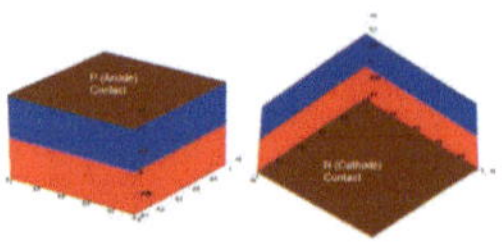	Step 2: Contact definitions for device simulation
Simple p-n junction diode	Device simulation
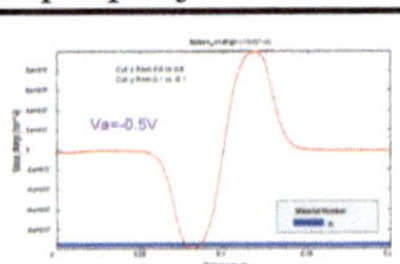	Step 3: Space charge vs. applied bias
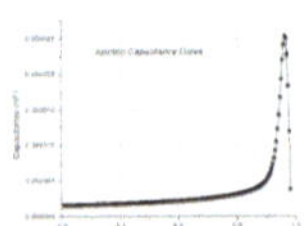	Step 4: Junction capacitance

Text Box 5.1 The Content of geo Files for the Simplest P-N Junction Diode

```
line x loc=   0.00000    spacing=  0.100000       tag=lft
line x loc=   1.00000    spacing=  0.100000       tag=rht

line y loc=   0.00000    spacing=  0.100000e-01 tag=top
line y loc=  0.100000    spacing=  0.100000e-01 tag=bot

region silicon xlo=lft xhi=rht ylo=top yhi=bot
bound exposed xlo=lft xhi=rht ylo=top yhi=top
bound backside xlo=lft xhi=rht ylo=bot yhi=bot
```

being the default value. Both n-type and p-type regions use a constant doping so for the simulation, no implant is necessary. A diffusion step is applied to activate all dopants.

Process Simulation Code

```
mode quasi3d
3d_mesh inf=geo

# initialize the substrate n region
init phosphorus conc=1e18 orient=111
struct outf=1_pn.str

# grow silicon epi layer for p region
deposit silicon thick=0.1 boron conc=1e18 meshlayer=50

# activate the dopants and export
diffuse time=1 temp=950

struct outf=2_pn.str
export outf=pn.aps xpsize=0.001
```

The `mode` command specifies that a quasi-3D will be used in this example. This will neglect the diffusion of dopants between mesh planes which is unimportant in this case.

The `3d_mesh` command loads mesh declaration statements from the MaskEditor GUI. This example has 2 cut planes and a basic mesh so 2 identical geo files (geo1 and geo2) are created to build the 3D structure. The content of the geo files is shown in Text Box 5.1. The `line` statements define the mesh spacing at various x-y locations and the thickness of the substrate is 0.1 um.

The `init` statement initializes the mesh substrate. The parameters indicate it is doped with phosphorus so it can act as the n-region of the diode.

The `deposit` statement does the epitaxy growth of the p+ silicon layer. Vertical mesh lines are inherited from the substrate and horizontal lines for the new layer are defined using the `meshlayer` parameter.

The `diffuse` command does a short (1 min) annealing step at a furnace temperature of 950°C to activate the dopants.

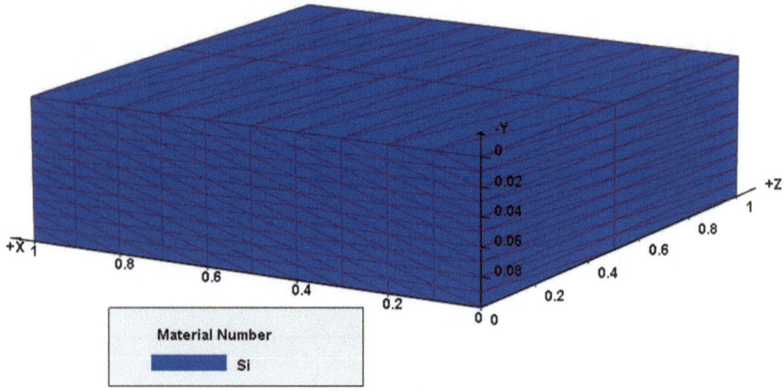

Fig. 5.2 Silicon substrate as the n+ region (1_pn.str)

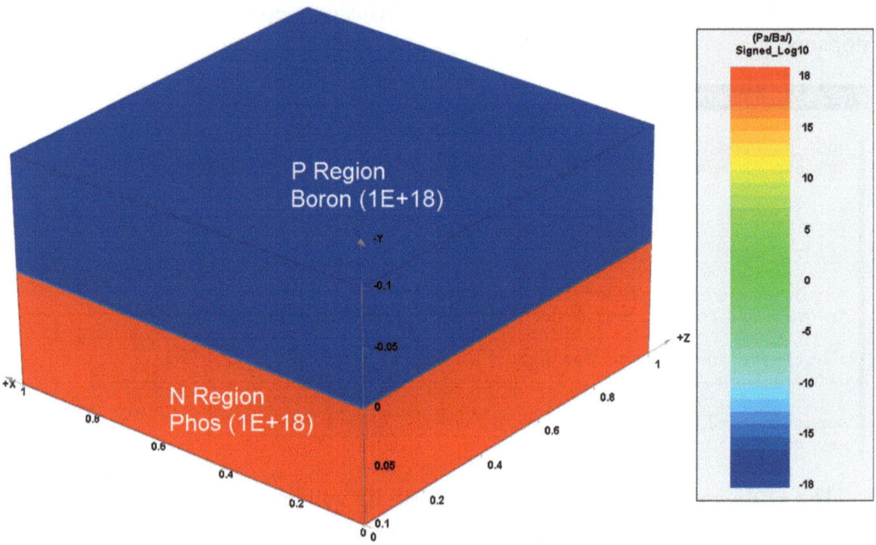

Fig. 5.3 Net doping chart of the finished structure (2_pn.str)

Figure 5.2 is the substrate (n-type region) of the p-n junction diode. The final net doping chart is given in Fig. 5.3.

5.1.3 Contact Definitions for Device Simulation

Contacts need to be defined to inform the device simulator of the location of the electrical boundaries. For this simple example, two contacts are defined, one on top and one at bottom: the doping of the layers is sufficient to provide a good quality ohmic contact so we skip the usual step of depositing metals. Figure 5.4 shows the contact definitions for simple p-n junction diode.

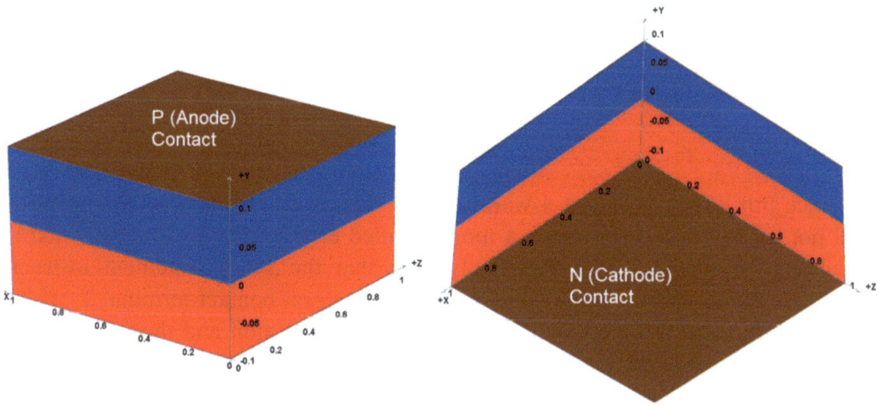

Fig. 5.4 Contact definitions for the simplest p-n junction diode

5.1.4 *Device Simulation*

Device simulation is performed for both DC bias condition and AC bias condition. The complete code for device simulation is copied below. Note that the contact_3d.sol contains contacts as well as material information for device simulation. Please refer to Chap. 4 for more information on how to set up these files.

Device Simulation Code

```
$-------Part 1: Welcome to Apsys 3D ---------------------------------
begin
convention positive_current_flow=inward

$------- Part 2: Input statement ------------------------------
$------- load mesh from process simulation output .aps file --------------
include file=zmesh.zst &&
  ignore1=load_mesh ignore2=output ignore3=export_3dgeo
load_mesh mesh_inf=pn.aps  suprem_import=yes

$------- Part 3: Output statement, additional physics & alias----
$------- You can change the name of output file for the device simulator -----
output sol_outf=pn.out
more_output impact_ionization=yes space_charge=yes

$------- Part 4: Load material and contact information --------
$ use ContactDesigner program to define contact
include file=contact_3d.sol

$------- Part 5: Main scan command ----------------------------
$ define Newton parameters
newton_par damping_step 2. var_tol=1.e-3 res_tol=1.e-3 &&
  max_iter=50 opt_iter=15 stop_iter=15

$ scan line #1--equilibrium sets all contacts to zero
equilibrium

$ define Newton parameters
newton_par damping_step=2. var_tol=1.e-2 res_tol=1.e-2 &&
  max_iter=50 opt_iter=15 stop_iter=15
```

```
$ scan line #2
scan var=voltage_1 value_to=-0.5 init_step=1.0e-5 min_step=1.e-6
max_step=1e-2
$ scan line #3
scan var=voltage_1 value_to=+0.5 init_step=1.0e-5 min_step=1.e-6
max_step=1e-2
$ end
```

Since this example is just a warm up, the details about the device simulation will not be covered at this time: a more detailed explanation will be provided in a subsequent section. We will simply note that the `scan` statement instructs the simulator to perform a DC bias scan of voltage on contact terminal 1, which is the anode. The voltage bias is increased in one direction and then reversed to show both forward and reverse bias conditions.

5.1.5 Space Charge vs. Applied Bias

We will now sweep the bias applied to the p-n junction to see how the space charge various according to different bias voltage. The anode to cathode voltage (V_a) sweeps from -0.5 to 0.5 V. The space charge region decreases as bias voltage increases, as can be seen from Fig. 5.5.

At equilibrium, the width of space charge region W can be written as [61, 62]:

$$W = \sqrt{\frac{2\epsilon V_0}{q}\left(\frac{1}{N_a} + \frac{1}{N_d}\right)} \tag{5.1}$$

where N_a and N_d are the concentration of accepters and donors, ϵ is the permittivity and V_0 is the built-in potential:

$$V_0 = \frac{kT}{q}\ln\left(\frac{N_a N_d}{n_i^2}\right) \tag{5.2}$$

For p-n junction with current flow, such as in this example, the V_0 should be replaced with the new barrier height: $V_0 - V$. We can see from this equation, as the applied voltage V increases from -0.5 to 0.5 V, W will decrease.

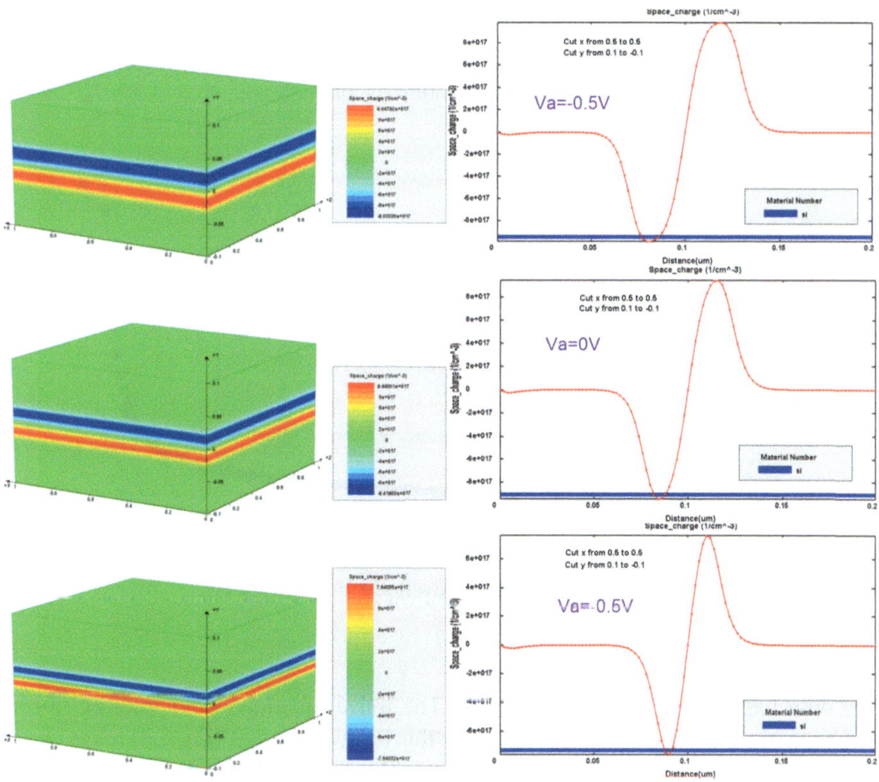

Fig. 5.5 Space charge charts and 2D center cuts (*top to bottom*) for different biases

5.1.6 Junction Capacitance

Variation of charge within a p-n junction will result in capacitance effect [63]. There are basically two types of capacitance associated with a junction [62]:

1. Junction capacitance. This is caused by the dipole in the transition region. Junction capacitance is a voltage variable capacitance.
2. Charge storage capacitance. Due to charge storage effect from the lagging behind of voltage as current changes.

Junction capacitance dominates the reverse bias condition, while charge storage capacitance dominates the forward bias condition.

The Junction capacitance can be calculated from the following equation [62]:

$$C_j = \frac{A}{2}\sqrt{\frac{2q\epsilon}{(V_0 - V)}\frac{N_d N_a}{N_d + N_a}} \tag{5.3}$$

where A is the junction area.

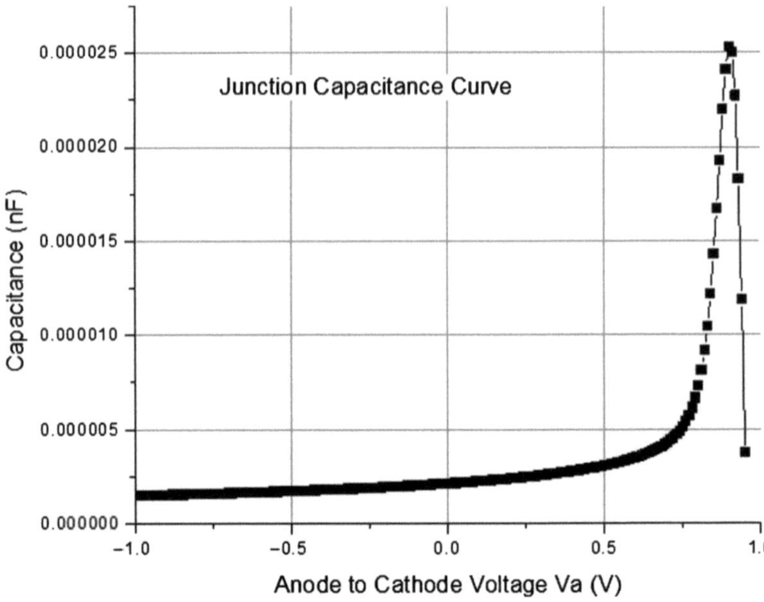

Fig. 5.6 C-V curve for the p-n junction diode

What happens in the limit of $V \rightarrow V_0$? The above equation would predict $C_j \rightarrow \infty$ which is clearly incorrect. The reason for this behavior is that (Eq. 5.3) relies on the following approximation of the depletion width:

$$W = \sqrt{\frac{2\epsilon(V_0 - V)}{q} \frac{N_a + N_d}{N_a N_d}} \tag{5.4}$$

Using the device simulator, we plot the Capacitance-Voltage (*C-V*) characteristics of the p-n junction diode at a frequency of 1 MHz. As shown in Fig. 5.6, the capacitance initially increases as predicted when $V \rightarrow V_0$ but then takes a sharp drop.

5.1.7 Simulation Data

Table 5.2 gives the simulation data for the simple p-n junction diode. For each example, the simulation data will be given with approximate process and device simulation time. Please note that these results are only intended to provide a rough estimate: the exact numbers depend a lot on the actual computer specifications.

Table 5.2 Simulation data for the simple p-n junction diode

	Process simulation	Device simulation (Va = 1.2 V)	Total mesh count	Total number of planes
Simple p-n junction diode	30 s	1 min	2,222	2
Computer Configuration: HP desktop with Intel i7-860/6G/1T/NVIDIA GeForce GTX260/Win7				

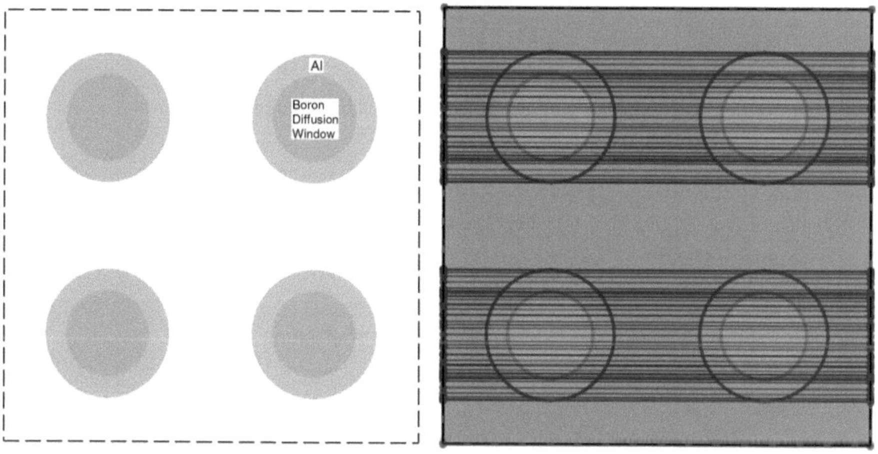

Fig. 5.7 Layout of 4 p-n junction diodes with two masks (*left*) and with cut lines (*right*)

5.2 A More Complicated P-N Junction Diode

In this section, we will simulate a more complicated p-n junction diode. The structure of the diode is based on the 4th edition of Dr. Ben G. Streetman's classic textbook [62], but will use a simplified process recipe to demonstrate the simulation process. This is a relatively "old" technology, but we will use it as an example here. The p-n junction diode we are going to simulate uses a selective diffusion process. The layout consists of several isolated p-n junctions on a single wafer but contains only two mask layers. Mask #1 is used to create openings in the oxide layer for boron diffusion windows. Mask #2 is used to etch deposited aluminum and create contacts for the p-type region. The doping, furnace temperature and diffuse time are all chosen arbitrarily and may not correspond to a real process. Since the bended-plane method is still under development at the time of writing, it will not be applied here. But please keep in mind that for this kind of structure with multiple circles, using bended planes will significantly reduce total mesh count and simulation time.

Figure 5.7 shows the mask layout for this process; the inner circle is the diffusion window for boron and the outer circle is the aluminum etching. It also illustrates the cut lines generated by MaskEditor.

Table 5.3 Overview of simulation steps for the p-n junction diode

P-N junction diode	Process simulation steps
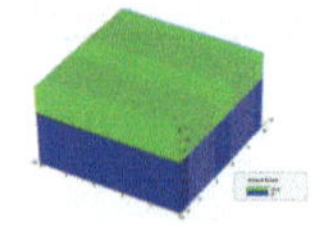	Step 1: Substrate and surface oxidation
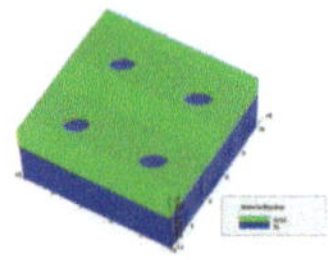	Step 2: Opening windows in the oxide layer
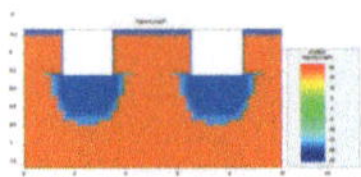	Step 3: Boron diffusion through the windows
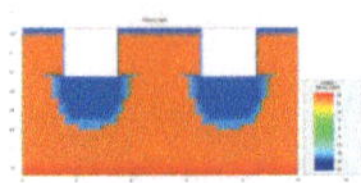	Step 4: Cathode ohmic contact n+ implant
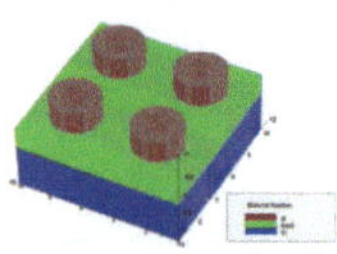	Step 5: Aluminum evaporation
P-N junction diode	**Contact definitions for device simulation**
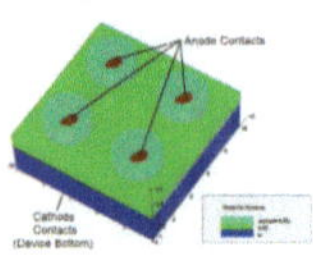	Step 6: Contact definitions for device simulation
P-N junction diode	**Device simulation**
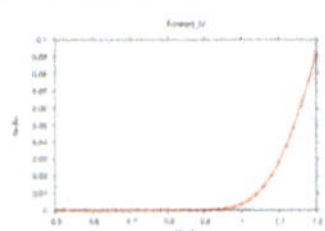	Step 7: Forward bias simulation results

5.2.1 *Overview of Simulation Steps*

The process steps used here do not necessarily correspond to a real device but this example has one particular feature not found elsewhere in this book and which is worth discussing: a boron diffusion step. Diffusion doping is an old technology that has been phased out in most commercial fabs. Newer ion implantation techniques are often preferred due to the increase control over the quantity of impurities introduced and the shape of the doping profile. However, many university labs struggle to make do with obsolete equipment and this technique is still occasionally used. We would therefore be remiss not to include at least one example of diffusion doping, if only because of its historical significance.

Since we only use the boron diffusion step to demonstrate the technique, we only use it once. For the back side n+ region, we will use a more typical phosphorus implantation. After the process simulation is done, contacts are defined for device simulation. In the device simulation section, we will discuss the band diagram of the p-n junction diode (Table 5.3).

5.2.2 *Substrate and Surface Oxidation*

A simple process is created to build the p-n junction diodes. The silicon sample is pre-doped with phosphorus (n-type) and boron is diffused through the openings of the oxide layer. The process steps are described below with the matching process simulation commands.

The silicon substrate is uniformly doped with phosphorus at a concentration of 1E+18 cm^{-3} and oxidized with 30 min wet oxidation at 1,100°C. The silicon surface orientation is chosen to be [111]. The output is saved to the file named 01_oxide.str. The cross section is drawn in Fig. 5.8 and the simulation result is shown in Fig. 5.9.

Process Simulation Code

```
mode three.dim
3d_mesh inf=geo

initialize phosphorus conc=1e18 orient=111
diffuse time=30 temp=1100 weto2
structure outfile=01_oxide.str
```

`mode three.dim` defines a full 3D simulation for this example. This is needed because of the circular shapes of the p-n junctions. As a comparison, a quasi 3D will also be run separately: timing results will be shown at the end of this chapter.

The `3d_mesh` command loads mesh declaration statements from the MaskEditor GUI; sample content of the geo files is shown below (Text Box 5.2). A uniform mesh in the lateral direction is used here so all the geo files should have the same content. Because of the circular shapes involved, many mesh planes are required; this can be improved in the future with better mesh generation techniques such as the bended planes method discussed in Chap. 4.

The `initialize` command processes all the mesh declaration statements and generates the initial substrate for the process simulation. An n-type substrate with a doping concentration of 1E+18 cm^{-3} and oriented in the [111] direction is used as the n region of the diode. An n+ implant will later be used at the bottom of the

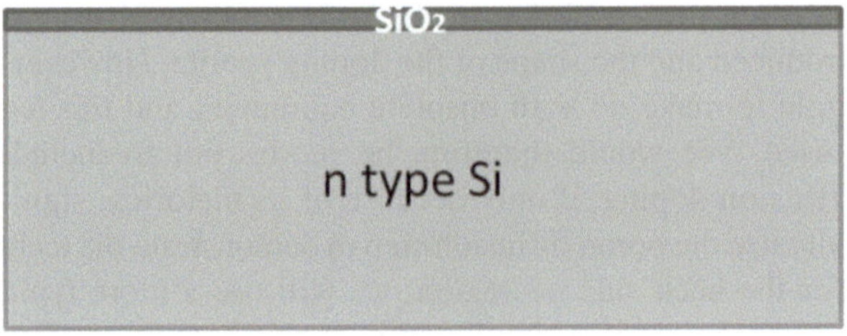

Fig. 5.8 Cross section of step 1: oxidize the Si sample

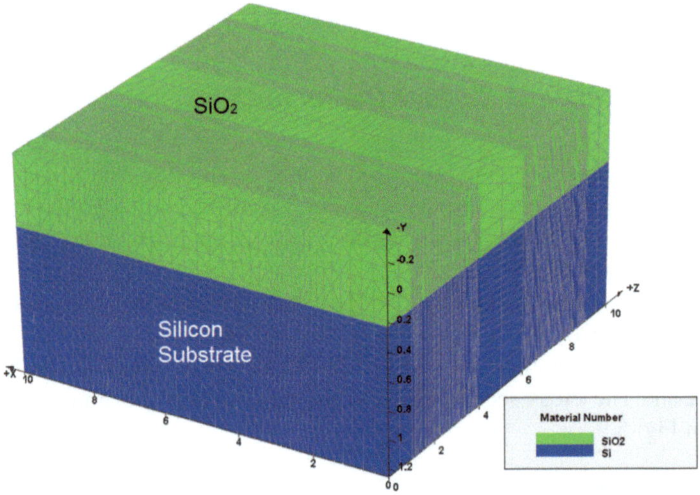

Fig. 5.9 Oxidation of the silicon sample (01_oxide.str)

Text Box 5.2 The Content of geo Files

```
line x loc=   0.00000   spacing=  0.347222       tag=lft
line x loc=   2.50000   spacing=  0.104167
line x loc=   5.00000   spacing=  0.347222
line x loc=   7.50000   spacing=  0.104167
line x loc=   10.0000   spacing=  0.347222       tag=rht

line y loc=   0.00000   spacing=  0.520833e-01   tag=top
line y loc=   0.75000   spacing=  0.520833e-01
line y loc=   1.25000   spacing=  0.173611       tag=bot

region silicon xlo=lft xhi=rht ylo=top yhi=bot
bound exposed xlo=lft xhi=rht ylo=top yhi=top
bound backside xlo=lft xhi=rht ylo=bot yhi=bot
```

substrate to ensure a good ohmic contact. This is postponed until the last part of simulation to avoid the excessive diffusion caused by the thermal cycle of the boron diffusion process.

The `diffuse` command is used here to perform the growth of a thermal oxide layer. A 30 min process with furnace temperature of 1,100°C is used under wet oxide condition. Wet oxide generally has less quality than dry oxide and is not suitable as a gate oxide. However, it has the advantage of being much faster to grow than dry oxide. In this case, the oxide is only used as a blocking layer for the boron diffusion and will remain after the fabrication finishes. Etch windows will be opened in the oxide for the boron diffusion.

The `structure` statement will write the mesh and solution information to a user-specified file (01_oxide.str). It is a good habit to apply a number to the beginning of the .str file name to organize the saved files and sort them in the order of the process step. This is especially true for lengthy simulations where dozens or even hundreds of process simulation structure files are generated. The simulation result after wet oxidation is shown in Fig. 5.9.

5.2.3 Opening Windows in the Oxide Layer

The next step is to open windows in the oxide layer for subsequent boron diffusion. First, a layer of photoresist (PR) 1.0 um thick is applied. The PR is exposed through mask #1 and after the exposed material is removed, HF is used to etch the oxide layer and create diffusion windows [62]. In order to give the reader a clear picture of what's happening in the mask step, the process steps of window opening is illustrated in Fig. 5.10.

Process Simulation Code

```
include file=diode1.msk
structure outfile=02_mask1.str
etch oxide depth=0.8 avoidmask
etch photoresist all
structure outfile=03_oxide_etch.str
```

The statement `include` instructs the process simulator to load the content of the diode1.msk file which was generated automatically by MaskEditor. This file contains the commands used for the PR deposition and its selective removal. The contents of the mask file are shown in Text Box 5.3 but are truncated for the sake of brevity since the file contains mask information for all 84 planes.

Figure 5.11 shows the mask layout. The polarity of this mask is negative and it uses the "general purpose" setting in MaskEditor rather than the simplified "etch" setting. The negative mask polarity means photoresist in the drawn area will be removed, leaving a window to etch oxide (Fig. 5.12).

The next step is to etch the oxide layer. Although HF etch is an isotropic process, we use an anisotropic dry etch here for simulation purposes. The statement `etch`

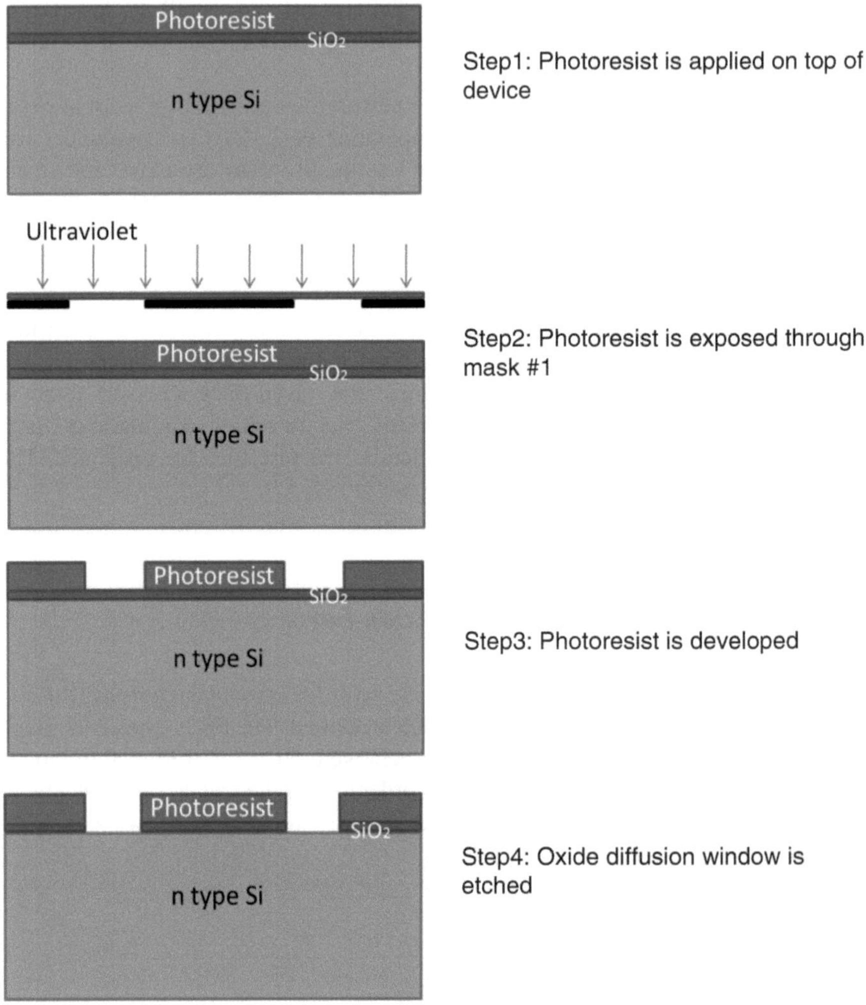

Fig. 5.10 Steps to etch SiO_2 diffusion windows

Text Box 5.3 Contents of Mask File diode1.msk (Truncated)

```
mask segm=1 thick=1. x1.from=0. x1.to=10.
..
mask segm=9 thick=1. x1.from=0. x1.to=2.34357 x2.from=2.65643 x2.to=7.34357
x3.from=7.65643 x3.to=10.
..
mask segm=35 thick=1. x1.from=0. x1.to=2.34357 x2.from=2.65643 x2.to=7.34357
x3.from=7.65643 x3.to=10.
..
mask segm=84 thick=1. x1.from=0. x1.to=10.
```

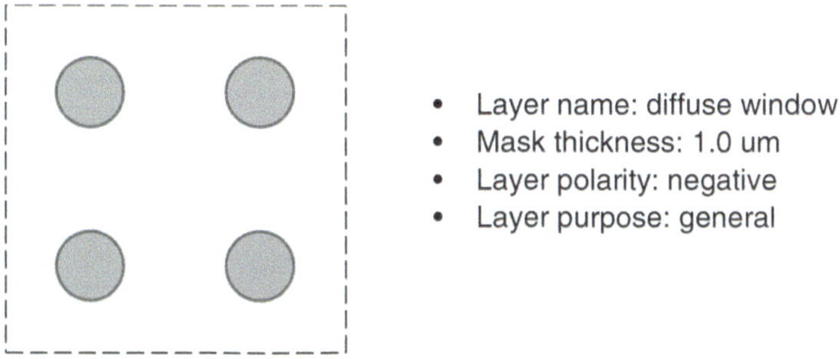

Fig. 5.11 Layout mask of oxide windows process step

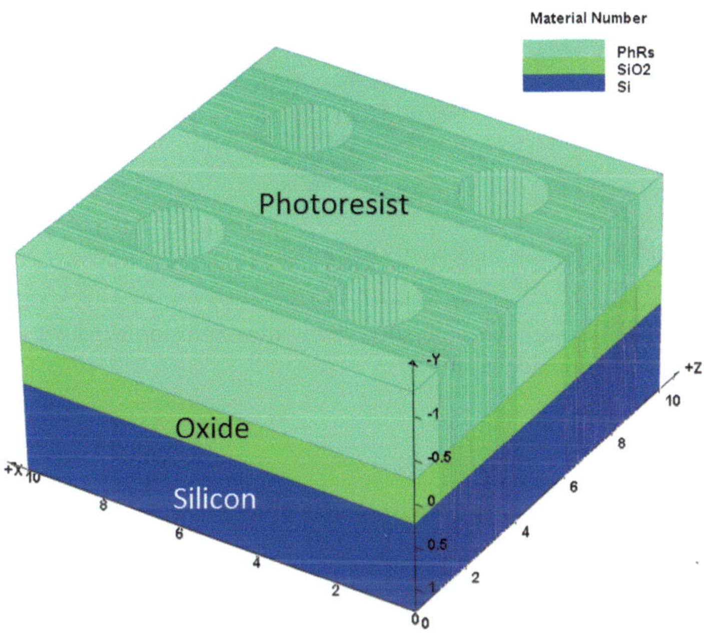

Fig. 5.12 After photoresist development of the oxide etch step (02_mask1.str)

oxide is used to instruct the simulator to perform an oxide etch with an etch depth of 0.8 um. Note that if the oxide layer is thinner than 0.8 um, the simulator will automatically stop at whatever thickness the oxide has because of the settings used in the `etch` command. The `avoidmask` parameter is somewhat redundant but guarantees that the area under the PR is not affected by the etch process.

We note that in this case, a general-purpose mask was used in MaskEditor so we must remove the left-over PR with `etch photoresist all` before continuing to the next step. Failure to do so will cause process simulator to print error messages.

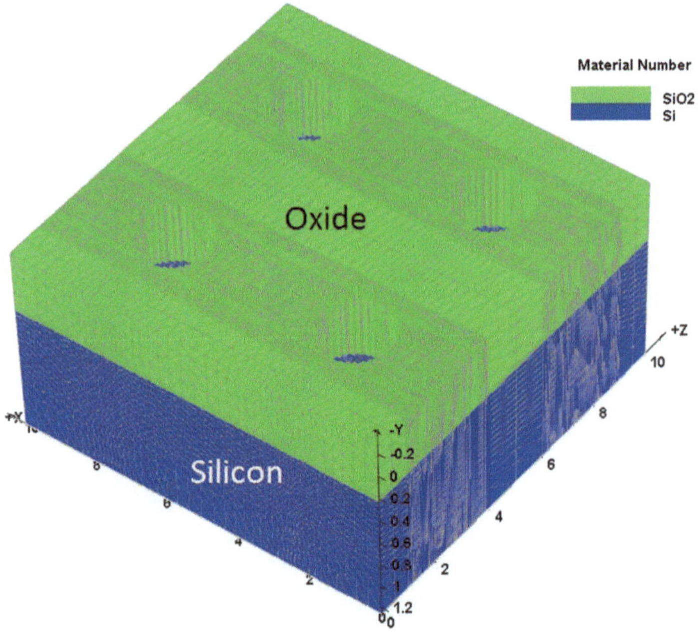

Fig. 5.13 Etch SiO_2 diffusion windows (03_oxide_etch.str)

Figure 5.13 illustrates the simulation result after diffusion windows are opened by selective etching oxide layer.

5.2.4 Boron Diffusion Through the Windows

Now that windows through the oxide layer have been opened, we can diffuse boron through them. The diffusion method to dope semiconductors is an old technology which has mostly been replaced by ion implantation in industry. However, many university clean rooms use older equipment and still rely on this technology (Fig. 5.14).

Process Simulation Code

```
diffuse time=90 temp=1000 boron gas.conc=1.e20
struct outf=04_p_region.str
```

The `diffuse` statement will cause boron to diffuse into the silicon through the oxide windows and create a p-well at the silicon surface. The diffusion time is set to be 90 min (not 90 s) and the furnace temperature is set at a constant temperature of 1,000°C.

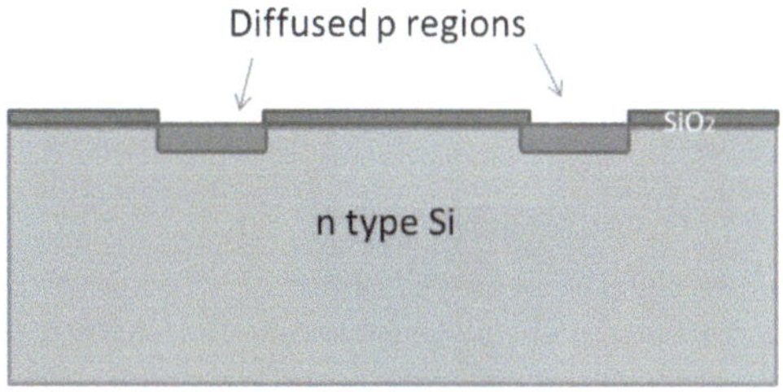

Fig. 5.14 Boron diffusion through oxide opening

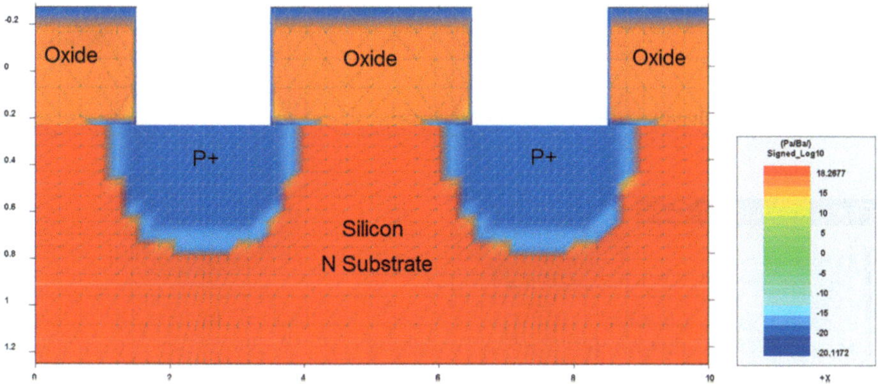

Fig. 5.15 2D cut (z = 2.75) net doping after boron diffusion (04_p_region.str)

Figure 5.15 is the net doping chart after boron dopant activation. From the 2D cut we can see that mesh is denser at the center of the p-wells. This was done intentionally to get a denser mesh in the center of the diode devices. Also, the top mesh is denser than the bottom because the junctions are located at the top half of the structure. Note that we can use `regrid` statement here to make the p-n junction smoother, at the cost of computing time.

5.2.5 Cathode Ohmic Contact n+ Implant

In order to create an ohmic contact at the cathode, the wafer bottom needs highly doped n region. We must flip over the wafer to perform bottom phosphorous implant. Note that implant is used here rather than using another diffusion step.

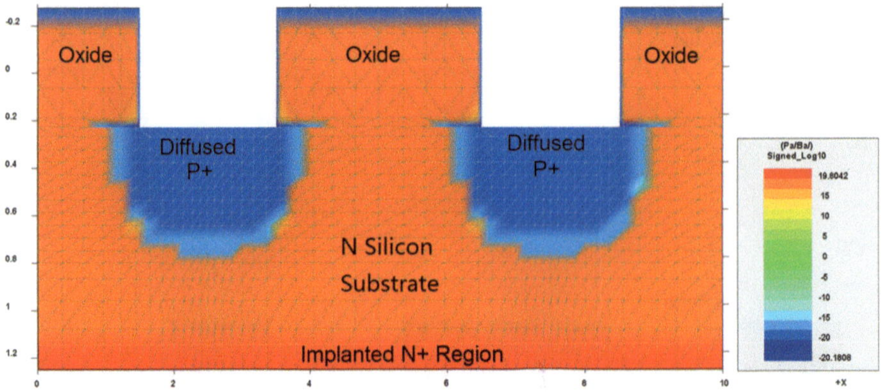

Fig. 5.16 2D net doping chart after n+ region anneal

Process Simulation Code

```
flip_y
implant phosphorus energy=20 dose=4e15
flip_y
diffuse time=1 temp=950
struct outf=04_n_region.str
```

The `flip_y` statement allows the user to flip the whole device in the y direction and do a coordinate transform of y → −y. This matches what is done in a real process where the wafer is flipped over. This statement is used again after the implantation is done to revert the coordinates back to normal.

The `implant phosphorus` command performs an ion implantation of phosphorus to create the backside ohmic contact. The implant dose is 4E+15 cm^{-2} and the ion energy is 20 keV. A quick thermal anneal (`diffuse`) is then used to activate the dopants of the newly implanted n+ region.

Note that the doping concentration of 1E+18 cm^{-3} specified for the substrate earlier is close to the limit of what is acceptable for a good ohmic contact so the extra doping on the backside helps. If this had been neglected, the device simulation might have trouble applying ohmic boundary conditions on the contact and a Schottky boundary would be required.

The 2D cut of the simulation result is shown in Fig. 5.16. Note the bottom of the device is heavily doped n+ region.

5.2.6 Aluminum Evaporation

Aluminum is evaporated onto the surface and etched away to form the p-contact (anode) layer. This step is illustrated in Fig. 5.17.

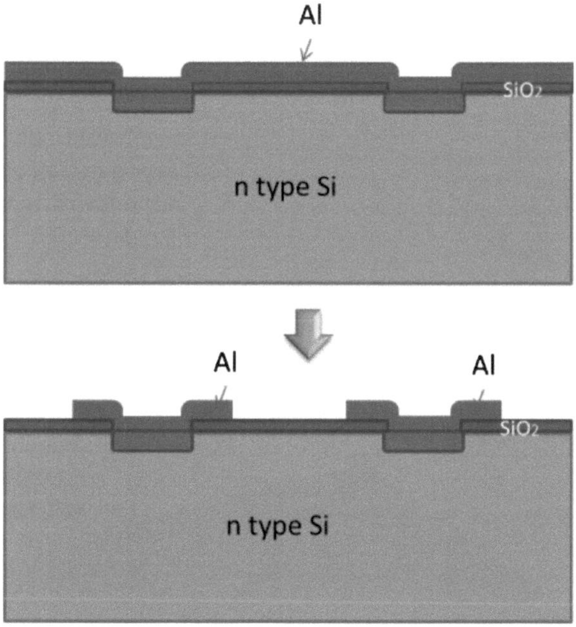

Fig. 5.17 Cross section of aluminum deposition and etch

Process Simualtion Code

```
deposit aluminum thick=0.5 meshlayer=3
struct outf=05_al_deposit.str
include file=diode2.msk
struct outf=06_al_etch.str
```

The process simulator treats evaporation as a deposition process, so the `deposit` command is used to deposit a layer of aluminum 0.5 um thick. The `meshlayer` parameter is used to insert extra mesh in this aluminum layer to improve the accuracy of the simulation. The simulation output after deposition is shown in Fig. 5.19.

As before, the `include` statement is used to process a mask file created by MaskEditor: its abbreviated contents are shown in Text Box 5.4. This time however, the mask used has an "etch" purpose so the aluminum is automatically removed to a depth of 0.5 um. This removes it completely except for the area defined by the second mask (positive mask polarity). The mask layout for this step is shown in Fig. 5.18.

Figure 5.20 shows the final structures after the aluminum etch. Figure 5.21 shows the net doping in 3D and for 2D cut after the aluminum etch.

The final step for the process simulation is to export the structure for device simulation.

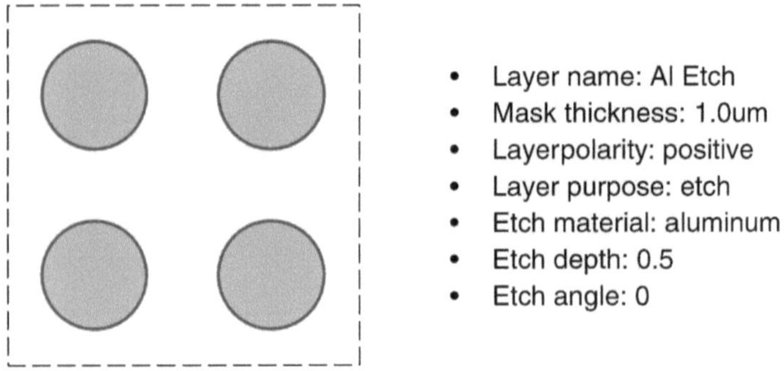

Fig. 5.18 After aluminum deposition (05_al_deposit.str)

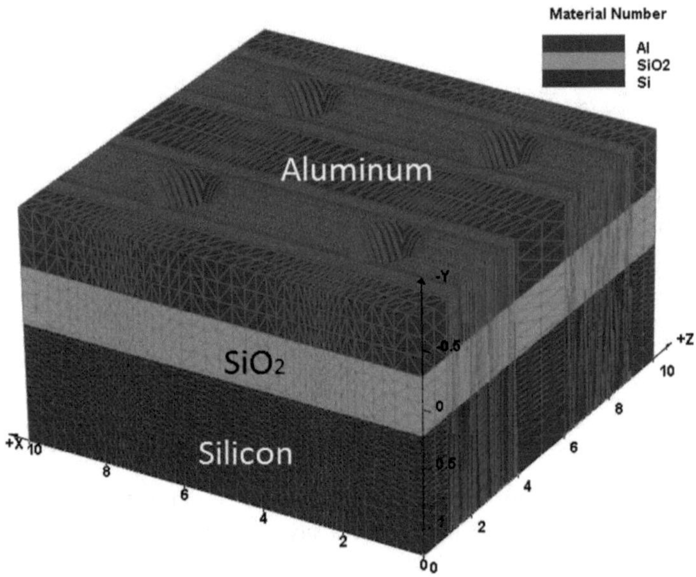

Fig. 5.19 Mask for aluminum contact etch

Text Box 5.4 The Contents of Mask File diode2.msk

```
etch dry segm=1 alumin thick=0.5
.........
mask segm=3 thick=1. x1.from=2.26535 x1.to=2.73465 x2.from=7.26535 x2.to=7.73465
etch segm=3 alumin avoidmask depth=0.5
etch segm=3 photoresist all
.........
etch dry segm=84 alumin thick=0.5
```

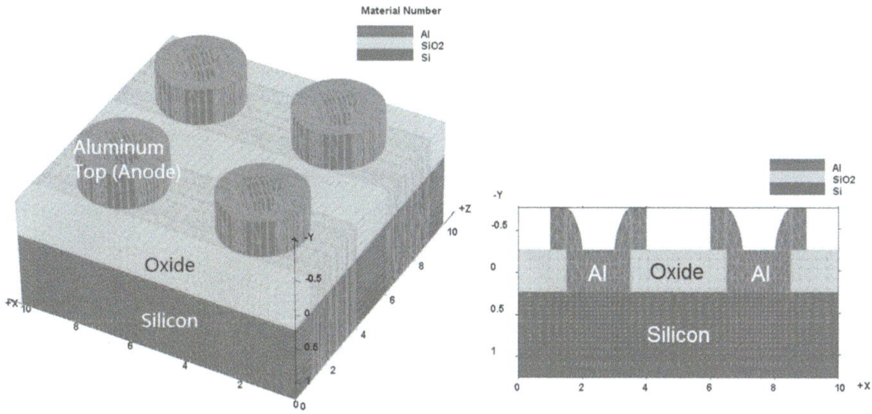

Fig. 5.20 After aluminum etch (06_al_etch.str)

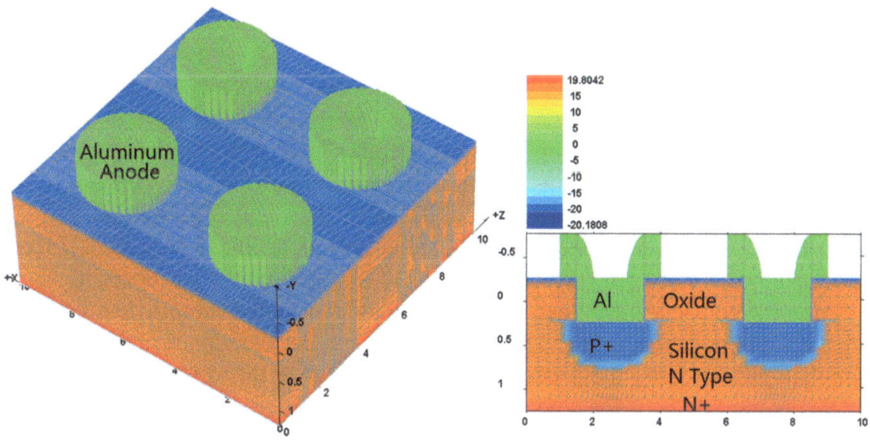

Fig. 5.21 Net doping after aluminum etch (06_al_etch.str)

Process Simulation Code

```
export outf=diode.aps xpsize=0.001
```

Here diode.aps will be used as the input file of the device simulation. The `xpsize` parameter is used for the mesh generation because the device simulator does not allow mesh points to be located at the exact boundary between two materials. Instead, two points on each side of the boundary are used; their spacing (in um) is determined by `xpsize`.

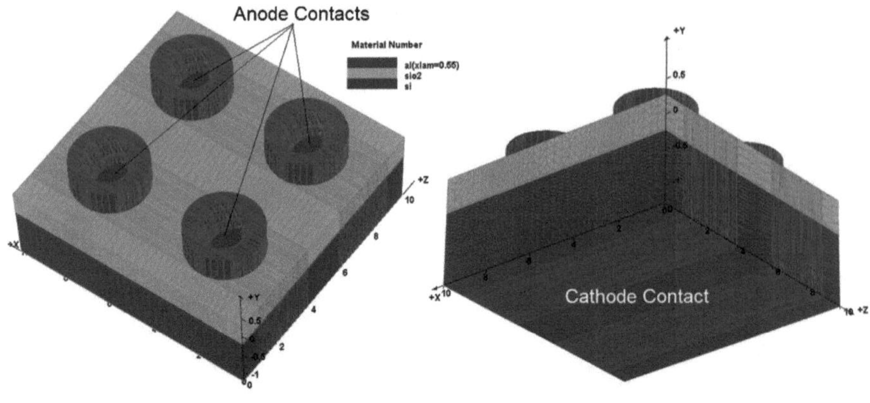

Fig. 5.22 Contact locations in p-n junction diode

5.2.7 Contact Definitions

Contacts are defined with the help of the embedded Contact Designer application. Contacts are defined by specifying x, y and z ranges as well as a contact number. We note that the top contacts (anode) have 4 declarations sharing a single number (#1); this means that these regions are electrically connected and will all share the same boundary conditions in the device simulation. On the other hand, the cathode contact (#2) at the bottom is declared only once since it covers the entire bottom surface. Figure 5.22 shows the contact locations.

After the contacts are defined and saved, a new file named contact_3d.sol will be created by ContactDesigner. This file is used by the default device simulation template generated by the process simulation export; refer to Chap. 4 for details on the interaction between device simulator and the process simulator.

5.2.8 Device Simulation of P-N Junction Diode

The exported file of the process simulation, diode.aps is the starting point of the device simulation; refer to Chap. 4 for details on the interaction between device simulation and process simulation. In this example, we will perform a simple simulation of diode forward biasing. All 4 p-n junctions will be forward- biased simultaneously: the top contacts are connected implicitly because they share the same contact number and thus apply the same boundary condition.

5.2.9 Device Simulation Setup

The device simulation code in the default template (main_3d.sol) is shown below.

Device Simulation Code

```
$-------Part 1: Welcome to Apsys 3D ---------------------------------
begin
convention positive_current_flow=inward

$------- Part 2: Input statement -----------------------------
$------- load mesh from PROCESS SIMULATION output .aps file --------------
include file=zmesh.zst &&
  ignore1=load_mesh ignore2=output ignore3=export_3dgeo
load_mesh mesh_inf=diode.aps  suprem_import=yes

$------- Part 3: Output statement ----------------------------
$------- You can change the name of output file for APSYS -----
output sol_outf=diode.out
$more_output impact_ionization=yes space_charge=yes

$------- Part 4: Load material and contact information ------
$ use contact designer program to define contact
include file=contact_3d.sol

$------- Part 5: Main scan command ---------------------------
newton_par damping_step=12. print_flag=3 res_tol=1.e-2 var_tol=1.e-2 &&
 max_iter=60 opt_iter=30 change_variable=no mf_solver=3

equilibrium

scan var=voltage_1 value_to=  3.0 print_step=0.3 &&
  init_step=0.5E-02 min_step=0.5E-05 max_step=0.2
```

Note that commenting out a line in the device simulator ("$") is a little bit different than in process simulator ("#"). Additionally, there is a restriction on the number of characters per line so the "&&" sign is used to separate lengthy statements into separate lines.

Most of the statements used in the template have already been explained in Chap. 4. We will concentrate on explaining only the additional physical models and bias commands of this example in this section. The additional physical models in part 3 are also commented out at this time and will be omitted.

As used here, the `convention` statement merely restates that current flowing into the device is positive. By default, the simulator treats current flowing out of the device and into a contact as being positive.

The `more_output` statement instructs the solver to save additional physical parameters not included in the default output. This can be used when the user is interested in certain parameters like the space charge and carrier mobility.

In part 5 of the template, we find the statements used to apply bias to the device. The starting point of any simulation is the `equilibrium` statement. This solves the Poisson equation exclusively and sets the voltage at all electrodes to 0. At thermal equilibrium, the net current is zero throughout the device so the current continuity equations are not solved and the quasi-fermi levels for the electrons and holes is flat. The default equilibrium temperature is set to 300 K but it can be adjusted with the `temperature` statement. Note that the equilibrium statement can be re-issued to reset the simulation and start a new series of bias scans: this can be useful when doing voltage sweeps for a family of *I*-*V* curves with different gate bias voltages, as will be seen in a later chapter.

The statement `scan` is used by the main simulator engine to activate the equation solver while one or more control variables are being changed (scanned). Usually this means the device electrode is biased to a certain voltage to calculate the electrical characteristics. For voltage/current control of the device simulators, the convention is that if the bias on an electrode is not explicitly changed by a `scan` statement, then the voltage value is implicitly forced to remain at its previous value. Since it is impossible to simultaneously impose both a voltage and a current boundary condition on an electrode, this means that electrodes under voltage control have a current which will be determined by the software.

For example, if a device has three electrodes and only one of them (say electrode number 3) is changed by a `scan` command. This means the other two (electrode numbers 1 and 2) keep the same voltage but have changing current. This is one of the key reasons why all simulations start with the equilibrium statement: it provides an initial voltage and current value for all electrodes (0 V and 0 A). This initial value is used as the starting guess of the solver when changing the scan variable.

In this particular diode example, the `scan` command controls voltage_1 (anode) which means that the voltage on electrode #1 is being changed while the voltage on electrode #2 stays at 0 V. Since both electrodes are under voltage control, the current will then be determined by the solver.

We make a quick note that current control can only be used on an electrode where a small current flow already exists. Since the initial guess to the Newton solver is extrapolated from previous solutions, the value of dV/dI plays a role in the numerical stability of the solver: small changes in the denominator of this derivative should produce only small changes in the numerator in order to get reliable convergence.

The simulation program can use the `scan` command to vary many other control variables such as the external circuit resistance and incident light power. It can also control several variables at once: for example a transient simulation will control a time variable and may also include a time-dependent bias such as a voltage pulse.

All scan commands progressively modify a variable from its initial value (i.e. the final value of the previous scan) to its target value. This is not done in smooth increments since it depends on the convergence of the non-linear Newton solver. In general, since the `scan` procedure involves perturbing a known good solution, smaller steps facilitate convergence and produce smoother output curves at the cost of extra simulation time. The steps can be controlled by various parameters of the scan command such as `init_step`, `print_step` and `value_to`, which are all illustrated in Fig. 5.23.

The `print_step` parameter instructs the solver to print output data at regular intervals and serves as a de facto limit on the maximum step size: an additional constraint can be set with the `max_step` parameter to limit the solver step without outputting data. The example of corresponds to the parameters `init_step=0.1`,

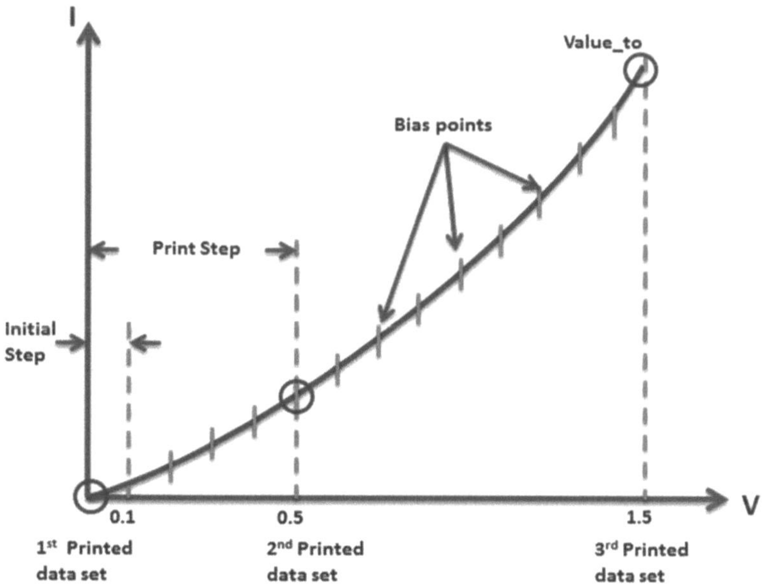

Fig. 5.23 Relationship between initial step, print step and value to

`print_step=0.5` and `value_to=1.5`. The first data set is the output from the `equilibrium` statement and the second data set is output when the first `print_step` value is reached. If the `print_step` value is small enough, additional data sets will be output before the end of the scan. Note that all `scan` statements output data upon reaching their final value, no matter the value of `print_step`.

The `min_step` parameter determines a bias step which will cause the solver to give up. This is used because in cases of non-convergence, the solver will attempt a smaller bias step to try to recover the solution: a minimum value is required to prevent the bias step from becoming infinitely small. Situations where this problem occurs usually indicate a problem in the device setup, simulation parameters or some other limitations of the program.

In addition to the bias step controls of the `scan` statement, the convergence of the simulation is affected by the `newton_par` statement; each use of this statement affects the `scan` statements which follow it. `newton_par` controls several aspects of the non-linear Newton solver including:

- `damping_step` which limits the correction vector length: smaller values normally yield slower but more reliable convergence. Larger values can cause oscillations in the solution guess.
- `res_tol` is the tolerance or lower limit of the Newton residuals beyond which the solution is considered converged: the larger the value, the easier it is to converge. But this will sacrifice the accuracy. Increased tolerance can also affect the quality of the extrapolated guess for the next bias step so it is sometimes counter-productive to use a larger value.

- `var_tol` is the tolerance or lower limit of the variable change between Newton iterations: smaller values indicate a stable solution. Typically, the convergence criterion of the Newton solver is that the solution be both stable and accurate: i.e. both `res_tol` and `var_tol` are lower than the specified values.
- `max_iter` is the maximum number of nonlinear iterations. If the required number of iterations is greater than this value, the solution is not found and the program will automatically reduce the bias step and start the nonlinear iteration again.
- `opt_iter` is an estimate of the optimal number of nonlinear iterations. The program adjusts its bias step by comparing the actual number of iterations needed to converge with this target value. Exceeding the threshold reduces the size of the next bias step and faster convergence is rewarded with a larger step size.
- `stop_iter` is an iteration number where the solver can make a decision to prematurely stop the non-linear iterations. It makes this decision based on the general trend of the Newton residual and the stability of the solution guess.
- `change_variable` is used to change the variables of the device simulator from the quasi-Fermi levels to the carrier densities. For many devices, minority carrier Fermi level fluctuation in some areas with low current can cause convergence problems; using the carrier densities directly reduces the effects of those fluctuations and improves the convergence. It is difficult to determine in advance which devices require this approach but some examples include buried heterostructure lasers with strong current blocking layers, JFET's and CCD's.
- `mf_solver` is a linear solver flag related to the multi-frontal sparse matrix solver. The default value is 3 which calls a parallel algorithm well suited to modern multi-core CPUs. A value of 4 calls a GPU-accelerated solver which scales better for larger 3D cases than "3"; however, it is slower for smaller cases due to the extra parallelization overhead.
- `print_flag` parameter controls how much information is printed during program execution.

5.2.10 Forward Bias Simulation Results

Figure 5.24 shows the *I-V* curve under forward biasing for this diode example.

Band diagrams can be plotted by selecting a 1D cut in the plotting GUI (Fig. 5.25). Figure 5.26 illustrates the location of band diagram cut line chosen from one of the top contacts, about 1 um is shown. Plotting the band diagram at various bias steps shows the ohmic boundary and a reduction in the electrostatic potential barrier at the junction with increasing anode voltage bias, as is illustrated in Fig. 5.26.

Note that at equilibrium, the Fermi level is very close to the valance band in the p-type region, indicating a high concentration of p-type dopants. The Fermi level is

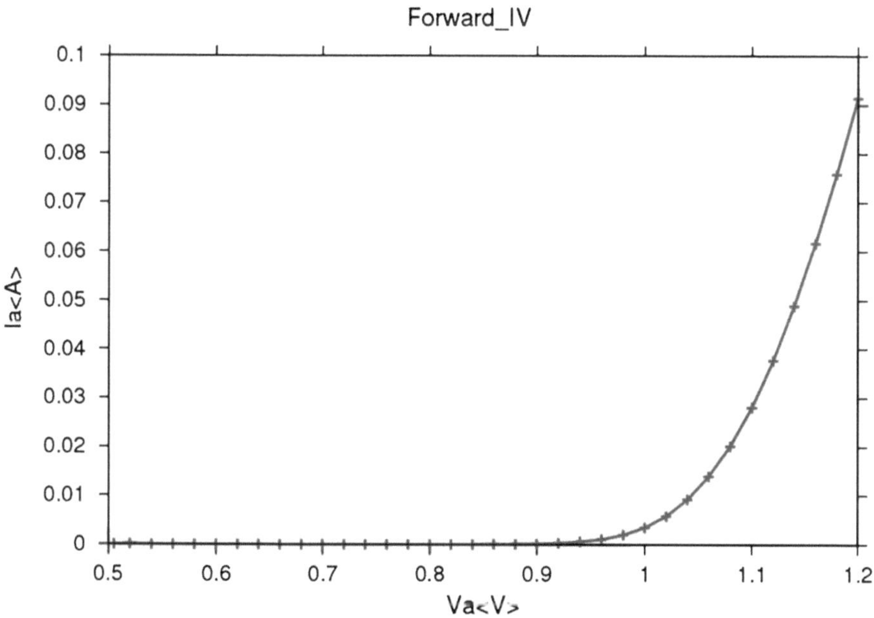

Fig. 5.24 Forward *I-V* curve of 3D diode simulation

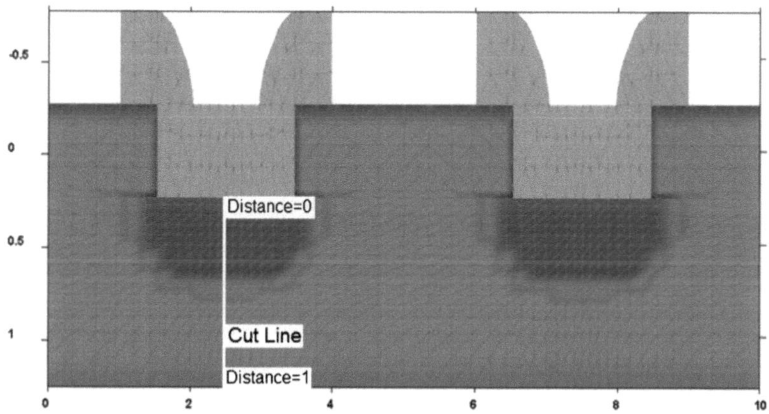

Fig. 5.25 The cut line location for band diagram plots

very close to the conduction band in the n region and even higher than that in the n+ region, indicating a very high doping concentration at the n+ region to make a good ohmic contact.

At equilibrium, the net current flow at the junction is zero. Drift current (caused by the electric field from space charge) is equal in magnitude and cancels out the

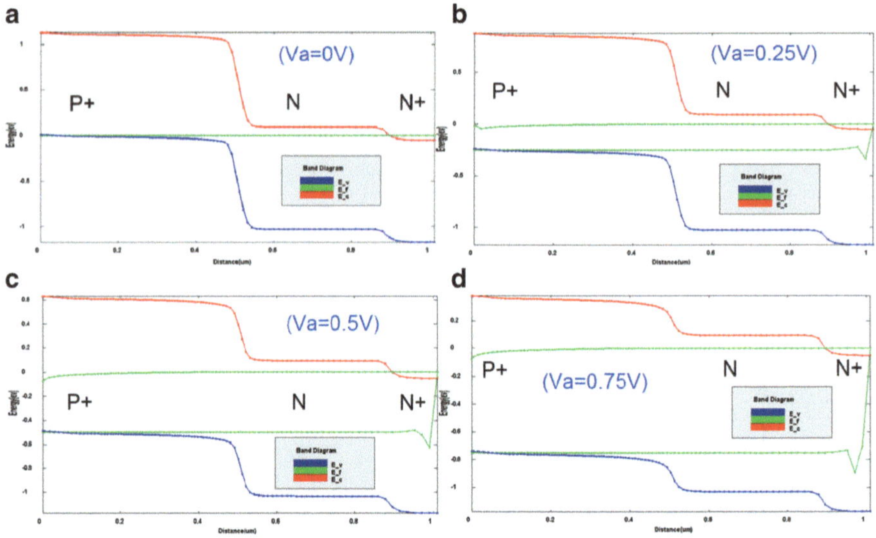

Fig. 5.26 Simulated band diagram of forward-biased diode (**a**)–(**d**)

diffusion current (caused by the concentration gradient at the junction) [61]. For electrons, this means:

$$J_n = J_n(drift) + J_n(diffusion) \tag{5.5}$$

$$= q\mu_n n\varepsilon + qD_n \frac{dn}{dx} \tag{5.6}$$

$$= \mu_n n \frac{dE_F}{dx} = 0 \tag{5.7}$$

Where ε is the electric field in the x-direction.

$$\varepsilon = \frac{1}{q}\frac{dE_i}{dx} \tag{5.8}$$

And from Einstein relation [61]:

$$D_n = kT\frac{\mu_n}{q} \tag{5.9}$$

With positive bias on the anode terminal (the 4 cylindrical aluminum contacts on top), the p-n junctions are forward biased, meaning the electron potential barrier at the junction is lowered by the forward voltage V_f from the equilibrium build-in potential V_o to the smaller value $V_o - V_f$ [62]. Forward bias increases the probability that a carrier can diffuse across the junction by the factor of $\exp(qV_f/KT)$ Diffuse current dominates the total current and we will start to see the large current flow from anode to cathode.

In Fig. 5.26, the separation of the energy bands is a direct function of the electrostatic potential barrier at the junction [62]. Shifting of the energy bands

Table 5.4 Simulation data for p-n junction diode

	Process simulation	Device simulation	Total mesh count	Total number of planes
P-N junction diode	Quasi 3D: 37 min Full3D (GPU): 2.5 h	100 min	92,522	84
Computer Configuration: HP desktop with Intel i7-860/6G/1T/NVIDIA GeForce GTX260/Win7				

under bias implies a separation of the Fermi levels on either side of the junctions. Under forward bias, the Fermi level on the n side E_{Fn} is above E_{Fp} by the energy qV_f.

5.2.11 Simulation Data

Table 5.4 gives the simulation data for the more complex p-n junction diode. Full 3D requires significant more time than quasi 3D. A GPU-accelerated simulation is used to speed up the full 3D simulation.

Chapter 6
MOSFET/CMOS Technology

6.1 3D Long Channel n-Type MOSFET Simulation

In this chapter, we will first simulate a very simple 3D device. A long channel n-type MOSFET (NMOS) is the simplest 3D simulation case for unipolar devices, as it is merely an extension of a 2D simulation. There is no variation along the z direction so only two planes are needed to perform a 3D simulation. The mask layout is fairly simple and a quasi-3D simulation is used. The process steps used in this example is based on the original Stanford SUPREM-IV.GS example 5 with minor modifications.

Note that due to symmetrical nature of MOSFET, we will first simulate half of the structure and use the `mirror` statement to complete the MOS structure. So unlike previous examples, the layout of this device represents half of the complete structure. The mask layers and MOSFET cross-section are shown in Fig. 6.1.

6.1.1 Overview of Simulation Steps

The first stage of simulation is the process simulation where we create the structure of the long channel MOSFET. This is followed by the contact definitions and electrical device simulation. An overview of the simulation steps is presented in Table 6.1.

6.1.2 Process Simulation of Long Channel MOSFET

Like in the previous chapter, the process simulation code will be divided into separate parts for clarification purpose. Most of the statements in the process simulation code will be explained in detail. Since Stanford's SUPREM-IV.GS is

S. Li and Y. Fu, *3D TCAD Simulation for Semiconductor Processes, Devices and Optoelectronics*, DOI 10.1007/978-1-4614-0481-1_6,

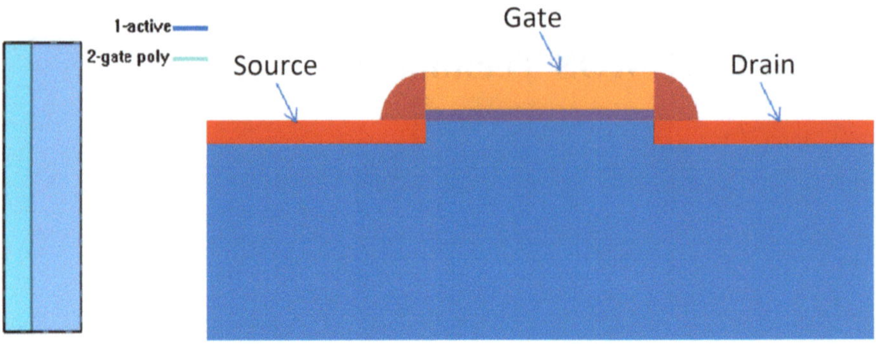

Fig. 6.1 Mask layout (half) and 2D view of the long channel MOSFET

Table 6.1 Overview of simulation steps for long channel MOSFET

Long channel MOSFET	Process simulation steps
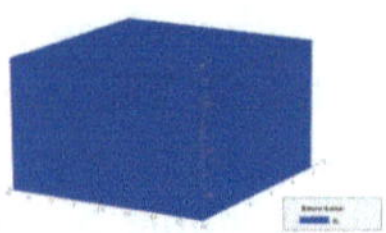	Step 1: Initial substrate
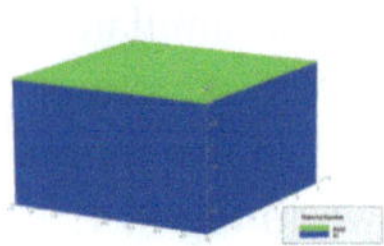	Step 2: Gate oxide deposition
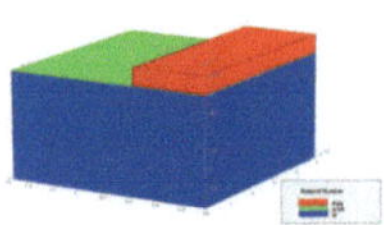	Step 3: Threshold voltage adjustment implant and gate poly
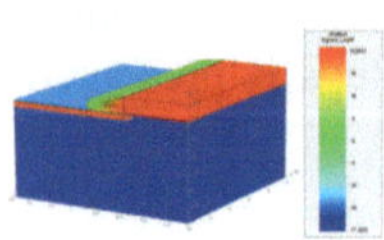	Step 4: LDD implant and nitride spacer
	Step 5: Source/drain implant

(continued)

Table 6.1 (continued)

Long channel MOSFET	Process simulation steps
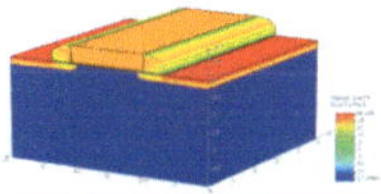	Step 6: Mirror structure and export to the device simulator
Long channel MOSFET	Contact definitions for device simulation
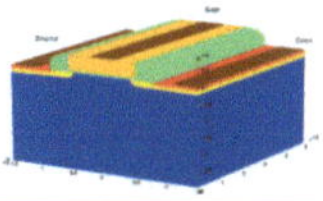	Step 7: Contact definitions for device simulation
Long channel MOSFET	Device simulation
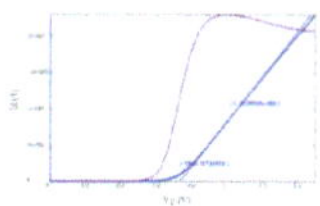	Step 8: Device simulation of threshold voltage
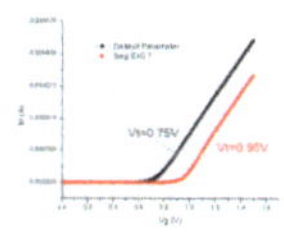	Step 9: Tuning material parameters
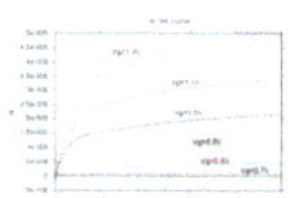	Step 10: I_D-V_D family of curves
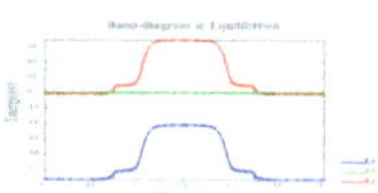	Step 11: Band diagram simulation
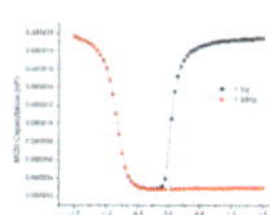	Step 12: MOS capacitor simulation

the original source for several commercially available simulators, the SUPREM statements used here are very similar to other simulators. Users of other simulators should consult their user manuals to find equivalent commands.

6.1.3 *Initial Substrate*

The initial substrate will be defined in this section.

Process Simulation Code

```
# 3D mode
mode quasi3d
3d_mesh infile=geo
init boron concentration=1.0e16 orient=100
structure outf=01_sub.str
```

First, in the `mode` statement, a quasi 3D model is chosen over full 3D to reduce simulation time. Since there is no variation in the z direction, this example could also have been done in 2D without any meaningful loss of information.

The `3d_mesh` statement is used to load commands contained in a number of input decks representing mesh lines for different planes. The `infile` parameter is the mesh command file name base: this is usually "geo" as files named geo*.in are automatically generated by the MaskEditor GUI program. Text Box 6.1 shows the content of geo1.in: for more information about geo files, please refer to Chap. 4. The default settings of `3d_mesh` will also load the zmesh.zst file which specifies locations of all the mesh planes.

The `init` statement instructs the simulator the substrate will be p-type silicon with a boron concentration of 1E + 16 cm^{-3}. The crystal orientation is chosen to be [100] because this results in fewer surface imperfections and better quality Si/SiO_2 interfaces [53].

The `structure outf=01_sub.str` will save the initialized structure to the working directory with file name of 01_sub. Figure 6.2 is the substrate of the long channel MOSFET. Note that the structure here is not proportionally scaled for better view.

6.1.4 *Gate Oxide Deposition*

In this section, gate oxide is deposited to the silicon surface.

Process Simulation Code

```
#deposit the gate oxide
deposit oxide thick=0.025
structure outf=02_gateoxide.str
```

Text Box 6.1 The Content of geo File geo1.in

```
line x loc=0.0 tag=lft spacing=0.03
line x loc=0.4          spacing=0.01
line x loc=0.45         spacing=0.01
line x loc=0.55         spacing=0.01
line x loc=0.75         spacing=0.01
line x loc=1.0         spacing=0.03
line x loc=1.4          spacing=0.03
line x loc=1.5 tag=rht spacing=0.03

line y loc=0.0 tag=top spacing=0.01
line y loc=0.1                 spacing=0.01
line y loc=0.25          spacing=0.01
line y loc=3.0 tag=bot    spacing=0.05

elimin x.dir xlo=0 xhi=1.5 ylo=0.25 yhi=3 ntimes=4
elimin y.dir xlo=0 xhi=1.5 ylo=0.25 yhi=3 ntimes=2
elimin y.dir xlo=0. xhi=0.4 ylo=0 yhi=0.25 ntimes=2
elimin y.dir xlo=1.0 xhi=1.5 ylo=0 yhi=0.25 ntimes=2
elimin x.dir xlo=0. xhi=0.4 ylo=0 yhi=0.25 ntimes=2
elimin x.dir xlo=1.0 xhi=1.5 ylo=0 yhi=0.25 ntimes=2

region silicon xlo=lft xhi=rht ylo=top yhi=bot
bound exposed xlo=lft xhi=rht ylo=top yhi=top
```

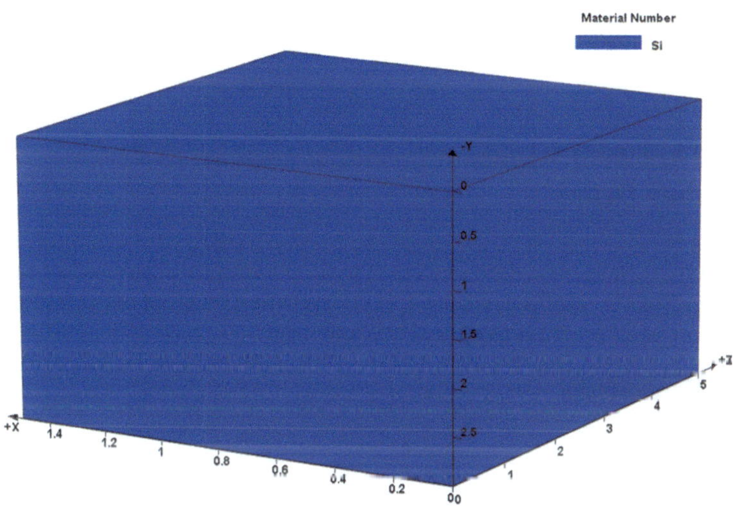

Fig. 6.2 Substrate of the long channel MOSFET (01_sub.str)

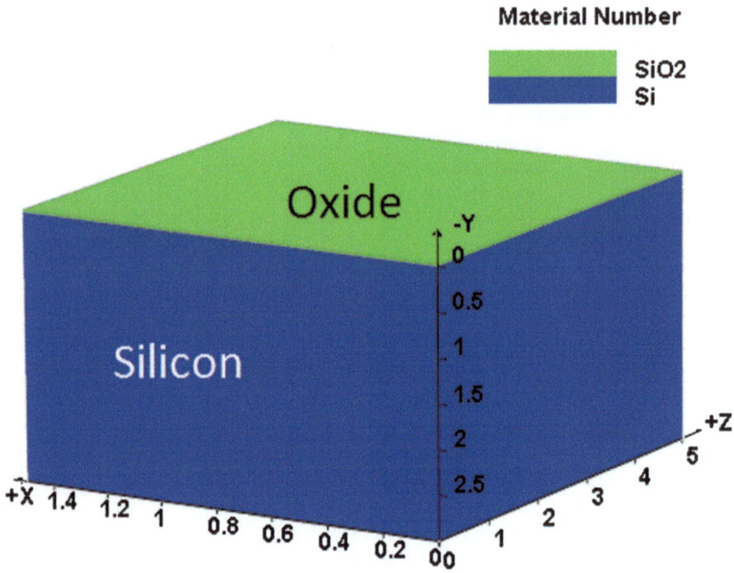

Fig. 6.3 Deposited gate oxide (02_gateoxide.str)

The `deposit` statement deposits 250 Å (0.025 um) of gate oxide. This oxide is deposited rather than thermally grown to simplify the MOSFET processes for demonstration purposes: thermally grown oxide will shift the silicon surface and make subsequent process steps more complicated. Several examples in this book use this simplification. Note that parameters relating to the quality of the oxide and the Si/SiO_2 (traps, fixed charges, etc.) are defined during the device simulation stage and are ignored during process simulation Fig. 6.3 is the simulation result of deposited gate oxide.

6.1.5 Threshold Voltage Adjustment Implant and Gate Poly Define

In this section, an implantation process is used to adjust the threshold voltage (V_{th}). This is followed by gate polysilicon (poly) deposition over the entire wafer surface with LPCVD and in-situ doping with phosphorous.

Process Simulation Code

```
#channel implant
implant boron dose=1.0e12 energy=15.0
structure outf=03_channelimplant.str
```

The `implant` statement will perform boron implantation with a dose of 1E + 12 cm^{-2} and energy of 15 keV. This is used to adjust the threshold voltage

Text Box 6.2 Content of mos2.Msk

```
mask segm=1 thick=1. x1.from=0. x1.to=0.55
etch segm=1 poly avoidmask depth=0.5
etch segm=1 photoresist all

mask segm=2 thick=1. x1.from=0. x1.to=0.55
etch segm=2 poly avoidmask depth=0.5
etch segm=2 photoresist all
```

by introducing precise quantities of impurities [62]. V_{th} adjustment step is important because existing process steps such as well implant may not be adequate to precisely control threshold voltage.

Process Simulation Code

```
#deposit the gate poly
deposit poly thick=0.500 meshlayer=10 phosphorus conc=1.0e19
structure outf=04_polydeposit.str

#anneal the device
diff time=10 temp=1000
structure outf=05_polyanneal.str

#etch the poly away
include file=mos2.msk
structure outf=06_polyetch.str

#anneal this step
diffuse time=30.0 temp=950
structure outf=07_polydiff.str
```

The `deposit` statement specifies a gate poly thickness of 0.5 um. The `meshlayer` parameter specifies the number of horizontal mesh lines used in this new layer: the vertical mesh lines are inherited from the previous process steps.

The polysilicon is in-situ doped with phosphorus at a doping concentration of 1E + 19 cm^{-3}. The `diff` (short for `diffuse`) statement then anneals the gate at a constant temperature of 1,000°C for 10 min.

The `include` statement reads in mask commands from the MaskEditor GUI and etch away the poly. The content of mos2.msk file is shown in Text Box 6.2. Note that since we are exploiting symmetry to save on computation time, only half of the device is modeled at this stage. This means we only need to etch away the polysilicon for $x > 0.55$.

After poly etching, the device is annealed again (`diffuse` statement) at 950°C for 30 min. Note that instead of a constant temperature anneal, a more realistic Rapid Thermal Annealing (RTA) process could also be used. This method involves rapidly ramping up and down the temperature to create spikes and plateaus Fig. 6.4 illustrates the simulation results after poly etch and anneal.

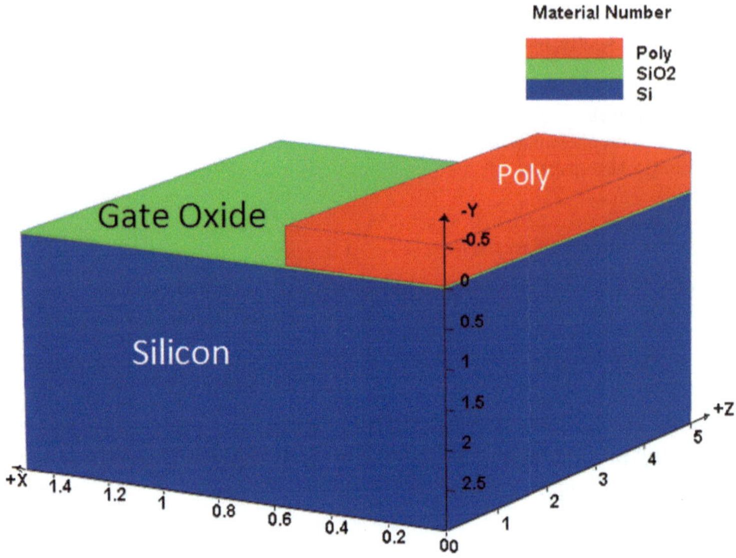

Fig. 6.4 After poly etch and poly anneal (07_polydiff.str)

6.1.6 LDD Implant and Nitride Spacer Etch

In this section a Lightly Doped Drain (LDD) is created through implantation.

LDD implant is usually performed before nitride (or oxide) spacer step. Device scaling has caused electric field to rise near the drain region. This high field can bring the so called "hot electron" problems. Hot electrons or hot holes are high energy carriers. These carriers can gain sufficient energy to surmount large energy barriers between Si conduction band and SiO_2 conduction band [53]. The result can be carriers injected into gate dielectrics, become trapped and eventually cause device reliability problems such as Time Dependent Dielectric Breakdown (TDDB).

LDD is created to cope with this problem. By introducing a graded doping profile (N+/N−/P), that allows drain voltage to drop over a large distance. In comparison with abrupt junction (N+/P), the graded doping profile reduces peak value of the electric field near the drain.

LDD usually has a shallower junction than source/drain implant; this is desired because a shallower junction can minimize the short channel effect.

In preparation for the source/drain implant of the next step, we also deposit and etch a nitride spacer and grow a thin screening oxide layer.

Process Simulation Code

```
#do the phosphorus LDD implant
implant phosphorus dose=1.0e13 energy=50.0
structure outfile=08_ldd_imp.str
```

The LDD implantation process uses phosphorus with a dose of 1E + 13 cm^{-2} and ion energy of 50 keV.

Process Simulation Code

```
#deposit the nitride spacer
deposit nitride thick=0.300 space=0.01

#etch the spacer back
etch dry nitride thick=0.30
structure outf=09_spacer.str

#after etch anneal
method vert fermi grid.ox=0.05
method no.triangle.flip=true
diffuse time=30 temp=950 flow=t O2=1
structure outf=10_dryo2_diff.str
```

The `deposit` command is used to create a layer of nitride 0.3 um thick. The `space` parameter can be used to control and smooth out the mesh on the curved surface of the nitride: this will affect the shape of the sidewall that will be left by the etching step that follows.

The `etch` command removes the nitride to a thickness of 0.3 um. Because dry etch is anisotropic, no lateral etch is performed and a sidewall will be left behind; its width will depend on the thickness of the original nitride layer.

The `fermi` parameter in the `method` command indicates that the defects are assumed to be a function of the fermi level only. The `vertical` model indicates that the oxide growth is entirely vertical. `grid.ox` is the desired spacing (in microns) between the grid points that will be added to the growing oxide layer. It is analogous to the `meshlayer` parameter of the `deposit` command and will affect the time steps.

`method no.triangle.flip=true`. This statement determines whether it is necessary to flip mesh triangles during the oxide growth. The flip check improves the quality of the mesh but can significantly increase the computation time. This setting disables the flip check.

The `diffuse` statement instructs the process simulator to do a 30 min, 950°C oxidation where the oxygen flow rate is defined. The `flow.control` option allows the user to set up the flow amount of different gas species that are present in the furnace during diffusion. Its default value is false but when set to true, the user can set up the flow rate of the following gas/stream species:

- H_2: the flow rate of hydrogen
- O_2: the flow rate of oxygen
- steam: the flow rate of H_2O
- HCL: the flow rate of HCL
- other: the flow rate of other user-defined species

The default value of these parameters is 0. The partial pressure of each species is calculated automatically using its flow rate and the total furnace pressure. The units of the flow rates are arbitrary since only the ratio is used to compute the partial pressure based on the total pressure. The program converts the flow rate of H_2 and O_2 into steam automatically (if applicable) which enables simultaneous dry and wet oxidations. Figure 6.5 shows the net doping after LDD and spacer etch anneal.

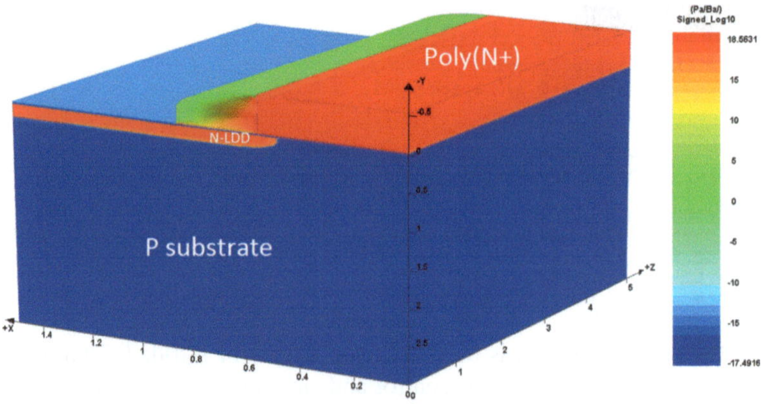

Fig. 6.5 After spacer etch anneal (10_dryo2_diff.str)

6.1.7 Source/Drain Implant

In this section, we perform n + implant for source/drain region.

Process Simulation Code

```
#implant the arsenic
implant arsenic dose=5.0e15 energy=80.0

#do the final anneal
diffuse time=20 temp=950
structure outf=11_sd.str
```

A blanket `implant` of arsenic is used to form the source/drain n+ region. The exposed gate poly is also doped in this step. Since arsenic is a heavier atom than phosphorus, arsenic implant can create a shallower junction, which is desired in modern technology.

Note that we use a "screening" oxide before source/drain implant. This is to prevent the channeling effect where ions oriented along certain crystal directions travel much further than intended [53]. An amorphous region such as oxide or a region damaged by heavy ion implantation can reduce this effect by introducing extra random collisions.

Another `diffuse` step is used to anneal the implanted arsenic with a 950°C furnace temperature for 20 min. Twenty minutes is reasonable here only because this is a long channel MOSFET. It would not be recommended for short channel MOSFET or CMOS technology as it would likely exceed the thermal budget for this step (Fig. 6.6).

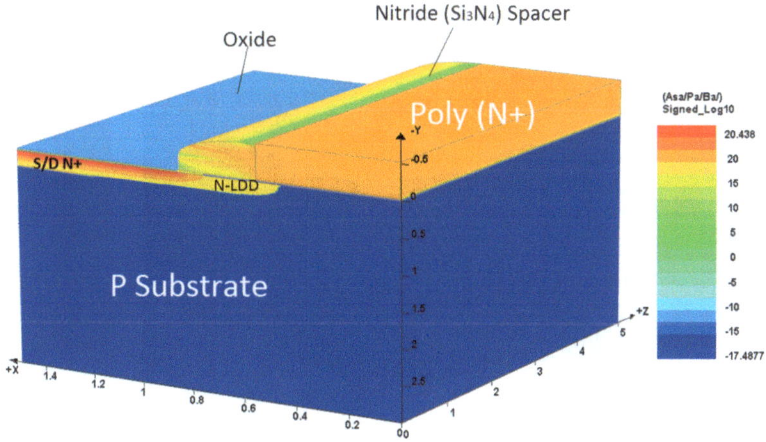

Fig. 6.6 After source/drain implant (11_sd.str)

6.1.8 Mirror Structure and Export to the Device Simulator

For all the process steps used until now, we have exploited the symmetry of the design to save on computation time. While the structure itself is symmetric, the voltage that will be applied on the source and drain contacts will be different. This will break the symmetry for the device simulation so we need to recover the full structure before exporting the structure to the device simulator.

Before the export though, we also need to remove the leftover oxide over the n+ source/drain to make the contact area.

Process Simulation Code

```
#make the contact hole
etch oxide right p1.x=0.85

#reflect the structure
structure mirror left
structure outf=12_sup.str
```

The `etch oxide` command is used to remove all oxide to the right of a certain point. The left/right distinction can be confusing when visualizing in 3D like the figures in this chapter. It is easier to imagine things using a 2D x–y cut so `right p1.x=0.85` means $x > 0.85$.

The `structure` command is used twice to generate the full structure and remove the symmetry we previously relied on. The first call uses the `mirror` option to create a mirror image of the structure to the left ($x < 0$ in the original coordinates). The second call to this command exports the full structure to a file so it can be visualized.

Process Simulation Code

```
#export to APSYS
export outf=sup.aps xpsize=0.001
```

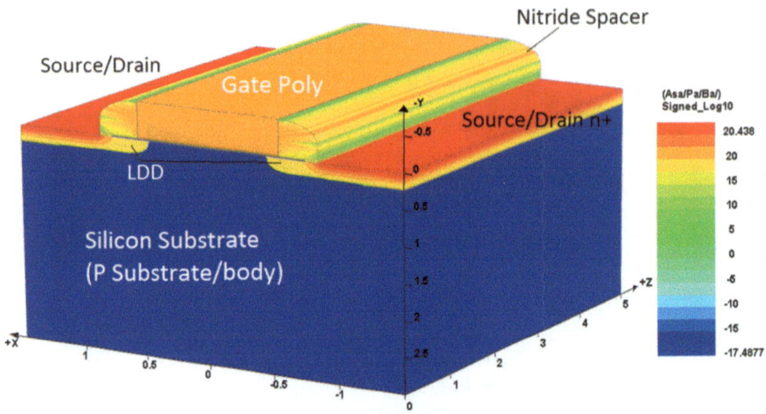

Fig. 6.7 Net doping chart of the long channel MOSFET

Finally, the device is exported for device simulation. The original process simulator mesh is used for the most part but points on material boundaries are duplicated and separated by a small gap controlled by the parameter `xpsize`. This is done to satisfy the convention used by the device simulator that mesh points must belong to only one type of material.

Figure 6.7 shows the final net doping chart. In the next section, we will continue to the device simulation.

6.1.9 Contact Definitions

The process simulation is now complete but contact regions must now be defined. This does not refer to metal or heavily-doped layers created during the process simulation but to equipotential boundary conditions used for device simulation. Using ContactDesigner, three contacts are defined, namely the source, gate and drain. Note that substrate contact can also be defined if the design calls for it. Please refer to previous chapters for details on how to define contacts using ContactDesigner.

Before we move on to device simulation code, we need to learn more about the file contact_3d.sol. From previous chapters we already know that this file is created by ContactDesigner, and should be included in the device simulator.sol file. The reader may be wondering what is included in the contact_3d.sol file. Text Box 6.3 shows the content of this file.

In the contact_3d.sol file, all materials used for simulation are loaded. Since each segment may have different materials to load, and to make the code more compact, loop function (`start_loop..end_loop`) can be used as well. The material is first mapped from process simulation to device simulation and then loaded in each segment. Contacts are defined segment by segment. Normally, the user doesn't need to change the content of this file; the modification of this file is reserved for advanced users.

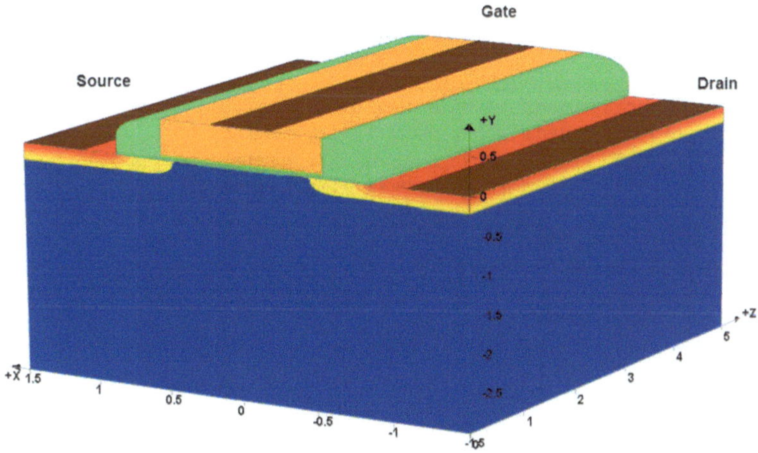

Fig. 6.8 Contact definitions for the long channel MOSFET

`begin_zmater` and `end_zmater`. All the material and contact definitions for each segment/plane should be within these statements.

`load_macro` activates a group of statements used to describe the properties of a given material. It associates a given material number with an entry in the material database. Material properties are then evaluated based on the supplied parameters and certain reserved keywords. In this case, a total of 4 types of material are loaded with material name and material number.

`define_material` statement used when a material number is re-used between z-segments. The first z-segment where this material occurs must use `load_macro` to load the macro parameters. The other z-segments use `define_material` to indicate `load_macro` has already been used.

`SUPREM_contact` is used to define electrodes on the mesh imported from process simulation, `num` statement allows user to specify the contact material number defined above. `xrange` and `yrange` is used to define the geometrical location of the contact. `side` is used to specify whether the upper, lower or inside of the material should be defined as contact. As an example, for source and drain, since the contact is placed on top of silicon surface, upper side is chosen. For polysilicon gate, the contact can also be defined as within the material itself. In this example, source contact is contact #1, gate contact is set to be contact #2 and drain contact is contact #3.

The `contact` statement defines the properties of an equipotential region belonging to an electrode or metal contact. It does not define any other properties associated with a metal contact such as optical absorption: this can be handled by using real metal layers in the device or other statements relevant to a particular model.

For ohmic contacts, the simulation program must determine a built-in voltage based on the flat band condition of semiconductor material in contact with the electrode. Therefore, the program will get confused if an ohmic contact is in contact with more than one semiconductor materials. In such a case, a solution is to use the

Text Box 6.3 Content of contact_3d.sol File

```
suprem_to_APSYS_material suprem_mater=3 APSYS_mater=1
suprem_to_APSYS_material suprem_mater=1 APSYS_mater=2
suprem_to_APSYS_material suprem_mater=4 APSYS_mater=3
suprem_to_APSYS_material suprem_mater=2 APSYS_mater=4

begin_zmater   zseg_num=1
suprem_property user_material_mapping = yes

load_macro name= si mater= 1
load_macro name= sio2 mater= 2
load_macro name= poly mater= 3
load_macro name= sin mater= 4

suprem_contact num=1 xrange=(-1.500000 -1.000000)   yrange=(-0.100000 &&
0.100000) touch_mater=1 side=upper
suprem_contact num=2 xrange=(-0.250000 0.250000)   yrange=(0.520000 0.530000) &&
touch_mater=3 side=upper
suprem_contact num=3 xrange=(1.000000 1.500000)   yrange=(-0.100000 0.100000) &&
touch_mater=1 side=upper

contact num=1
contact num=2
contact num=3
end_zmater

begin_zmater   zseg_num=2
suprem_property user_material_mapping = yes

load_macro name= si mater= 1
load_macro name= sio2 mater= 2
load_macro name= poly mater= 3
load_macro name= sin mater= 4

define_material mater=1
define_material mater=2
define_material mater=3
define_material mater=4

suprem_contact num=1 xrange=(-1.500000 -1.000000)   yrange=(-0.100000 &&
0.100000) touch_mater=1 side=upper
suprem_contact num=2 xrange=(-0.250000 0.250000)   yrange=(0.520000 0.530000) &&
touch_mater=3 side=upper
suprem_contact num=3 xrange=(1.000000 1.500000)   yrange=(-0.100000 0.100000) &&
touch_mater=1 side=upper

contact num=1
contact num=2
contact num=3
end_zmater
```

parameters started with "touch_" to force the electrode to use the properties of a specific semiconductor.

In thermal simulations, additional parameters should be turned on for the `contact` statement, in order to guarantee that the correct thermal boundary conditions are applied to the simulation. See the self-heating example in Sect. 7.3.15. The default value (a 300 K heat sink) may not always correspond to the desired settings.

6.1.10 Device Simulation of Threshold Voltage

In this example, we will simulate the threshold voltage (V_{th}) and use the plotting tool to automatically extract the V_{th}.

Device Simulation Code

```
$-------Part 1: Welcome to Apsys 3D ----------------------------------
begin
convention positive_current_flow=inward

$------- Part 2: Input statement ------------------------------
include file=zmesh.zst &&
  ignore1=load_mesh ignore2=output ignore3=export_3dgeo
load_mesh mesh_inf=sup.aps  suprem_import=yes

$------- Part 3: Output statement, additional physics & alias----
output sol_outf=Vth.out
mobility_xy mater=1 &&
  elec_field_model=lombardi  hole_field_model=lombardi

material mater=1 &&
  el_vel_model=[beta]  hole_vel_model=[beta]

define_alias alias=%Vg name=voltage_2
define_alias alias=%Vd name=voltage_3
define_alias alias=%Id name=current_3

$------- Part 4: Load material and contact information ------
include file=contact_3d.sol

$------- Part 5: Main scan command ---------------------------
newton_par damping_step=5. max_iter=100 print_flag=3

equilibrium

newton_par damping_step=12. print_flag=3 res_tol=1.e-1 var_tol=1.e-1 &&
  max_iter=20 opt_iter=15 stop_iter=12

scan var=voltage_3 value_to=0.1 init_step=0.001 max_step=0.02
min_step=1.e-9

scan var=voltage_2 value_to=1.5 init_step=0.01 max_step=0.01 min_step=1.e-9
stop
```

The `mobility_xy` command is used to create to an anisotropic mobility model for the carriers. Here, we define a vertical field-dependent mobility for the MOSFET channel. For more information about mobility models, including the Lombardi model, please refer to the previous chapters.

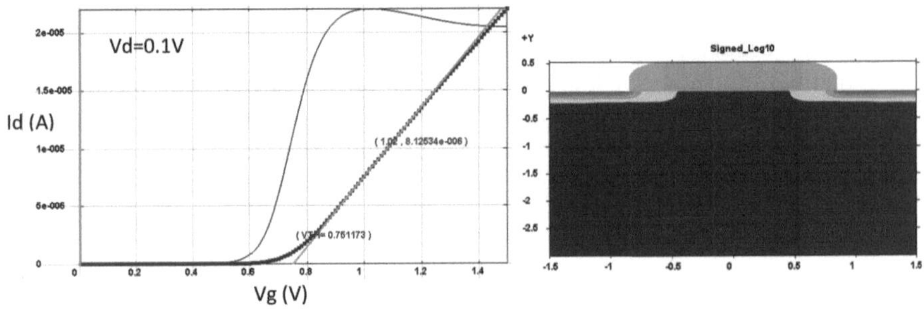

Fig. 6.9 Threshold voltage extraction and device 2D cut showing net doping chart

The `define_alias` commands allow user to define user-friendly names for electrode currents and voltages rather than relying on the contact numbers used internally by the device simulator.

The last part of device simulator .sol file defines the scan parameters and settings for the non-linear Newton solver; refer to the previous chapters for help on how to setup these values. We remind the reader that the `equilibrium` statement sets the initial state of the simulation: zero net current and zero voltage on all the contacts. This must be issued before any `scan` command alters the bias on a contact.

In this example, the voltage on contact #3, or the drain is ramped up to 0.1 V after equilibrium, this is followed by applying a 1.5 V on contact #2, or the gate terminal. The I_D-V_G curve will be plotted after the simulation is done.

The plotting GUI used to plot output curves, can automatically extract the threshold voltage from a MOSFET. In this example, the threshold voltage is extracted to be about 0.75 V. While there are different ways to extract threshold voltage, here we adopt the peak G_{m} (peak transconductance) method:

1. Plot I_{D}-V_{G} at low V_{D} (say 0.1 volt)
2. Plot the slope of I_{D}-V_{G} to get conductance curve.
3. Select V_{G} at maximum conductance (or slope) and obtain the (`Vg_max_slope, Id_max_slope`) point on the I_{D}-V_{G} curve.
4. Draw a tangent line through (`Vg_max_slope, Id_max_slope`) and let it intercept voltage axis of the I_{D}-V_{G} plot. This intercept is defined as V_{th}.

Figure 6.9 is the threshold extraction of the NMOS example as well as 2D cut of the structure with net doping shown.

Note that in theory, the threshold voltage is calculated using the formula below [62]:

$$V_T = \phi_{ms} - \frac{Q_i}{C_i} - \frac{Q_d}{C_i} + 2\phi_f \tag{6.1}$$

where Q_i is the interface charge, Q_d is the depletion charge, ϕ_{ms} is the work function potential difference between gate and semiconductor:

$$\phi_{ms} = \phi_m - \phi_s \tag{6.2}$$

ϕ_f is the Fermi level:

$$\phi_f = \frac{kT}{q} \ln\left[\frac{N_A}{n_i}\right] \tag{6.3}$$

In industry, several practical methods are adopted; the peak G_m method used in this example is only one of them. Different extraction methods will yield different V_{th} results [64]. Sometimes threshold voltage can even be extracted by a predefined drain current and corresponding gate voltage is defined as threshold voltage. Interested reader can refer to the literature [64] for more information.

In terms of device design, the most important parameters that govern the threshold voltage are:

- Gate oxide thickness, which is usually not easy to modify once a particular technology, has been chosen. The thicker the gate oxide, the higher the threshold voltage. In today's technology scaling, thinner gate oxide is desired. Unfortunately, thinner gate oxide also brings larger gate leakage current. To cope with this problem, high-k materials are widely used for advanced CMOS technologies. High-k means higher dielectric constant, which can yield the same gate capacitance with thicker dielectric.
- Doping concentration underneath the gate oxide. The doping concentration plays a major role and is usually the easiest parameter to alter to achieve the desired threshold voltage. Threshold voltage adjustment steps, which exist in many technologies, allow engineers to fine tune the channel doping.
- Interface charges. This is the most difficult aspect to control in practice. Many efforts have been made to reduce the influence of these charges, but they remain unpredictable.

6.1.11 Tuning Material Parameters

TCAD engineers often need to calibrate some specific parameters to accurately predict the real process. This can be because the physical models have some simplifications or assumptions that prevent an accurate reflection of the real world. It can also be because of variations in the test results from fab to fab, wafer to wafer and even from die to die.

One of such parameters that engineers frequently tune is the segregation coefficient. The segregation phenomenon, put simply, is a physical preference of dopants for either oxide or silicon. The segregation coefficient is defined as [65]:

$$m = \frac{equi.\,conc.\,in\,Si}{equi.\,conc.\,in\,SiO_2} \tag{6.4}$$

For Boron, m is less than 1. The silicon surface is depleted of boron as boron atoms segregate preferentially into the oxide. For phosphorus and arsenic, m is greater than 1, which will cause dopant atoms to "pile up" at the surface of the silicon substrate [65].

Here we will show how to tune the segregation coefficient in SUPREM. We will choose boron as an example for the long channel MOSFET. The default values of segregation coefficients for boron are listed in the code below.

Process Simulation Code

```
boron silicon /oxide Seg.0=1126.0 Seg.E=0.91 Trn.E=0.0
```

The segregation model is computed using the following model [9]:

$$\overrightarrow{J_{\text{seg}}} \cdot \vec{n} = T_{\text{r}}\left(C_1 - \frac{C_2}{M_{12}}\right) \tag{6.5}$$

where $\overrightarrow{J_{\text{seg}}} \cdot \vec{n}$ is the normal flux of dopants flowing from material 1 to material 2 across the interface. C_1 and C_2 are the concentrations in material 1 and 2 respectively. T_{r} is the transport velocity which takes into account the rate of dopants crossing the interface. M_{12} is the segregation term taking into account the variation of the threshold solubility of dopants in the two neighboring materials. For each diffusing dopant A, we have:

$$M_{12} = \frac{TS_1(A)}{TS_2(A)} \tag{6.6}$$

where $TS_1(A)$ is the threshold solubility of the dopant A in material 1. $TS_2(A)$ is the threshold solubility of the dopant A in material 2. M_{12} and T_{r} are calculated using a standard Arrhenius relationship. The process simulator allows the user to set these parameters using impurity name as command with its parameters `Seg.0`, `Seg.E`, `Trn.0`, and `Trn.E`. The meanings of these parameters are listed below [9]. The materials are ordered as: the first material in the command is material 1 and the second material is the material 2. The two materials should be separated by forward slash.

- `Seg.0` Pre-exponential constant for segregation
- `Seg.E` Activation energy for segregation
- `Trn.0` Pre-exponential constant for transport
- `Trn.E` Activation energy for transport

Two cases are compared, one with the default setting (`Seg.E=0.91`), the other with the modified setting (`Seg.E=0.7`). All other parameters stay the same, and the modification is just for boron. This change shifts the threshold voltage by 0.2 V because with `Seg.E=0.7`, more boron resides in silicon than the default case. Figure 6.10 illustrates the simulation result.

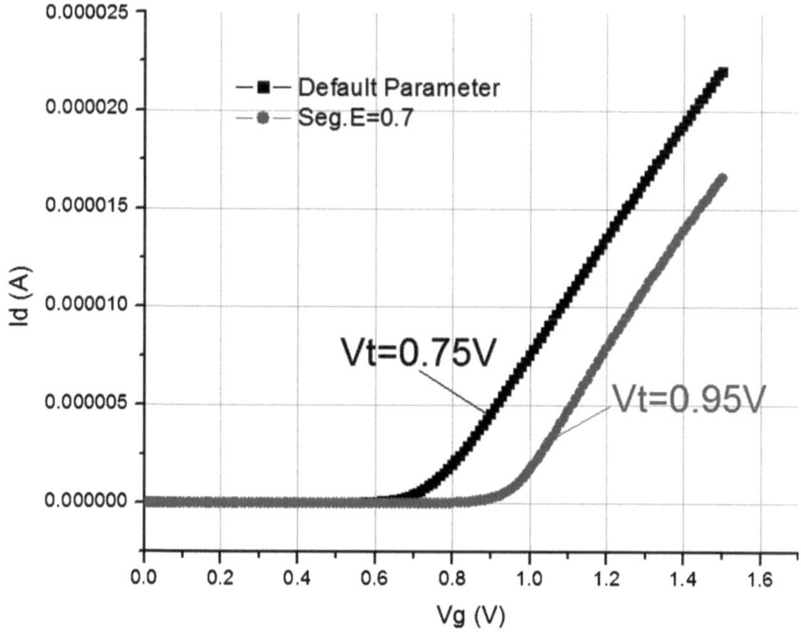

Fig. 6.10 Comparison of threshold voltage for segregation parameter calibration

6.1.12 I_D-V_D *Family of Curves*

Shown in Fig. 6.11 is the I_D-V_D family of curves with impact ionization parameter turned off. A typical MOSFET has three regions of operation: linear, saturation and breakdown. Breakdown region is not shown here. In semiconductor theory and compact modeling, the drain current I_D is calculated using the following formula, which includes both linear and saturation regions [62].

$$I_D = \frac{\overline{\mu_n} Z C_i}{L}\left\{\left(V_G - V_{FB} - 2\varphi_f - \frac{1}{2}V_D\right)V_D - \frac{2}{3}\frac{\sqrt{2\epsilon_s q N_a}}{C_i}\left[\left(V_D + 2\varphi_f\right)^{\frac{3}{2}} - \left(2\varphi_f\right)^{\frac{3}{2}}\right]\right\} \quad (6.7)$$

where V_{FB} is the flat band voltage, which is the gate voltage required to compensate for work function differences between the gate and substrate, and for any electrical charges that may be present in the gate oxide [53]

$$V_{FB} = \Phi_{ms} - \frac{Q_i}{C_i} \quad (6.8)$$

From the equations above, we can see that drain current is a function of both drain voltage V_D and gate voltage V_G.

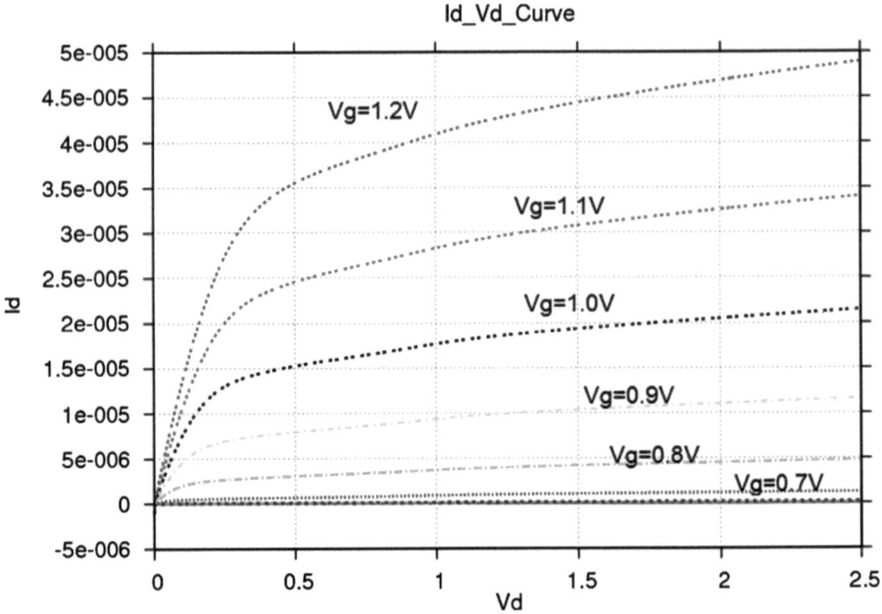

Fig. 6.11 I_D-V_D curves for long channel MOSFET

6.1.13 Band Diagram Simulation

Band diagram of the device is illustrated with 1D cut. Figure 6.12 is the band diagrams view along the channel at equilibrium and $V_D = 2$ V (the cut is close to the surface at $y = -0.01$ um).

6.1.14 MOS Capacitor Simulation

With some modification to the process simulation and device simulation, we can further simplify the structure to simulate the MOS Capacitance C-V characteristics as shown in Fig. 6.13. We leave it as an exercise to the reader to modify the process simulation to change the structure from a MOSFET to a three layered (poly-oxide-silicon) MOS capacitor.

The gate is biased from -1.5 to 1.5 V while the body/substrate is grounded. Note that the body p-type body doping has been changed to 1E + 16 cm^{-3}. The Cgb vs. bias curve is plotted in Fig. 6.13. For both high frequency (1 MHz) and low frequency (1 Hz) simulation, the MOS cap C-V curve illustrates the typical characteristics found in many semiconductor textbooks such as Sze [61].

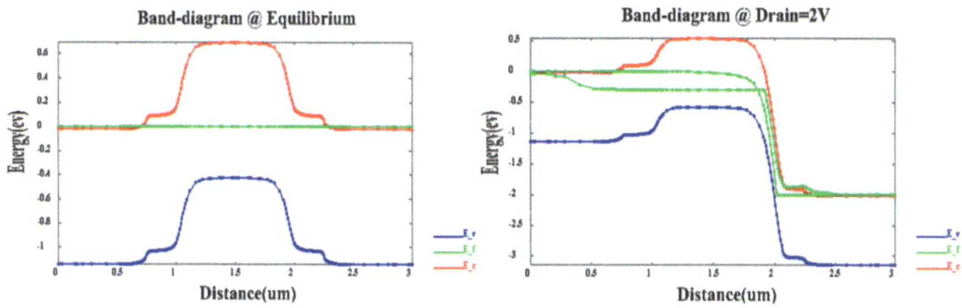

Fig. 6.12 Band-diagrams at equilibrium and at $V_D = 2$ V (surface cut @ $y = -0.01$ um)

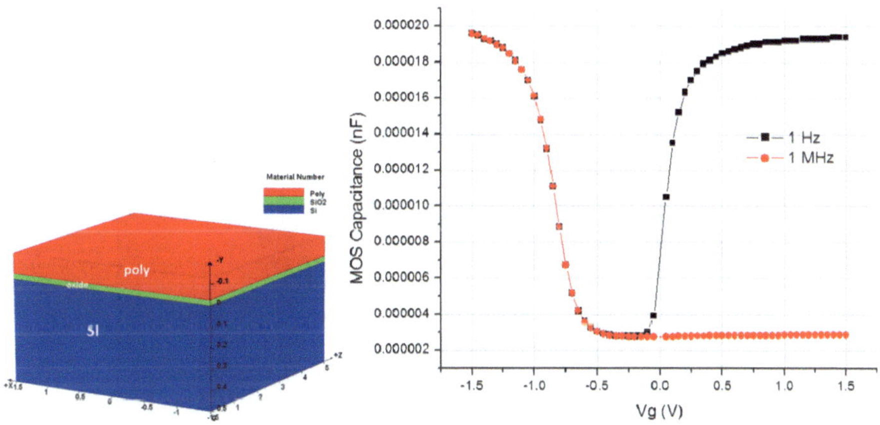

Fig. 6.13 MOS CAP simulation structure and simulation result

Table 6.2 Simulation data for long channel MOSFET

	Process simulation	Device simulation ($V_D = 2$ V)	Total mesh count	Total number of planes
Long channel MOSFET	3 min	2 min	7,634	2
Computer configuration: HP desktop with Intel i7-860/6G/1T/NVIDIA GeForce GTX260/Win7				

6.1.15 Simulation Data

Table 6.2 gives the simulation data for the long channel MOSFET.

6.2 CMOS Technology Process Flow

CMOS (Complimentary Metal Oxide Semiconductor) technology is the backbone of today's integrated circuit industry. Both n-type and p-type MOSFETs are used to realize logic functions. In comparison with transistor-transistor logic (TTL) built by

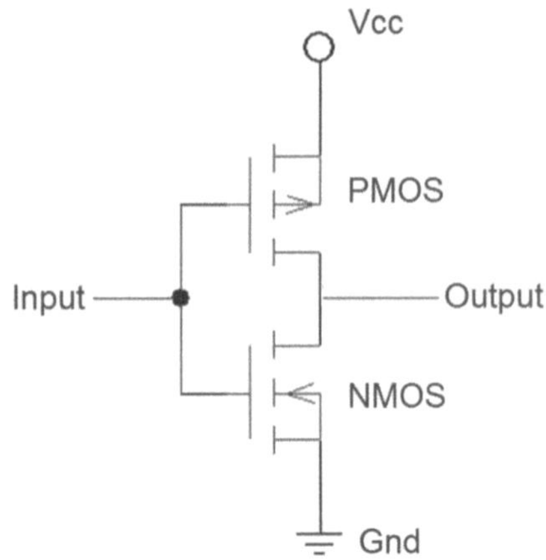

Fig. 6.14 A simple CMOS inverter circuit

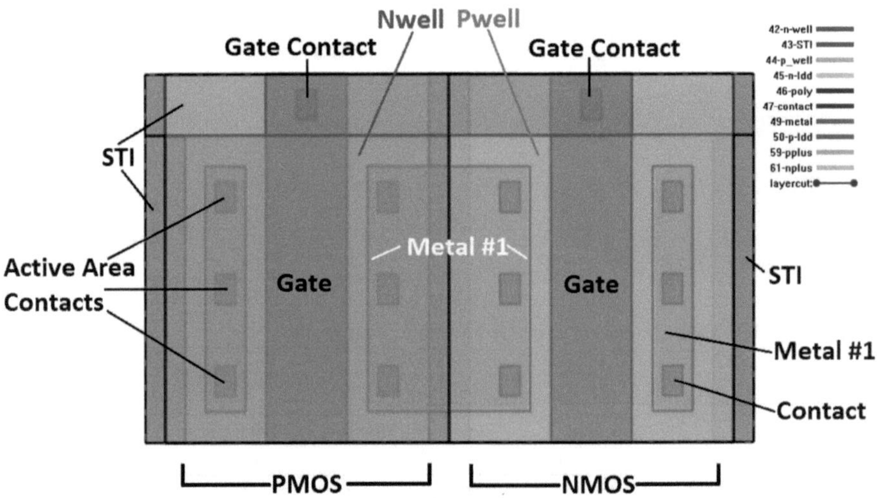

Fig. 6.15 Mask layout of CMOS Inverter

BJTs, CMOS has much less power dissipation due to insulated gate control [66]. This section will run a process simulation to create a structure similar to a typical CMOS inverter circuit such as the one shown in Fig. 6.14.

In this section, a CMOS technology process simulation is carried out according to the process steps found in Plummer's book [53]. The mask layout shown in Fig. 6.15 incorporates the important mask steps of CMOS technology process. STI (Shallow Trench Isolation) mask layer is used instead of active layer (device area outside STI). The following section will show a simplified 3D process simulation for CMOS process flow based on the layout information. Both FEOL (Front End of the Line) and BEOL (Back End of the Line) are included in this simulation.

The parameters chosen and process conditions are arbitrary, and do not reflect any real processes. Users can use these settings as a starting point for their 3D process simulations.

In this example, both NMOS and PMOS are simulated. The technology node is still in the long channel regime, meaning that short channel and quantum effects are not considered. Both p-type and n-type devices are surrounded by the STI region. Poly gate contacts are over the STI to prevent the possible damage to the channel region when forming contacts on poly.

6.2.1 *Overview of Simulation Steps*

This example will deal exclusively with the process simulation for CMOS technology. Contact definitions and device simulation will not be performed. Table 6.3 is the overview of simulation steps for CMOS technology.

6.2.2 *Substrate and Initialization*

The substrate is chosen according to different applications. Major considerations include wafer type (bulk or SOI), substrate type, resistivity and orientation [53]. SOI (silicon on insulator) wafers will be considerably more expensive than bulk wafers. For modern CMOS technologies, a p-type substrate is usually chosen with twin well implant. While it's true that NMOS can be built without introducing p well, in practice, it is always easier for process control to build NMOS within a p well [53]. Like always, the process file will be divided into several parts for explanation.

Process Simulation Code

```
mater_define material_label=BPSG macro_name=BPSG
mater_define material_label=W macro_name=W
mater_define material_label=TiN macro_name=TiN

mode quasi3d
3d_mesh infile=geo

# p substrate
initial boron conc=1.0e15 orient=100
structure outf=01_sub.str
```

mater_define is used to introduce user-defined materials to the process simulation. In this example, BPSG, Tungsten (W) and TiN are defined for use in the backend process. BPSG is used for the ILD (interlayer dielectric) while Tungsten and TiN are used for the contacts and vias. Additionally, a link is made between

Table 6.3 Overview of simulation steps for CMOS technology

<table>
<tr><th>CMOS technology</th><th>Process simulation steps</th></tr>
<tr><td>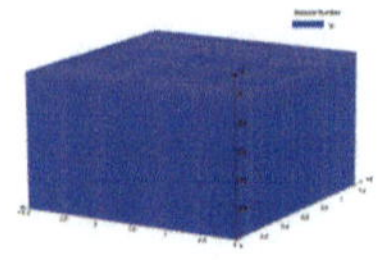</td><td>Step 1: Substrate and initialization</td></tr>
<tr><td>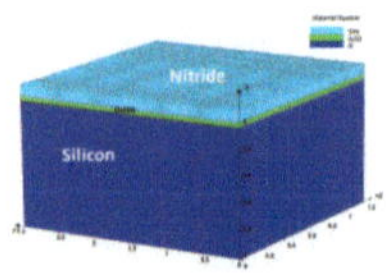</td><td>Step 2: STI pad oxide and nitride formation</td></tr>
<tr><td>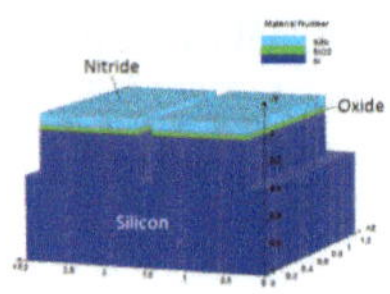</td><td>Step 3: STI plasma etch</td></tr>
<tr><td>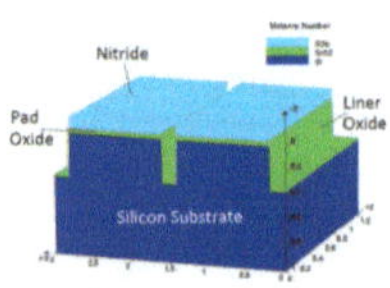</td><td>Step 4: STI liner oxide growth</td></tr>
<tr><td>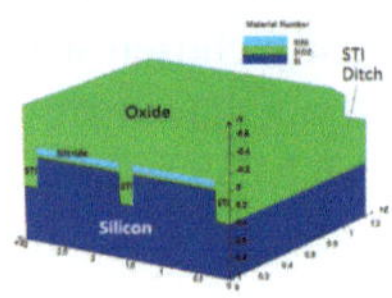</td><td>Step 5: STI oxide fill</td></tr>
<tr><td>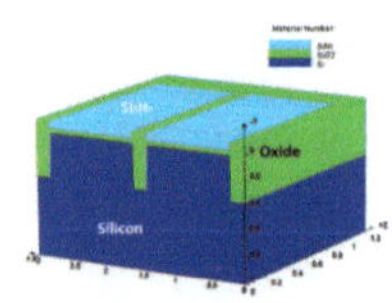</td><td>Step 6: Chemical mechanical polish (CMP) and chemical strip of nitride</td></tr>
<tr><td>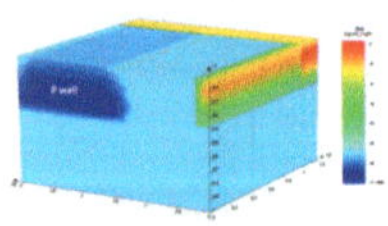</td><td>Step 7: P well and n well implant</td></tr>
</table>

(continued)

Table 6.3 (continued)

CMOS technology	Process simulation steps
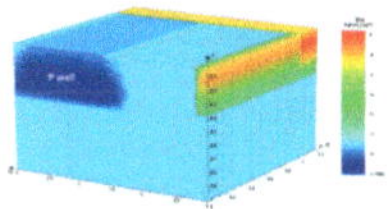	Step 8: Threshold voltage (V_{th}) adjustment steps
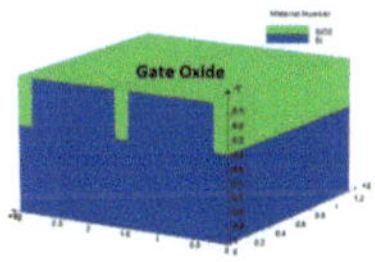	Step 9: Gate oxide growth and gate poly deposition
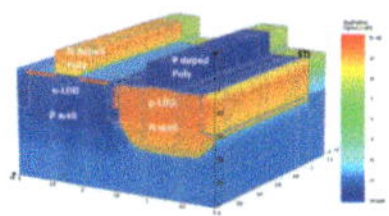	Step 10: Lightly Doped Drain (LDD) implant
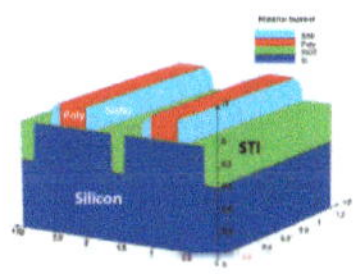	Step 11: Nitride spacer
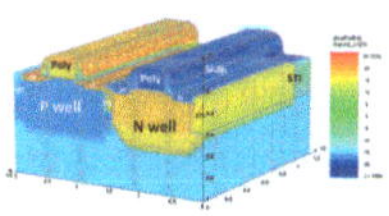	Step 12: Drain and source implant
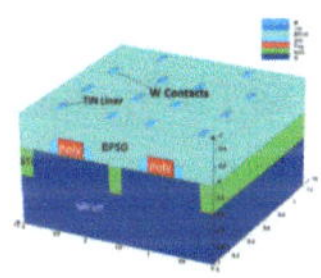	Step 13: Contacts placement
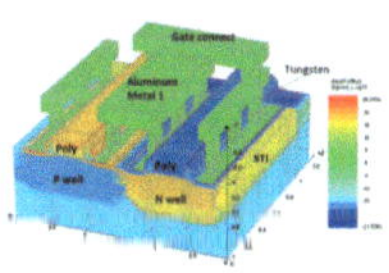	Step 14: Metal layer placement

Text Box 6.4 Content of geo1.in File

```
line x loc=    0.00000     spacing=  0.120536       tag=lft
line x loc=  0.550000     spacing=  0.180804e-01
line x loc=  0.800000     spacing=  0.120536
line x loc=    1.05000     spacing=  0.180804e-01
line x loc=    1.50000     spacing=  0.120536
line x loc=    1.95000     spacing=  0.180804e-01
line x loc=    2.20000     spacing=  0.120536
line x loc=    2.45000     spacing=  0.180804e-01
line x loc=    3.00000     spacing=  0.120536       tag=rht

line y loc=    0.00000     spacing=  0.731707e-02 tag=top
line y loc=    1.00000     spacing=  0.146341       tag=bot

region silicon xlo=lft xhi=rht ylo=top yhi=bot
bound exposed xlo=lft xhi=rht ylo=top yhi=top
bound backside xlo=lft xhi=rht ylo=bot yhi=bot
```

these user-defined materials and the material parameter macros used to represent them in the device simulation if required.

Like the previous long-channel MOSFET example, we use `mode quasi3d` to simplify the model and save on computation time. This is reasonable because for this particular example, no z-direction diffusion for the dopants necessary.

The `3d_mesh` statement is used to load 3D mesh declaration statements that have been automatically created by the MaskEditor GUI; refer to previous chapters for details. Text Box 6.4 shows the content of geo1.in as an example of the commands being loaded.

The `initial` command creates a boron doped substrate with a constant doping concentration of $1E + 15\ cm^{-3}$. The surface orientation is chosen as [100] since it has a lower interface trap density for both NMOS and PMOS.

The `structure` command is used to save the data structure to a user defined file. The simulation result is show in Fig. 6.16. We note in passing that SUPREM allows shorthand for commands and parameters (`outf` instead of `outfile`) except when there is a risk of confusion.

6.2.3 STI Pad Oxide and Nitride Formation

In today's technology, active devices are integrated side by side in a common silicon substrate [53]. In order to electrically isolate these devices, fairly thick SiO_2 is usually grown or deposited between the devices. Oxide is a perfect insulator to provide the

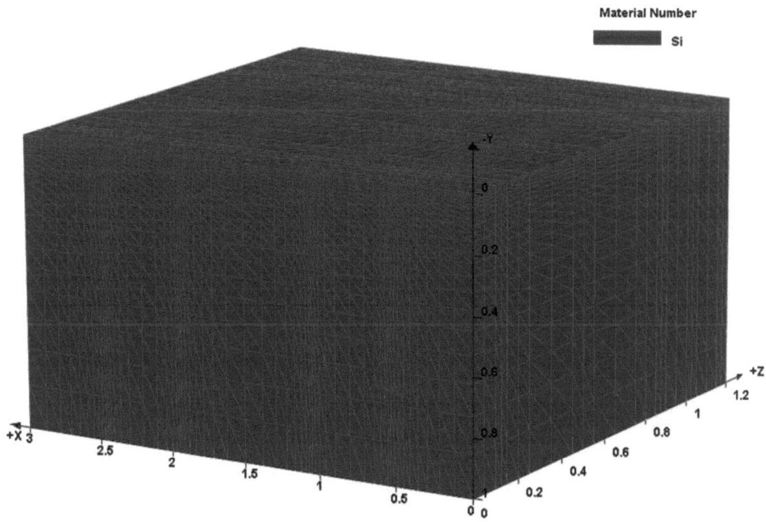

Fig. 6.16 Substrate formation (01_sub.str)

necessary isolation. Since LOCOS (Localized Oxidation of Silicon) is relatively old for CMOS technology, this book will focus primarily on STI (Shallow Trench Isolation) technology. Note that LOCOS is still widely used in analog/power technologies.

There are many advantages of STI over LOCOS including completely eliminating the "bird's beak" effect. It takes up less silicon space (which enables designers to achieve higher density) and produces a flat surface that is easier for lithography.

The first step of a STI process is to grow pad oxide. A nitride (Si_3N_4) layer is then deposited. This layer is usually highly stressed and produces a large compressive stress in the underlying silicon substrate; this can lead to defect generation. The main purpose of the pad oxide under the nitride layer is to help relieve the stress [53]. The nitride layer itself is used as an etch stop for later chemical mechanical polish (CMP).

Process Simulation Code

```
# Wet Oxide
diffuse temp=900 time=15 weto2

# deposit thin layer of Si3N4
deposit nitride thick=0.08 temp=800
structure outf=02_pad.str
```

The `diffuse` statement specifies a wet oxidation step at a constant furnace temperature of 900°C for 15 min.

The nitride is then deposited with a thickness of 0.08 um at temperature of 800°C. In practice, this deposition is done with Low Pressure Chemical Vapor Deposition (LPCVD) to achieve better uniformity. The resulting structure data is saved to file "02_pad.str" as seen in Fig. 6.17.

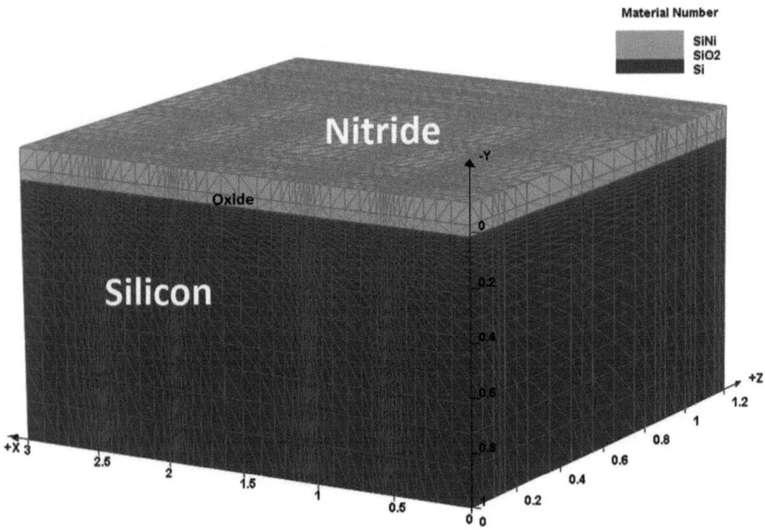

Fig. 6.17 Pad oxide and nitride (02_pad.str)

6.2.4 STI Plasma Etch

Etching in the process simulator is always geometrical. Plasma etch is treated as dry etch in the simulator. As stated previously, the mask file (cmos.gds43.msk) created automatically by MaskEditor contains all the simulation commands necessary for the etch process.

In manufacturing, trench etching sometimes can be quite sophisticated. In order to prevent undercutting of adjacent active regions, trenches need to be vertical. However, a perfect vertical trench wall will cause voids from oxide fill. A small slope is usually applied for trench walls in practice. Also, the corners need to be rounded to prevent sharp tips, since sharp corners will induce electrical effects later on.

In this simulation, the trench walls are etched with an angle of 2° to mimic the real process.

Process Simulation Code

```
# STI formation
include file=cmos.gds43.msk
struct outf=03_STI_etch.str
```

The `include` statement will load the mask file (cmos.gds43.msk) into the simulation program: the layout for the STI etching is shown in Fig. 6.17. Note that the mask layer polarity is negative, which means the drawn area will be etched away. In this mask file, three steps of process are included. First is deposition of photoresist, this is followed by mask define. Finally plasma etching of nitride, oxide

Text Box 6.5 The Content of the Mask File for Segment #2

```
mask segm=2 thick=1.3 x1.from=0.2 x1.to=1.4 x2.from=1.6 x2.to=2.8 x2.left.theta=2.
x2.right.theta=2.
etch segm=2 nitride avoidmask depth=0.35
etch segm=2 oxide avoidmask depth=0.35
etch segm=2 silicon avoidmask depth=0.35
etch segm=2 photoresist all
```

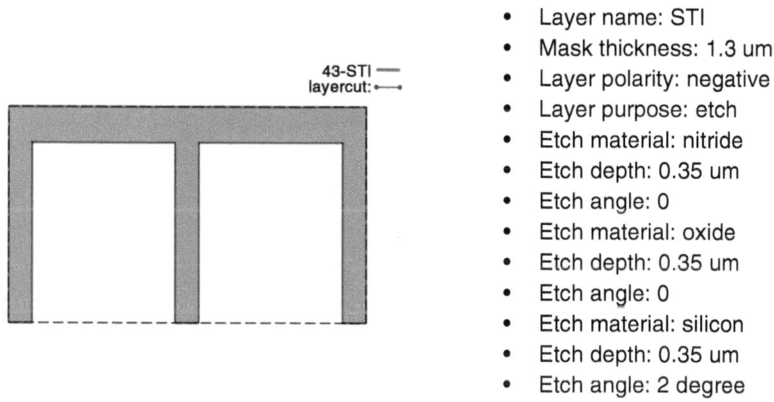

Fig. 6.18 MaskEditor layout for STI layer

and silicon layers is performed in sequence to create the groves. All of these are made possible by specifying the layer cut properties in the MaskEditor. MaskEditor will then automatically generate the mask file with command lines like the one seen below in Text Box 6.5, which includes the mask and etch statements for segment / plane #2. The statement `x2.left.theta=2` instructs the simulator to perform angled etch Fig. 6.18 shows the STI layer layout while 6.19 is the simulation result after STI etch.

6.2.5 STI Liner Oxide Growth

After plasma etching, a thin STI liner is grown on trench sidewalls and bottoms. The thickness of the STI liner is usually in the range of 10–20 nm [53].

The STI liner will ensure a better Si/SiO_2 interface with lower charge densities. This step will also help round the STI corners from the viscoelastic flow properties

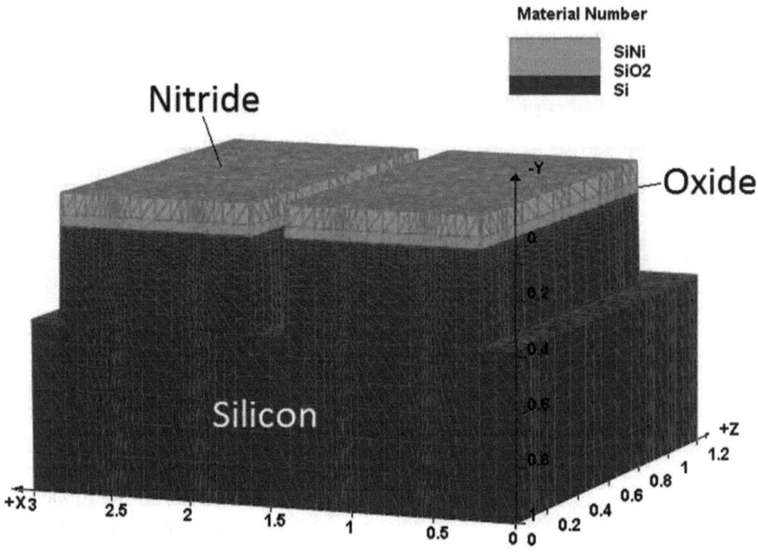

Fig. 6.19 After plasma etch of STI step (03_STI_etch.str)

of SiO_2 at high temperatures. For power LDMOS devices, the STI liner may influence the hot carrier injection, when STI is used in the drain drift region in these devices.

In this example, a dry thermal oxide liner is grown.

Process Simulation Code

```
# STI liner oxide
method viscous grid.oxide=0.02
diffuse temp=1100 time=2 dryo2
structure outf=04_STI_liner.str
```

The `method` command is used to let the user modify solver or model settings. Here, the `viscous` parameter tells SUPREM to treat the oxide as an incompressible viscous fluid [9]. As before, `grid.oxide` is used to control the mesh generation for the new oxide layer during growth.

Dry oxidation (`diffuse dryo2`) at a constant temperature of 1,100°C for 2 min is used to create the liner oxide of the STI side walls. The resulting structure is shown in Fig. 6.20.

6.2.6 *STI Oxide Fill*

A thick SiO_2 layer is deposited by Chemical Vapor Deposition (CVD). In practice, to reduce the filling problems like void, High Density Plasma (HDP) is generally applied.

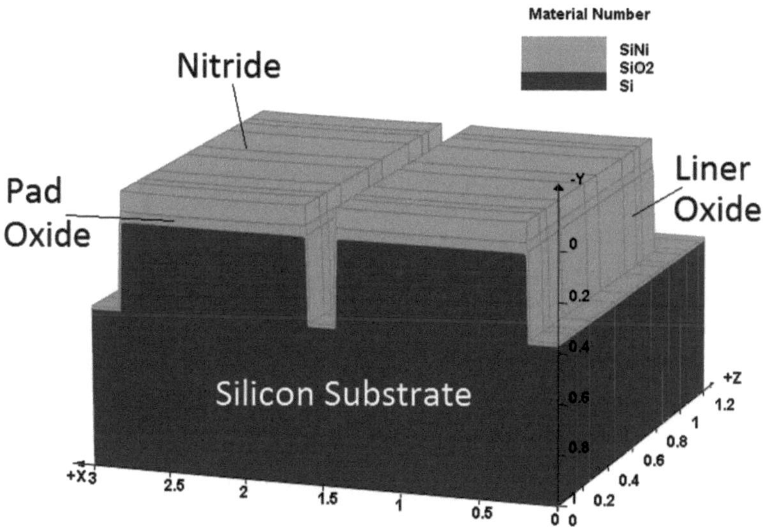

Fig. 6.20 STI liner oxide growth (04_STI_liner.str)

Of course, the process simulation will not distinguish different deposition method used in the real world. Note that a slope is usually necessary for void-free deposition. This example uses a trench wall etch angle of 2°.

Process Simulation Code

```
# STI HDP fill
deposit oxide thick=0.5 meshlayer=4
struct outf=05_STI_fill.str
```

High Density Plasma (HDP) CVD is used to `deposit` the fill-in oxide to a thickness of 0.5 um. The parameter `meshlayer` specifies the number of mesh lines added to this deposited material: higher numbers increase both the accuracy and simulation time Fig. 6.21 illustrates the simulation result.

6.2.7 *Chemical Mechanical Polish (CMP) and Chemical Strip of Nitride*

After High Density Plasma (HDP) CVD fill, the wafer will undergo a polishing step called Chemical Mechanical Polish (CMP) using high-pH silica slurry. Excess SiO_2 is polished off leaving a planar surface. The nitride layer acts as a polishing stop and is then chemically removed using hot phosphoric acid.

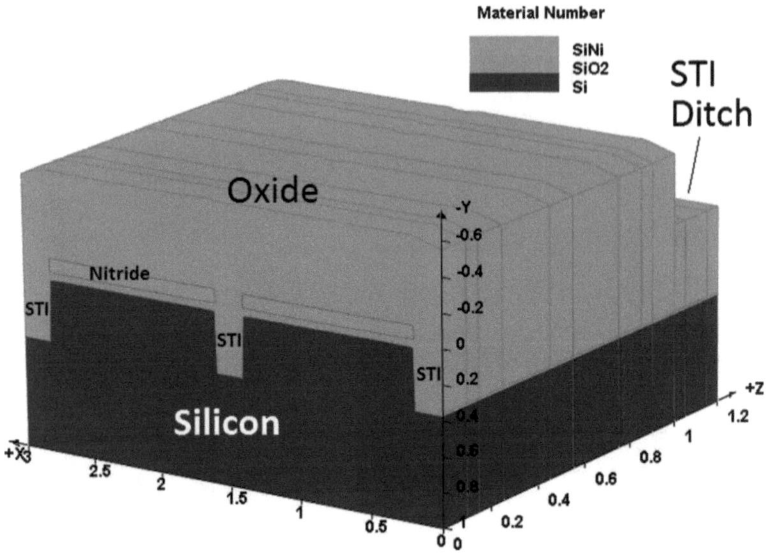

Fig. 6.21 STI HDP fill (05_STI.fill.str)

Process Simulation Code

```
# CMP stops at nitride
etch start x=0 y=-2
etch continue x=0 y=-0.08
etch continue x=3 y=-0.08
etch done x=3 y=-2
structure outf=06_STI_CMP.str
```

Here, CMP is modeled as a geometric etching process: the `etch start ....` `continue ....done` commands are used in combination to define an arbitrary complex region that will be etched away. The $x-y$ coordinates are easy to visualize in a 2D-cut but we note that the process simulator varies from the usual convention in that $x > 0$ in on the right but $y > 0$ is towards the bottom: attentive readers may have noted that the 3D plots have a $-y$ label on the vertical axis. The reference line $y = 0$ is set to be at the initial surface of the substrate so $y < 0$ values are needed to etch away deposited layers above that line. Coordinates are given in microns, as per the usual SUPREM convention.

We also note the absence of the `segm` statement. This means that the etch command operates in the same manner on all planes. The simulation result after the CMP step is shown in Fig. 6.22.

Process Simulation Code

```
# oxide nitride chemical strip
etch nitride all
etch start x=0 y=-2
etch continue x=0 y=0.025
etch continue x=3 y=0.025
etch done x=3 y=-2
structure outf=07_ON_strip.str
```

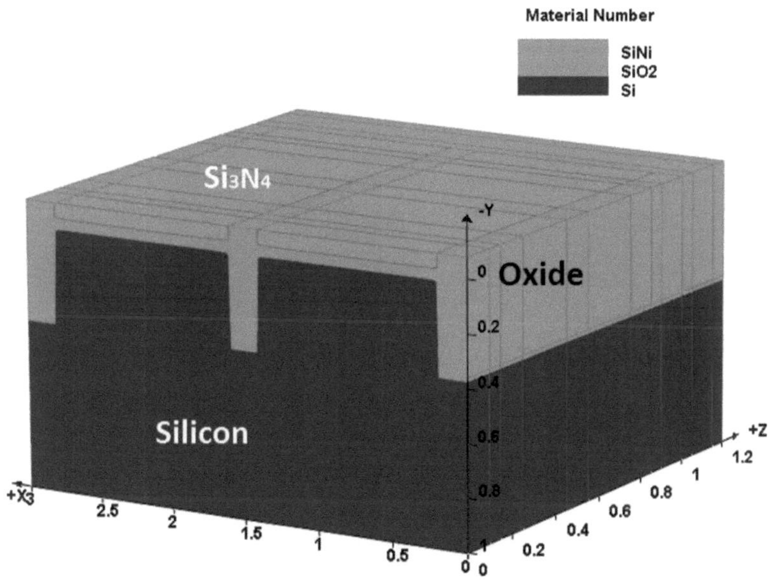

Fig. 6.22 CMP stops at nitride layer (06_STI_CMP.str)

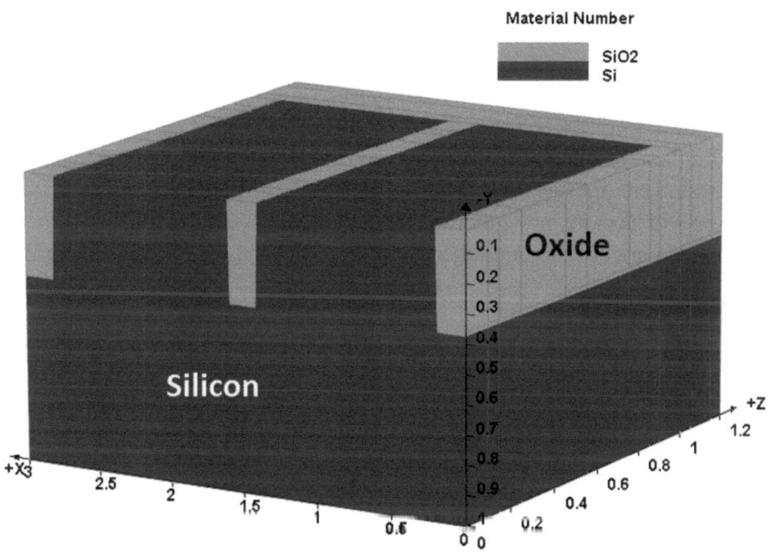

Fig. 6.23 Chemical strip of remaining oxide and nitride layer (07_ON_strip.str)

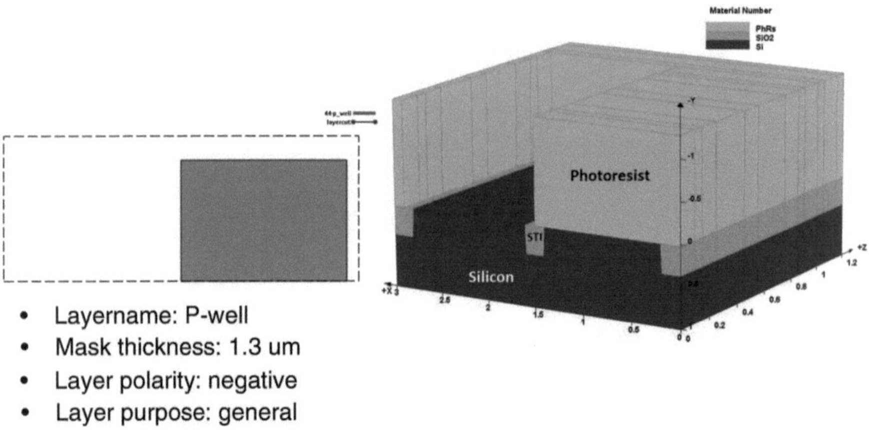

Fig. 6.24 P well mask and photoresist of the p-well implant (08a_pwell_mask.str)

Another `etch` command is used to remove the remaining nitride. Once again, a geometric etch is used to model the effects of the real etching process (phosphoric acid). Figure 6.23 shows the simulation result after nitride is removed.

6.2.8 P Well and N Well Implant

Twin well implant technology is used to optimize both NMOS and PMOS at the same time. After each implant, a Rapid Thermal Anneal (RTA) is generally performed to repair the implant damage and activate the dopants. For the sake of simplicity, constant temperature anneals are used here. Thermal diffusion for well drive in is necessary even though there will be many more thermal steps that follow. This is because these thermal steps usually do not have enough thermal budgets for the well drive-in purpose.

Process Simulation Code

```
# p well implant
include file=cmos.gds44.msk
structure outf=08a_pwell_mask.str
implant boron dose=6e12 energy=80
etch photoresist all
diffuse time=20/60 temp=1000 nitrogen
structure outf=08_pwell.str
```

The `include` statement will load the mask file cmos.gds44.msk which is shown in Fig. 6.24. Note that this is a negative layer, meaning that the drawn area is where the photoresist is removed and the implantation is allowed to proceed. A large thickness of photoresist is used to protect the other areas of the device. Please be aware that the origin of the *x*-axis of the device structure view is on the right hand side, while on the mask layout, the origin of the *x*-axis in on the left hand side. So the structure view mirrors the mask layout.

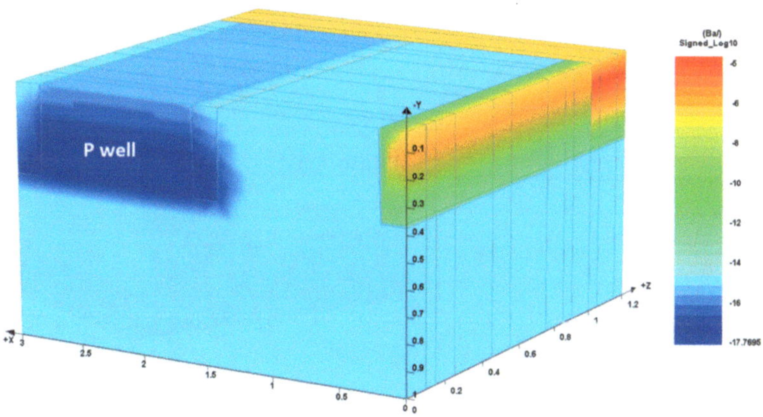

Fig. 6.25 Net doping chart after p-well implant (bottom: 08_pwell.str)

The `implant` statement performs a p well implant with a boron dose of 6E + 12 cm^{-2} and ion energy of 80 keV. After implantation an `etch photoresist all` command is issued to strip off all the photoresist before moving on to the diffusion cycle.

The `diffuse` command starts a diffusion cycle in a nitrogen environment (no oxidation). The diffusion time is given in minutes so 20/60 is used to specify 20 s. This time is especially short since the diffusion step in the subsequent n well implant will operate on both dopants.

The simulation result of net doping chart after p well implant is given in Fig. 6.25.

Process Simulation Code

```
# n well implant
include file=cmos.gds42.msk
structure outf=09a_nwell_mask.str
implant phosphorus dose=5e12 energy=100
etch photoresist all
diffuse time=90 temp=1000 nitrogen
regrid log10.change=5.0 refine
structure outf=09_nwell.str
```

The n well implant commands follow the same general steps as those for the p well. The main difference is the species used for implantation (phosphorus) and the ion energy (100 keV). The diffusion time is set to 90 min and drives in the dopants from both the n well and p well implantations.

After diffusion, a `regrid` command is used to refine the mesh and obtain a smoother doping profile and denser mesh near the p-n junction. The parameters for the regrid procedure impose a maximum change (on a logarithmic scale) in the doping concentration between neighboring mesh points. Figure 6.26 is the mask layout from MaskEditor.

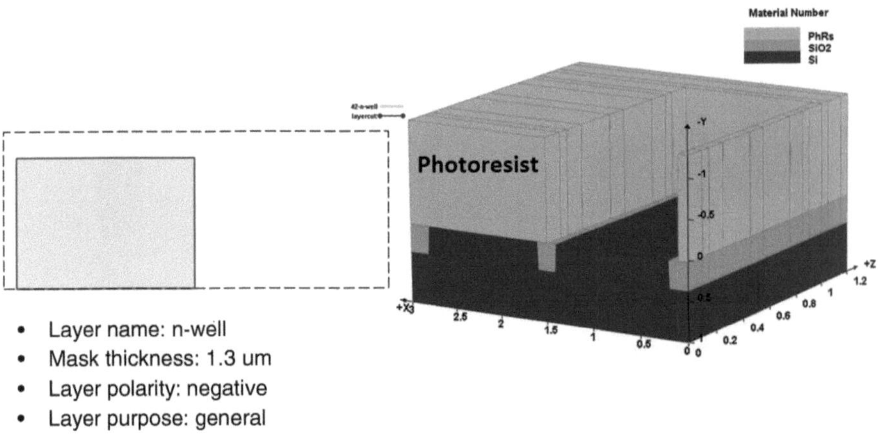

Fig. 6.26 Mask layout for the n well implant layer and photoresist (09a_nwell_mask.str)

6.2.9 Threshold Voltage (V_{th}) Adjustment Steps

Note that a threshold voltage adjustment step is often necessary to optimize the threshold voltage of both NMOS and PMOS. Equation 6.1 can be written in another way:

$$V_{th} = V_{FB} + 2\phi_F + \frac{\sqrt{2\varepsilon_s q N_A (2\phi_F)}}{C_{ox}} \Phi_{ms} + \frac{qQ_I}{C_{ox}} \tag{6.9}$$

where Q_I is the implant dose. This equation assumes that the entire implant dose is located close to the surface region, which is often a reasonable approximation [53]. This means Q_I should be adjusted to accommodate the desired threshold voltage. This can be achieved by V_{th} adjustment implant steps. Since in terms of process simulation steps, the V_{th} adjustment step is similar to n well and p well implant, and use the same mask sets. The detailed explanation is neglected here.

Process Simulation Code

```
# vt adjustment pwell
include file=cmos.gds44.msk
structure outf=09a_Vt_adj_pwell_mask.str
implant boron dose=2e12 energy=20
etch photoresist all
diffuse time=1 temp=950 nitrogen
structure outf=09a_Vt_adj_pwell.str

# vt adjustment nwell
include file=cmos.gds42.msk
structure outf=09b_Vt_adj_nwell_mask.str
implant phosphorus dose=1e12 energy=40
etch photoresist all
diffuse time=1 temp=950 nitrogen
#regrid log10.change=5.0 refine
structure outf=09b_Vt_adj_nwell.str
```

6.2.10 *Gate Oxide Growth and Gate Poly Deposition*

It is well-known that gate oxide needs to be of high quality so a chemical cleanup is generally used. This removes any remaining oxide and leaves behind a clean silicon surface for dry oxide growth. Gate poly is deposited after the oxide is grown using LPCVD. Because of poly contact issues, the gate pattern is defined in such a way that at least part of the gate poly extends onto the field oxide (STI region). Generally, the gate poly contact should not be placed on top of the active region as the stress may make the underlying gate oxide and channel region less reliable. The poly doping is done either through in-situ doping or blanket implant after poly deposition.

In situ doping of polysilicon is usually performed by simply including a dopant gas, usually diborane (B_2H_6) or phosphine (PH_3) in the CVD process. The in situ doping can increases (boron) or decreases (phosphorous) the deposition speed of poly depending on the dopant type [67].

Process Simulation Code

```
# Gate oxide
etch start x=0 y=-1
etch continue x=0 y=0.025
etch continue x=3 y=0.025
etch done x=3 y=-1

# re-grow dry oxide
diffuse temp=850 time=45 dryo2
structure outf=10_gate_oxide.str
```

`etch start, continue, done` will remove existing material to prepare the surface to grow gate oxide. This requires high quality thermal oxide so a `diffuse dryo2` step is used. Based on the time (45 min) and furnace temperature (850°C), the gate oxide thickness is about 8 nm Fig. 6.27 is the simulation result of gate oxide growth.

Process Simulation Code

```
# Gate Poly deposition
deposit poly thick=0.2 meshlayer=2
diffuse time=20/60 temp=1100 nitrogen
include file=cmos.gds46.msk
structure outf=11_gate_poly.str
```

The first step in the creation of the gate poly is to uniformly `deposit` a layer of polysilicon 0.2 um thick. While it is possible to in-situ dope the polysilicon with either n or p-type, it is not used here. In this demo, a separate implantation of the n and p-type gate poly is applied. NMOS will have n+ doped poly while PMOS will have p+ doped poly. After deposition, a short anneal step is used; this would also be required to activate the poly dopants had we chosen to use in-situ doping.

The `include` command loads a mask file generated by the MaskEditor GUI. Rather than a conventional mask, this particular file has etch-purpose setting which

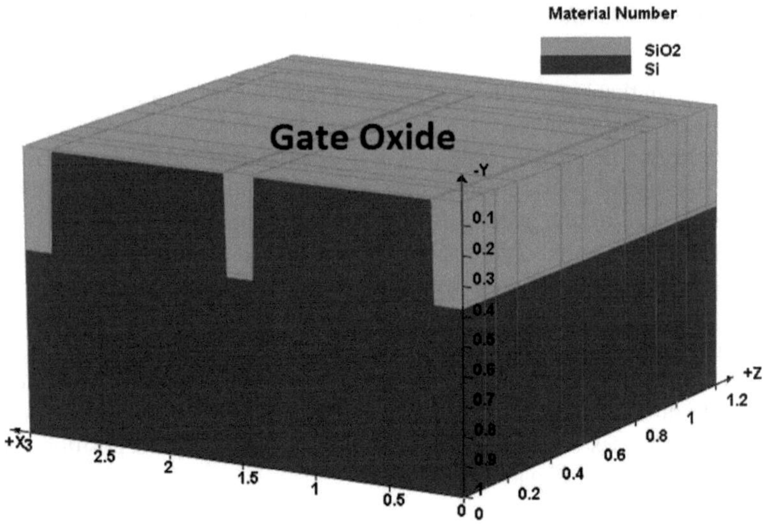

Fig. 6.27 Gate dry oxide growth (10_gate_oxide.str)

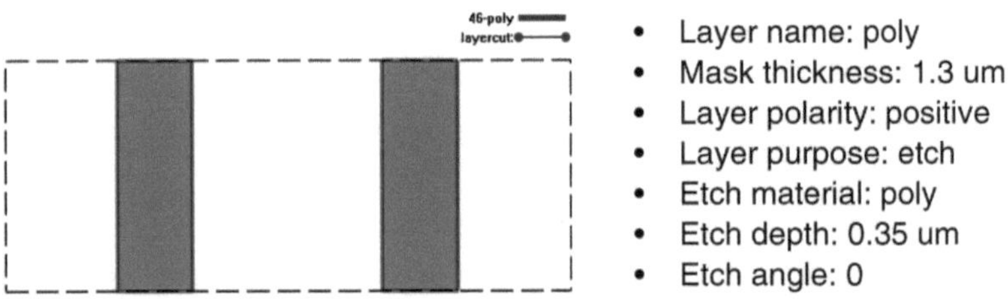

Fig. 6.28 Mask layout of poly gate

Text Box 6.6 The Content of cmos.gds46.msk for Segment 1

```
mask segm=1 thick=1.3 x1.from=0.6 x1.to=1. x2.from=2. x2.to=2.4
etch segm=1 poly avoidmask depth=0.35
etch segm=1 photoresist all
```

combines various process steps into a single command: please refer to earlier chapters for details on mask purposes. This etch mask has a positive polarity so the drawn area will be protected and everything else is etched away, as can be seen in Fig. 6.28. The content of this mask file (cmos.gds46.msk) for segment #1 is shown in Text Box 6.6 and the simulation result is shown in Fig. 6.29.

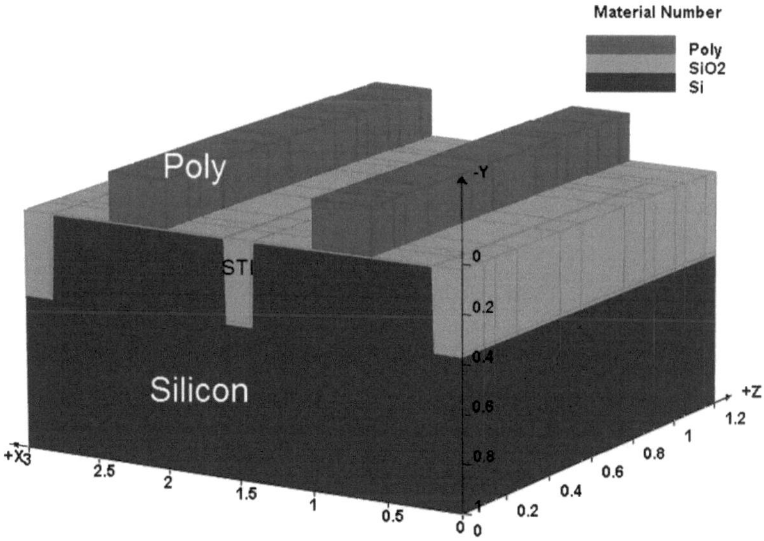

Fig. 6.29 Gate poly deposition (11_gate_poly.str)

6.2.11 Lightly Doped Drain (LDD) Implant

As discussed in Sect. 6.1.6, with the ever decreasing channel length of CMOS technology, if voltage is not properly scaled down as feature size scales, the electric field in the channel will increase dramatically. The LDD concept was invented to grade the doping in the drain region and to produce an N+/N/P profile between drain and the channel in NMOS and vice versa for PMOS [53].

The LDD also acts like a shallow junction that connects the deeper source and drain. This shallow junction works well to cope with the drain induced barrier lowering effect, usually found in short channel devices [53]. In practice, another implant step called a halo implant is frequently used for advanced CMOS process. This helps alleviate the short channel effect found in sub-micron CMOS processes.

Process Simulation Code

```
# p-LDD implant
include file=cmos.gds50.msk
structure outf=12a_pldd_mask.str
implant bf2 dose=8e12 energy=7
etch photoresist all
diffuse time=30/60 temp=1000 nitrogen
structure outf=12_pldd.str
```

Figure 6.30 shows the mask layout loaded by the `include` statement. Note that a negative mask is applied; meaning that in the drawn area, the photoresist is removed and the implant can take place. We remind the user to be careful in comparing the 2D and 3D plots because of the direction of the axes.

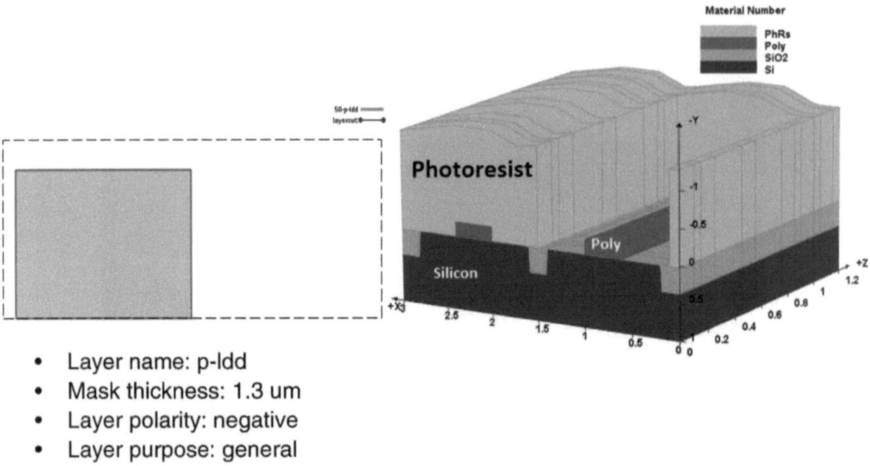

Fig. 6.30 The mask layout for p-LDD layer

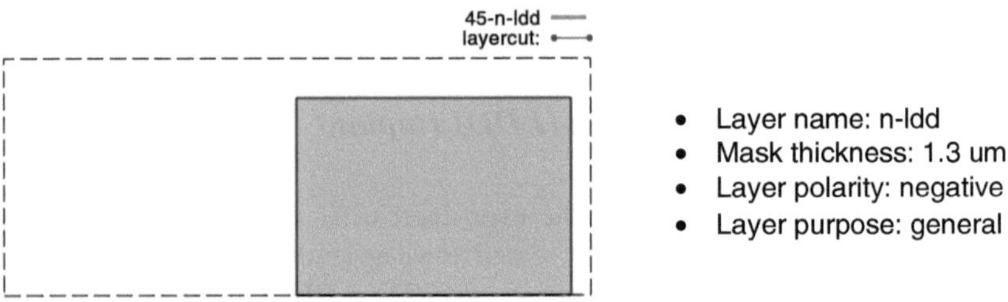

Fig. 6.31 The mask layout for n LDD layer

The `implant` for LDD should be relatively shallow so we use BF_2 with a dose of 8E + 12 cm^{-2} and ion energy of 7 keV for the p-LDD. As with previous well implants, we etch away the photoresist and do a quick diffusion (20 s) before moving on to the n-LDD implant Fig. 6.31 shows the mask layout for n-LDD.

Because a shallow p+ junction is not easily obtained using boron implantation, the molecular species BF_2 is used instead. The dissociation of BF_2 upon its first atomic scattering event gives a lower-energy boron atom [68]. Note that the energy used in this demo is fairly low which also helps to create a shallow junction.

Process Simulation Code

```
# n-LDD implant
include file=cmos.gds45.msk
structure outf=13a_nldd_mask.str
implant arsenic dose=9e12 energy=15
etch photoresist all
diffuse time=20/60 temp=1000 nitrogen
regrid log10.change=7.0 refine
structure outf=13_nldd.str
```

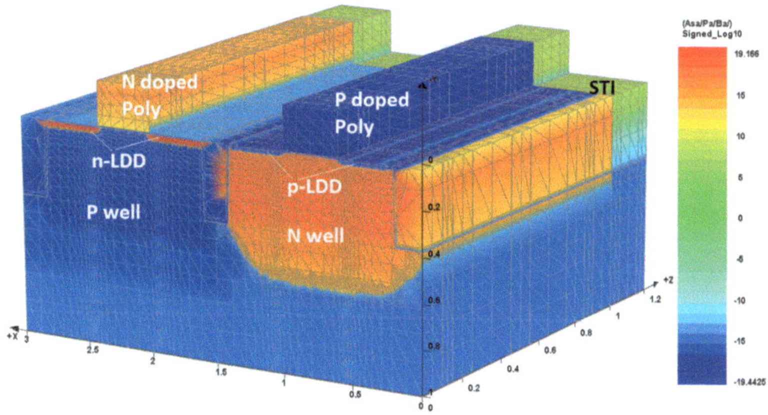

Fig. 6.32 Net doping plot after p-LDD and n-LDD implant anneal (13_nldd.str)

The commands for this step are roughly the same as for the p-LDD implant. Arsenic is used to create a shallow LDD region. Note that since the poly is used as implant mask, this step will dope the poly gate as well. Once again, the photoresist should be removed prior to annealing.

The `diffuse` command for this step is very quick: a designer should always keep in mind the thermal budget for the entire process since annealing steps always operate on all the dopants that were introduced in previous process steps. Instead of a constant temperature anneal, RTA can also be used for a more realistic process step.

The `regrid` command has the same function as previous used and increases the mesh density near the p-n junctions by imposing a limit on the maximum change in doping concentration between neighboring mesh points. The structure after n LDD implant is shown in Fig. 6.32.

6.2.12 *Nitride Spacer*

Now we will deposit a thick layer of dielectric, typically SiO_2 or Si_3N_4 on top of the device. This layer is usually named as "spacer" because after the subsequent anisotropic etch, it will leave a sidewall along the poly edge like a spacer. The thickness of the deposited layer will determine the width of the sidewall spacer, and should be carefully designed. For some technologies, this layer is also used as a source of stress.

In this example, nitride spacer is deposited, and then anisotropically etched leaving spacers around the gate poly. The thickness of the nitride layer is chosen to be thin because we don't want the spacer to occupy too much space in the device active area.

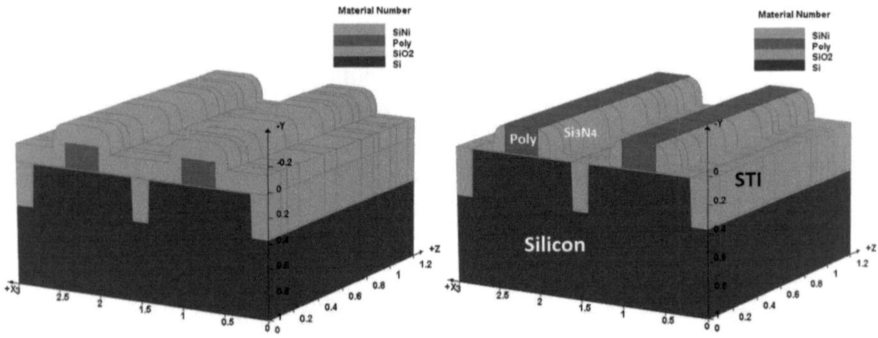

Fig. 6.33 Nitride spacer deposit (*left*: 14a_spacer_depo.str) and after spacer etch (*right*: 14_spacer.str)

Process Simulation Code

```
# nitride spacer
deposit nitride thick=0.15 meshlayer=2 space=0.05
structure outf=14a_spacer_depo.str
etch nitride dry thick=0.155
structure outf=14_spacer.str
```

As in previous examples, the spacer is created by using a pair of `deposit` and `etch` statements. The `space` and `meshlayer` parameters are used to control the smoothness of the surface and mesh allocation for this layer. Figure 6.33 shows simulation results after nitride spacer deposition and after spacer etch.

6.2.13 Drain and Source Implant

Source/Drain implant are high dose implant, they are usually implanted through a thin layer of oxide called screen oxide. The purpose of screening oxide is to prevent channeling and minimize the incorporation of trace impurities [53]. Implant channeling is caused by the implanted ions have a velocity vector line up with the substrate crystal structure. Ions can go deeper into the silicon substrate without encountering silicon atom, which is undesirable when shallow implantation is desired. A thin amorphous oxide layer can help randomize the directions of implanted ions and minimize this channeling effect [53].

Note that for source and drain implants, heavier ions (e.g. arsenic instead of phosphorus) are preferred to create shallower junctions. While boron is much lighter than arsenic, p+ implants often use much lower ion energies than n+ implants so the same effect can be achieved.

In practice, after source and drain implant, there will be a silicidation process, typically Titanium or Cobalt or some other metal is deposited by sputtering to make a better contact with low contact resistance. For example, if Titanium is deposited,

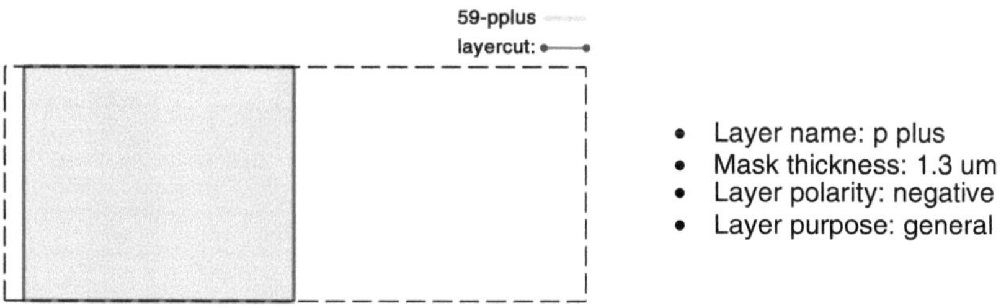

Fig. 6.34 Mask layout of PMOS source/drain implant

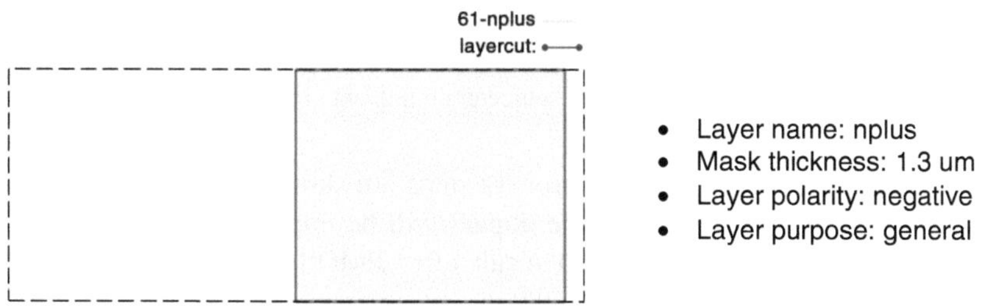

Fig. 6.35 The mask layout of n MOS source/drain implant

and the device subsequently goes through a short thermal cycle, the deposited Ti will react with Si to form $TiSi_2$, which is a fairly good contact material. This silicidation step is neglected in this simulation.

Process Simulation Code

```
# source and drain implant
include file=cmos.gds59.msk
structure outf=15a_psd_mask.str
implant bf2 dose=1e15 energy=10
etch photoresist all
diffuse time=5/60 temp=950 nitrogen
structure outf=15_psd.str

# source and drain implant
include file=cmos.gds61.msk
structure outf=16a_nsd_mask.str
implant arsenic dose=1e15 energy=25
etch photoresist all
diffuse time=5/60 temp=950 nitrogen
etch oxide dry thick=0.008
structure outf=16_nsd.str
```

The process commands used here are very similar to those of the LDD and well implants and will be skipped. We note once again the very quick diffusion time used here to limit the diffusion of the dopants from previous implants. After implantation, the screening oxide layer is removed with an `etch` command. The photoresist is also removed before every `diffuse` step.

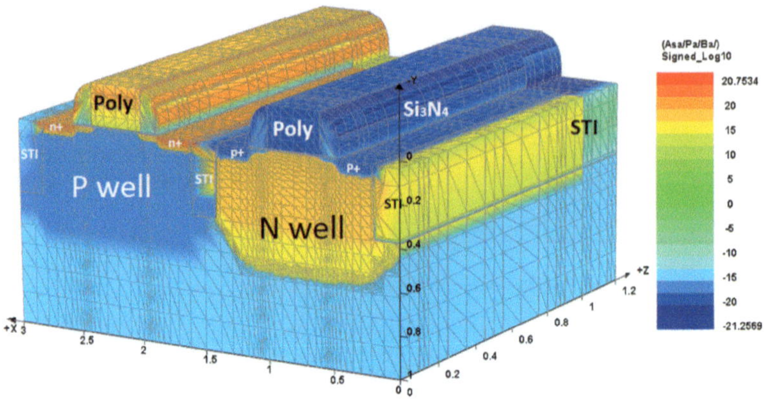

Fig. 6.36 Net doping concentration after source/drain implant (16_nsd.str)

The mask layers used in these process steps are shown in Figs. 6.34 and 6.35. The layer polarity is negative so the dopant will be implanted through the drawn area. Figure 6.36 is the simulation result after PMOS and NMOS source drain implant.

6.2.14 Contacts Placement

By now, we have finished the Front End of the Line (FEOL) process steps. The next step is to be devoted to Back End of the Line (BEOL). First, Interlayer Dielectric (ILD) needs to be deposited and contacts will be placed.

Normally, there are many materials that can be used as interlayer dielectric: a thick SiO_2 layer is an obvious choice. In practice, this oxide is often doped with phosphorus and boron and is known as BPSG (boronphosphosilicate glass). Phosphorous provides some protection against mobile ion impurities like Na^+, which can cause instability problems in MOSFET [53]. Boron is used to lower the temperature at which the dielectric glass flows: this fluidity helps smoothen out the surface for further processes.

CMP is often used in conjunction with the reflowing of deposited BPSG. CMP, or Chemical Mechanical Polish, is a process of smoothing surfaces with the combination of chemical and mechanical forces. It can be thought of as a hybrid of chemical etching and free abrasive polishing [69].

Contact holes are then etched through the ILD layer. In order to fill the contact holes with metal contacts, a thin layer of TiN is first deposited by sputtering or CVD. This layer provides good adhesion to SiO_2. Then, a blanket Tungsten (W) layer is deposited on top.

The final step is to use CMP again to prepare for metal deposition.

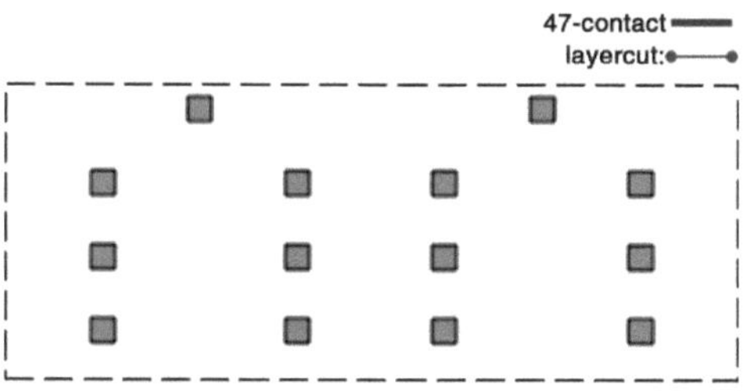

- Layer name: contact
- Mask thickness: 1.3 um
- Layer polarity: negative
- Layer purpose: etch
- Etch material: BPSG
- Etch depth: 0.8 um
- Etch angle: 0

Fig. 6.37 The mask layout of the contact layer

Process Simulation Code

```
# contact holes
deposit BPSG thick=0.5 meshlayer=3
structure outf=17a_BPSG.str
option etch.typ=2
etch start x=0 y=-2
etch continue x=0 y=-0.4
etch continue x=3 y=-0.4
etch done x=3 y=-2
structure outf=17b_BPSG.str
include file=cmos.gds47.msk
structure outf=17_contacts.str

# contacts fill
deposit TiN thick=0.01 meshlayer=2
structure outf=18a_TiN_fill.str
deposit W thick=1.2 meshlayer=2
structure outf=18a_W_fill.str
etch start x=0 y=-2
etch continue x=0 y=-0.4
etch continue x=3 y=-0.4
etch done x=3 y=-2
structure outf=18_contacts_fill.str
```

`deposit BPSG` deposits the Interlayer Dielectric (ILD): a single layer of BPSG is used to simplify the simulation process. Real-world ILDs are usually very complicated and contains several layers which meet competing design goals. A low-K ILD reduces crosstalk, while a high-K gate dielectric allows the designer to increase the gate oxide thickness and thus reduce the leakage current.

The `etch start, continue, done` commands define a geometric etch region that mimics the CMP process used in real manufacturing. CMP is often necessary to flatten the surface for subsequent process steps.

The `include` statement loads a mask file from the MaskEditor GUI which will be used to etch away the contact holes. Figure 6.37 shows the mask layout of the contact layer. The simulation result is shown in Fig. 6.38.

After contact holes are opened, a thin layer of TiN is deposited followed by a tungsten (W) layer which fills the gap; the first layer provides better adherence to the SiO_2 layer. After another CMP etch, the final structure is shown in Fig. 6.39.

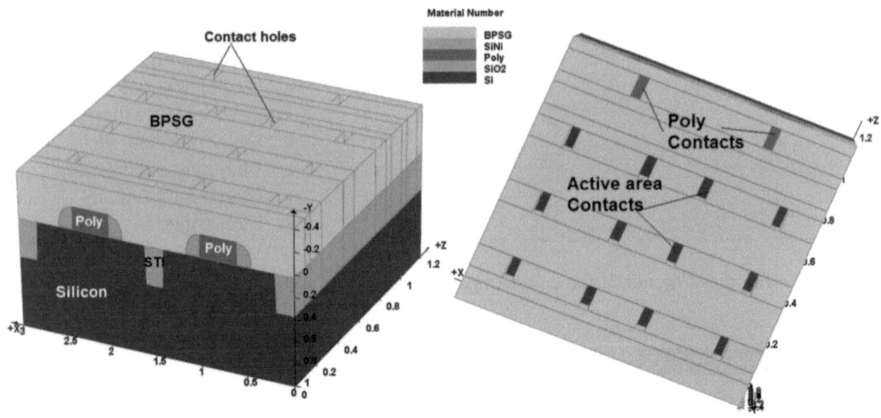

Fig. 6.38 Contact hole etch (17_contacts.str)

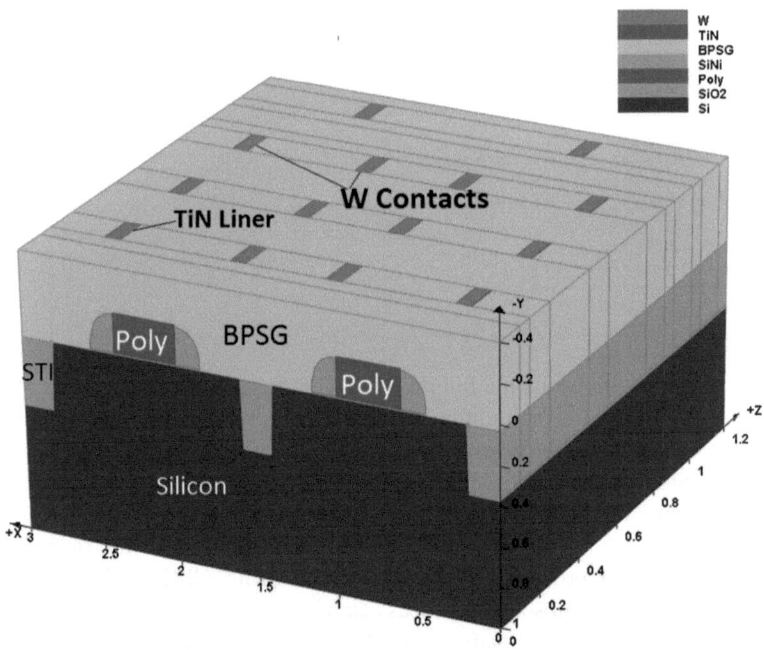

Fig. 6.39 Contact holes fill (18_contacts_fill.str)

6.2.15 *Metal Layer Placement*

Finally metals are deposited and patterned on top of the contacts. In real manufacturing, multiple metal layers are involved and copper (Cu) is commonly

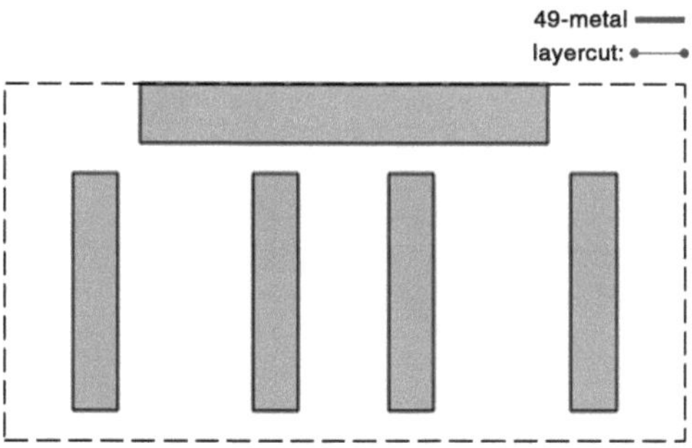

- Layer name: metal
- Mask thickness: 1.3 um
- Layer polarity: positive
- Layer purpose: etch
- Etch material: aluminum
- Etch depth: 0.2 um
- Etch angle: 0

Fig. 6.40 The mask layout of metal layer

used instead of aluminum like in this simulation. This provides lower sheet resistance and is less susceptible to electron migration.

Process Simulation Code

```
# metal layers
deposit aluminum thick=0.2 meshlayer=2
include file=cmos.gds49.msk
structure outf=19_metal.str
etch BPSG all
structure outf=20 final.str
```

The aluminum layer is deposited on top of the ILD and contact layer in the usual manner. It is then etched away using another etch-purpose mask designed in the MaskEditor GUI. The mask layout can be seen in Fig. 6.40; the positive mask polarity means that the drawn area will be protected while the blank regions will be etched away. Figure 6.41 is the structure after metal patterning.

The final command is `etch BPSG all` which removes the BPSG. This is merely a simulation trick to help visualize the internal structure of the contacts and is not a real process step. The net doping chart is shown in Fig. 6.42 with the BPSG stripped. In this figure, the NMOS can be seen in the range $x = (1.5\ 3.0)$ while the PMOS is in the range $x = (0\ 1.5)$. As with our original inverter circuit, the gates are connected. Note due to the symmetrical nature of both NMOS and PMOS devices, the source and drain terminals are interchangeable.

6.2.16 Simulation Data

Table 6.4 gives simulation data for the CMOS technology. The `regrid` command used to increase the mesh at p-n junctions significantly increases total mesh count.

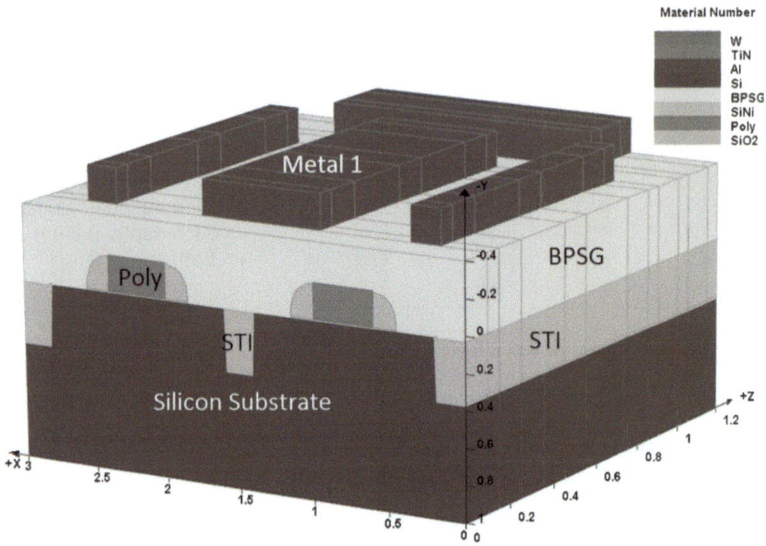

Fig. 6.41 CMOS final structure after metal layer (18_metal.str)

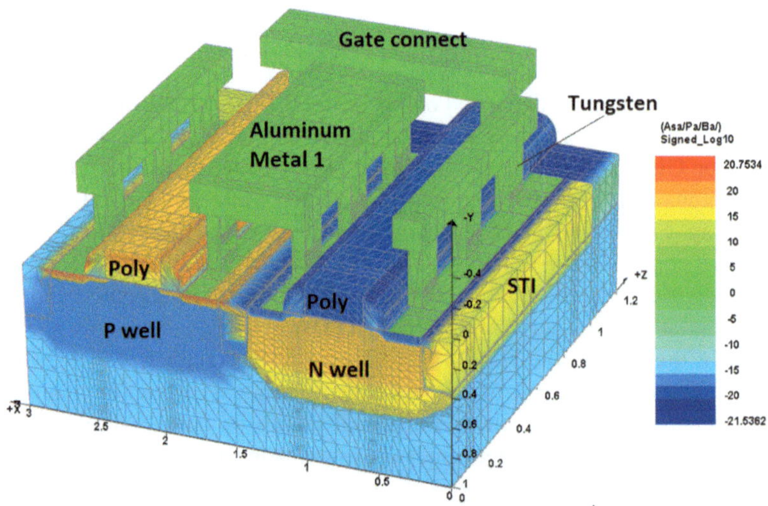

Fig. 6.42 CMOS final structure with TEOS removed and showing net doping

Table 6.4 Simulation data for CMOS technology

	Process simulation	Device simulation	Total mesh count	Total number of planes
CMOS Technology	60 min	N/A	68,626 (regrid) 36,010 (no regrid)	24

Computer configuration: HP desktop with Intel i7-860/6G/1T/NVIDIA GeForce GTX260/Win7

6.3 Nano MOSFET Device Simulation Add-on

This is a special section reserved for advanced users. Since we already have a complete CMOS process flow, the nano MOSFET process will be neglected here. But please keep in mind that for nano MOSFET devices, a halo implant should be added.

6.3.1 Including Quantization Effect in MOSFET

When under inversion condition, the oxide and silicon channel form a quantum well (QW) with the oxide having a high potential barrier. Advanced MOSFET models usually include quantization effects in the QW. Another effect equally important is the split of the quantum levels originated from the various band valleys in the QW. A common approach in mobility calculation is to consider scattering of carriers from these various band valleys. Therefore, the inclusion of quantization effect is a basic requirement in advanced TCAD for CMOS.

There are various approaches to include quantum mechanical (QM) effects into the device simulation. This section describes an approach specific to the device simulator from Crosslight which directly uses the solution from Schrödinger wave equations to construct the 2D density of states (DOS). The Schrödinger wave equation is solved self consistently with all other equations in the drift-diffusion model.

The following commands are used to set up quantum mechanical simulation:

Device Simulation Code

```
$ ------ special treatment to set up quantum for nano-MOSFET---------
$ to run quantum, part of the gate oxide must be renamed to be a cx-Macro
$ for quantum model to work
$
$ to make a material quantum, define it as "active"
get_active_layer name=cx-Si mater=1
active_reg mater=1 thickness=0.01
renumber_mater xrange=(-0.02 0.02) yrange=(-0.001 0.003) &&
orig_mater=2 new_mater=6
load_macro name=s-sio2 mater=6
get_active_layer name=cx-SiO2 mater=6

$ we also define a complex-region for quantum
begin_complex layer_num=2 column_num=1 use_xy_range=yes &&
cx_qw_side=bottom qw_thick=1.000000e-002 &&
x_start_complex=-0.02 &&
y_start_complex=-0.1
complex_region thickness=0.1 x_size=0.04 mater=1
complex_region thickness=0.002 x_size=0.04 mater=6
end_complex
self_consistent
bulk_treatment type=p
modify_qw right_mesh=0.002 confine_left=yes
```

In the device simulator, the quantum model is activated by calling an active macro. The term "active" originates from the original usage of this model in the optically active regions of semiconductor lasers. Here, it is used to define regions

Text Box 6.7 Simulation Reports for the Macro

```
[100] Si-interface with
s-sio2 cx-SiO2
mass_para= 0.190000000000000
mass_perp= 0.916000000000000
Cond. Band Valley Split = 2 4
```

which will be connected together and where the Schrödinger equation will be solved. The active macro contains additional material data which is not included in the standard bulk material macro data. For example, it contains details of the six *X* band valleys of silicon as well as information on the LH and HH masses (Luttinger parameters). In a bulk macro, these details are absent and averaged effective masses are used for the conduction and valence bands.

This case requires the use of a complex MQW model so the relevant macros use a "cx-" prefix: the silicon layer uses cx-Si and the oxide layer uses cx-SiO_2. The command `begin_complex` and `end_complex` defines the layers which belong to the QW. An infinite potential wall some distance from the outer barrier is used as the boundary condition of the Schrödinger equation. We also note that it is important that the oxide be treated as a wide bandgap semiconductor rather than a pure insulator so small amounts of current are allowed to flow in the drift-diffusion model.

In this example, we use the fully self-consistent model of carrier density explained in earlier sections so the shape of the well is affected by the external potential. To simplify the convergence and because this is a channel designed for a 2D electron gas (2DEG), the quantization of the holes is deliberately ignored (`bulk_treatment type=p`).

When running the simulator, a 2/4 band valley split is reported as shown in Text Box 6.7. The masses parallel to the interface are much lighter than those perpendicular to the interface. So it is clear that all six band valleys have anisotropic masses. Two of the valleys ($[100]$ and $[\bar{1}00]$) are oriented so their light mass is parallel to the interface and the other four have their masses switched since they lie in a perpendicular direction.

The quantization is determined by the motion of carriers perpendicular to the interface so the perpendicular mass will determine the position of the confined energy levels. However, the dispersion relation depends on the mass parallel to the interface. So the overall 2D DOS is a complicated function of the quasi-Fermi level, quantum level spacing and the effective mass parallel to the interface.

The simulator also reports the following solution at equilibrium (see Text Box 6.8):

The band valley c1 has heavier mass and has more levels. As the gate voltage increases, the QW is more confined and more levels are found by the solver.

Text Box 6.8 Solution at Equilibrium Reported by the Simulator

```
---Updating MQW potential and states ...
MQW Reg.# 1 Average Conc. (n&p)= 0.2016E+22 0.7875E+23
Cmplx.# 1 Levels: c1= 7 c2= 3 HH= 0 LH= 0
and at Vg=0.55:
---Updating MQW potential and states ...
MQW Reg.# 1 Average Conc. (n&p)= 0.3664E+24 0.3050E+22
Cmplx.# 1 Levels: c1= 12 c2= 5 HH= 0 LH= 0
```

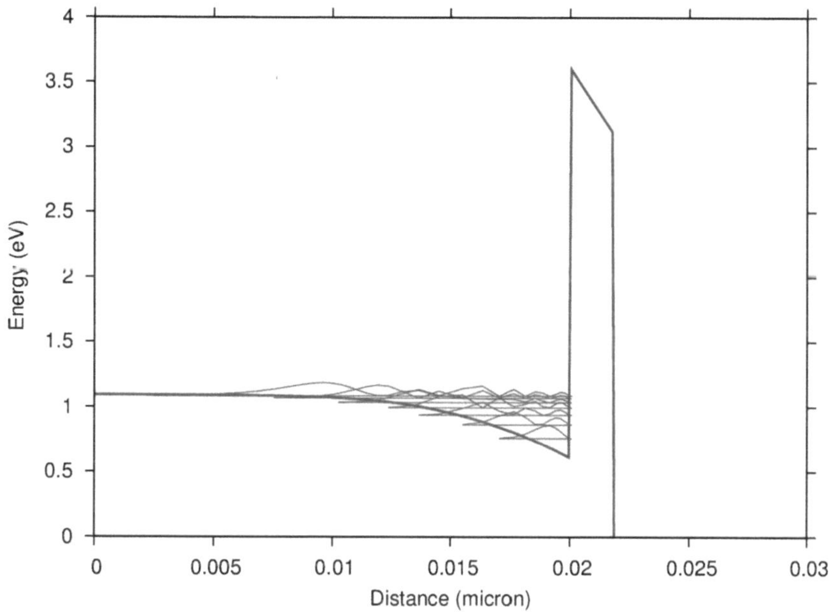

Fig. 6.43 Subband levels for c1 valley

The subband levels for c1 and c2 valleys are shown in Figs. 6.43 and 6.44, respectively. The I_D–V_G of this Nano-MOSFET of 40 nm is shown in Fig. 6.45.

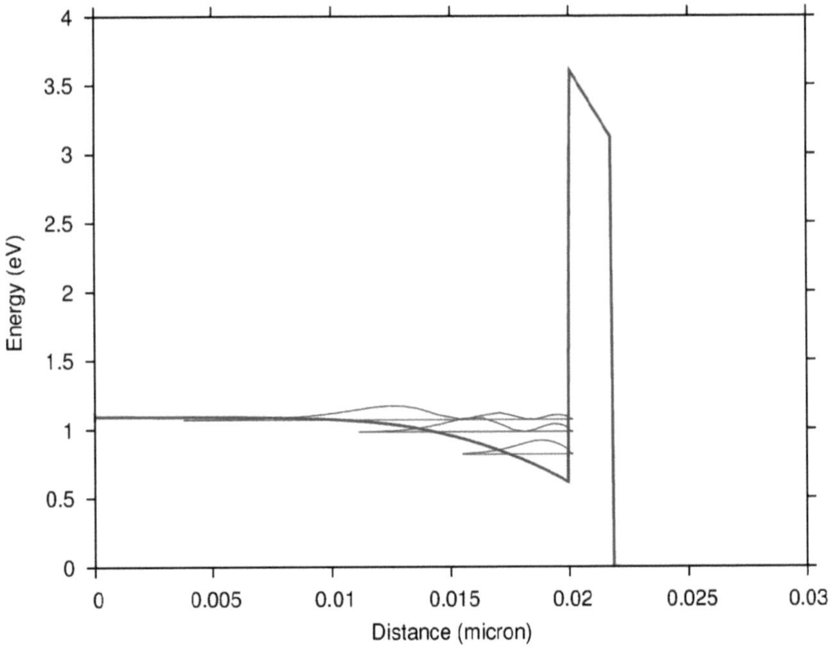

Fig. 6.44 Subband levels for c2 valley

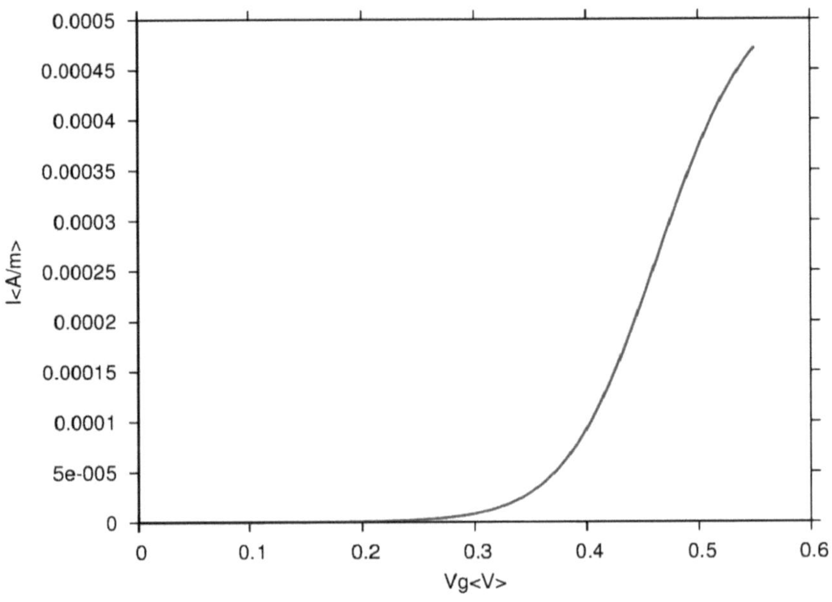

Fig. 6.45 The ID-VG curve for the nano MOSFET

Chapter 7
Smart Power Technology and Power Semiconductor Devices

7.1 Smart Power IC Technology

Smart power IC has gained popularity in the analog world. A typical power IC includes power devices, analog part, digital part and some passive devices. Figure 7.1 illustrates the interrelationship between these components. Traditionally, power ICs and digital ICs have been considered to be two very different technologies and were manufactured using different processes.

Technology has now enabled the integration of power and digital ICs and has given birth to a new category of ICs: "Smart Power" ICs which combine the "muscles" and "brain" functions into a single chip. Integrating Power, Analog, Digital and Passive components on the same chip is an art rather than technology! Without proper isolation, integration causes severe component cross-talk and compromises chip performance.

Smart power technologies are usually several generations behind CMOS technology in terms of feature and wafer sizes. They do not use the smallest feature sizes achievable or largest wafer sizes like advanced CMOS technology does. This is because power devices generally have a much larger pitch size, channel length and channel width than logic devices.

Even though logic gate density is an important consideration, it just is not cost effective to use the latest technology for smart power applications. A comparison between CMOS technology and smart power technology as of the year 2009 is listed in Table 7.1.

7.1.1 Things to Consider in Typical Smart Power Technology

Smart power technology requires optimization of multiple device technology parameters. Table 7.2 lists 18 things to consider according to different categories (e.g. power, analog, etc.). Of course, in real manufacturing, there are many other considerations besides the ones listed below.

S. Li and Y. Fu, *3D TCAD Simulation for Semiconductor Processes, Devices and Optoelectronics*, DOI 10.1007/978-1-4614-0481-1_7,

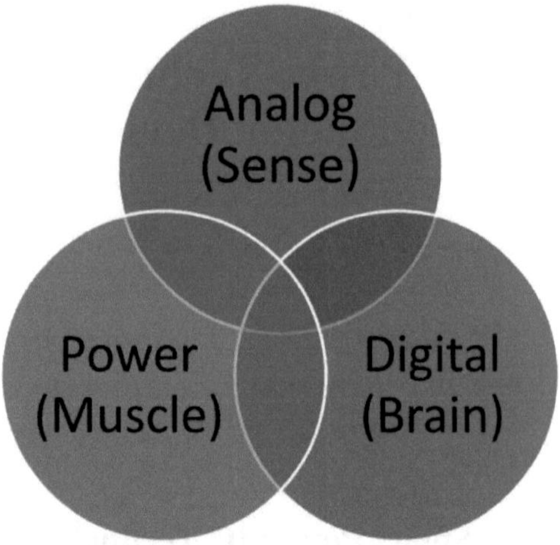

Fig. 7.1 Smart power IC technology

Table 7.1 Comparison between CMOS technology and smart power technology

	CMOS technology	Smart power technology
Feature size	90 nm and below	0.13 um and above
Wafer size	Mostly 12 in., some 8 in.	8 in. and below
Device types	NMOS and PMOS	MOSFETs, diodes, capacitors, resistors, etc.
Isolation	Mostly STI	STI, LOCOS, DTI, P-N junction

Table 7.2 Things to consider in a typical smart power technology

Categories	Things to consider
Power	1. Breakdown voltage (high side/low side)
	2. On-state resistance (R_{on})
	3. Energy capability
	4. Safe Operating Area (SOA)
Analog	5. MOS: G_m, R_{out}, matching, off state current, noise
	6. BJT: gain, Early voltage, matching, noise
	7. Resistors: voltage coefficient, temperature coefficient, matching, noise, PSRR
	8. Capacitors: linearity, matching, noise
	9. Diodes: breakdown voltage, temperature coefficient of BV, isolation
	10. Voltage and frequency trim –fuse/Anti-fuse/NVM
Logic	11. Gate density
	12. Cost due to added complexity of state of the art CMOS process
	13. Noise isolation
System level	14. Substrate injection isolation
	15. Embedded parasitic BJTs
	16. Noise immunity
	17. Heat dissipation
	18. High voltage ESD

Table 7.3 Typical devices in a smart power technology

Categories	Devices
Digital	NMOS and PMOS
Analog	NMOS and PMOS, BJTs, diodes, anti-fuse, NVM, etc.
Power	n-LDMOS, p-LDMOS, NPN, PNP, LIGBT, power diodes, etc.
Capacitors	MOS capacitors, double-poly capacitors, metal-insulator-metal capacitors, etc.
Resistors	Diffused resistors, poly resistors, etc.
ESD Devices	ggnmos, SCR, etc.

7.1.2 Devices in a Typical Smart Power Technology

Table 7.3 gives list of devices used in smart power technology. The requirements on these devices and the choice of technology depend on the intended application. For example, cell phones require a breakdown voltage for LDMOS of less than 20 V but for automotive application, values as high as 80 V may be required. Resistors are chosen based on their temperature and voltage dependence. For example, diffused resistors suffer from high voltage coefficients while poly resistors do not change much with the applied voltage. Capacitors are chosen according to the rated breakdown voltage and capacitance value. MOS capacitors can achieve higher capacitance but suffer from low breakdown voltage while double poly (poly1-insulator-poly2) capacitors have high breakdown voltage but much lower capacitance. Anti-fuse or Non-volatile Memory (NVM) is used for trimming purposes in analog ICs.

7.2 Isolation Methods

In a smart power IC, the isolation method plays an important role in limiting how the integrated devices influence each other. There are four isolation methods in common use, namely: P-N Junction isolation, LOCOS (Local Oxidation of Silicon), STI (Shallow Trench Isolation) and DTI (Deep Trench Isolation). Table 7.4 is a comparison of these isolation methods.

7.2.1 P-N Junction Isolation

An implant chain is used for the simulation of p-n junction isolation (code is not shown here). High energy implant may damage the silicon lattice and may cause dislocations and reliability problems. P-N Junction isolation is not used very often in smart power IC with technology node less than 0.25 um since it takes up too much space. However, this technology is still commonly used for High Voltage IC (HVIC). Figure 7.2 is an illustration of the p-n junction isolation.

Table 7.4 Comparison of different isolation methods

Method	Leakage	Process cost	Silicon area cost	Feature size example	Comments
Junction Isolation	High	Low	Large	0.5 um or above	• Simple but consumes a lot of space • Simple process but lots of cross-talk
LOCOS	High	Low	Large	0.35 um or above	• large size, uneven surface, suitable for older technologies
STI	Medium	Medium	Small	CMOS 90 nm or below	• Standard for CMOS and analog components • Edge effect, HCI reliability issue for LDMOS • Inferior to DTI for power devices
STI + DTI	Low	High	Small	BCD 0.18 um	• Suitable for all devices but especially important for power devices • Expensive and increases the difficulty for process integration • Potential reliability issues including dislocation and clustering

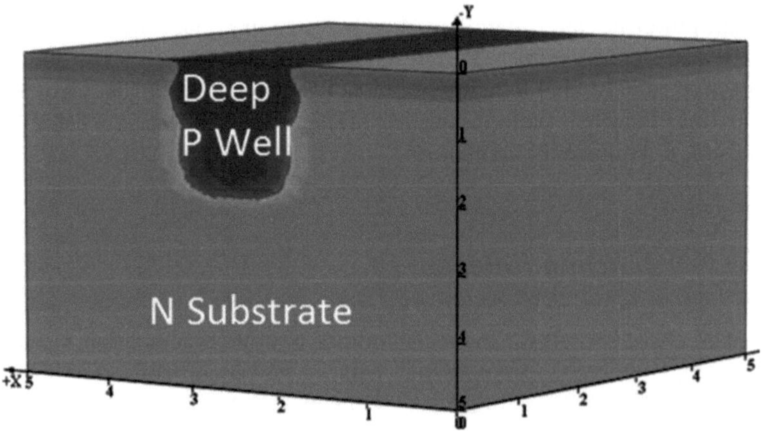

Fig. 7.2 P-N junction isolation

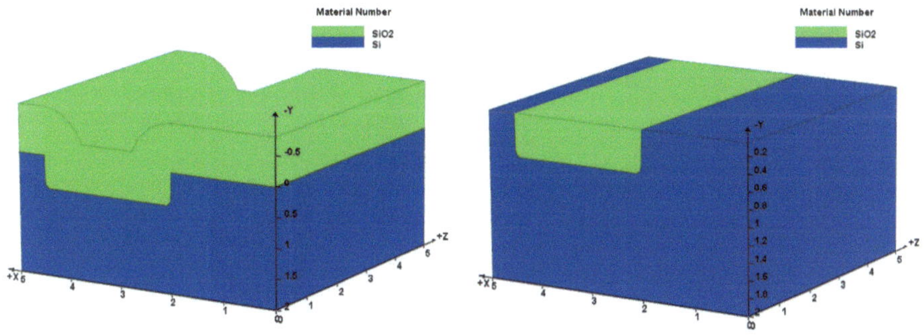

Fig. 7.3 Shallow trench isolation (STI) before CMP (*left*) and after CMP (*right*)

7.2.2 Shallow Trench Isolation (STI)

Shallow Trench Isolation (STI) is used extensively and is the main isolation method in CMOS technologies. The process was described in detail in previous chapters dealing with MOSFET and CMOS but can be resumed simply as:

- Selective etching of silicon to a depth of 0.3–0.5 um, depending on the technology
- Growth of liner oxide to improve Si/SiO_2 interface
- Deposition of filler oxide to refill the etched wells
- Chemical Mechanical Polish (CMP) to etch away excess oxide and leave a flat surface, as shown in Fig. 7.3

In terms of process modeling, the etch pattern can be obtained from the mask layout or can be based on SEM images of real devices. A geometric etch method can be used to selectively remove mesh triangles in the affected area instead of modeling the chemical/mechanical processes.

7.2.3 Deep Trench Isolation (DTI)

In terms of the real process step, STI and DTI are quite different and the exact method used varies from company to company. However, when it comes to process modeling, deep trench isolation is quite similar to STI, only with deeper trench etch and fill. Interested readers can modify the STI examples to model DTI. A small variation in the process is that the filler oxide can be replaced by polysilicon such as in [70]

7.3 Racetrack LDMOS

LDMOS with a racetrack shaped gate is popular in the power IC industry. Example can be found in many research papers [71]. In this section a demo LDMOS is built with an arbitrary process technology flow to show important process steps such as

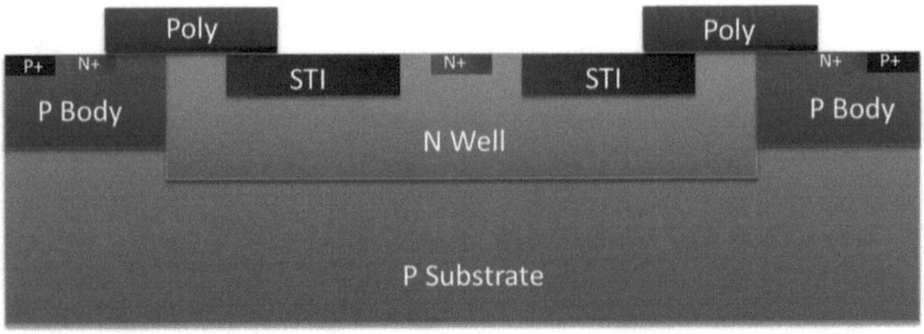

Fig. 7.4 Cross-section view of a typical LDMOS

Fig. 7.5 Mask layout of racetrack gate LDMOS

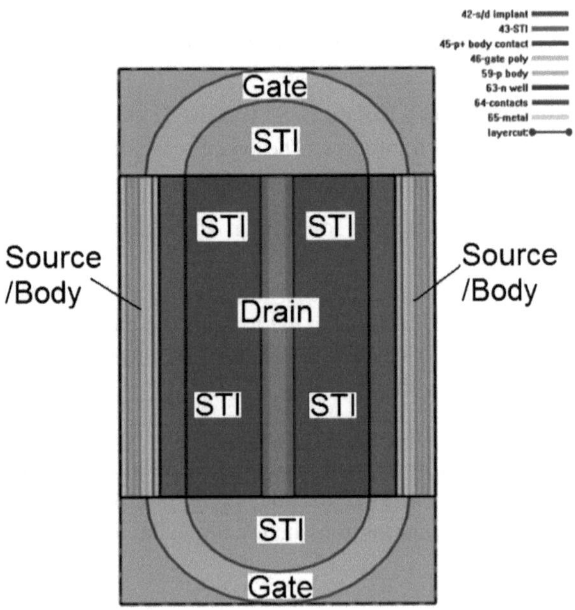

etching, implantation, diffusion, deposition, etc.... Usually, a drain with the racetrack structure of an LDMOS is surrounded by the source connected to the bulk for power ICs; this has many advantages, such as electrically isolating each device and reducing edge effects on the LDMOS [71]. Here a simplified racetrack LDMOS with STI in the drift region is simulated. All the process parameters are arbitrarily defined. Figure 7.4 is the cross-section view of a typical LDMOS.

Figure 7.5 is the MaskEditor layout mask for racetrack-gated LDMOS. 8 layers are used in this example. We remind the reader this is a simplified tutorial: the process steps are not calibrated and the design is not optimized with respect to breakdown voltage, on-state resistance (R_{on}) or any other parameters. Readers should use this example as a starting point for their own devices and calibrate process steps using their own fab results.

7.3.1 Overview of Simulation Steps

We will use the process simulator to create the structure of the racetrack LDMOS. This is followed by contact definitions and device simulation. An overview of simulation steps is presented in Table 7.5.

7.3.2 Substrate

Boron doped p-type substrate is chosen with a constant doping concentration of 1E + 14 cm^{-3}. The silicon sample orientation is [100] which is optimal for MOSFET fabrication. The process simulation input code is listed below:

Process Simulation Code

```
mode quasi3d
3d_mesh inf=geo
init boron conc=1e14 orient=100
```

`mode quasi3d` simplifies the process simulation by turning off the dopant diffusion in the z direction. Only the process steps that are relevant to the LDMOS fabrication are used in this simulation. We neglect some of the real process steps such as the buried layer and RESURF (Reduced Surface Field) technology.

`3d_mesh` loads mesh declaration statements generated by the MaskEditor GUI. This device has a curved gate located within the STI and a rectangular-shaped active region where the current flows. Since the current flow is the important part of the device simulation, a dense lateral mesh is used in the active region and a sparse lateral mesh is used for the STI region. In the z direction though, most of the cut planes [40] are located in the STI to approximate the curved region: the active region only requires a minimum of 2 mesh planes to properly define the volume. Figure 7.6 shows the STI and active regions locations on the mask layout. Figure 7.7 compares the lateral mesh definitions in the STI and active regions.

The `init` command processes all the mesh statements and creates the initial structure for the process simulation. The silicon substrate is p-type with a boron concentration of 1E + 14 cm^{-3} and the [100] orientation.

7.3.3 N Well Implant

The goal of this process step is to implant the n well which will later be used as the drain drift region connecting the channel to the drain.

Table 7.5 Overview of simulation steps for racetrack LDMOS

Racetrack LDMOS	Process simulation steps
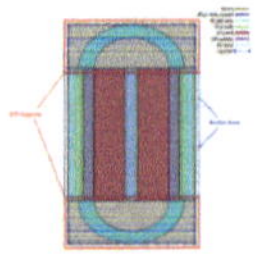	Step 1: Substrate
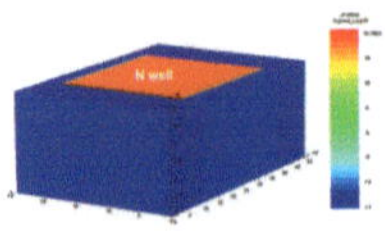	Step 2: N well implant
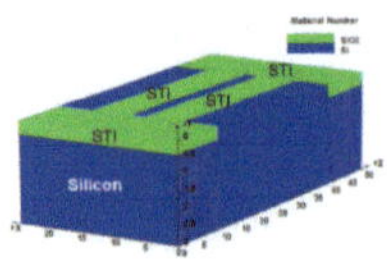	Step 3: STI formation
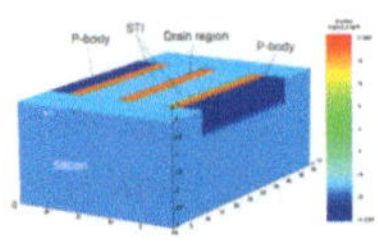	Step 4: P body implant
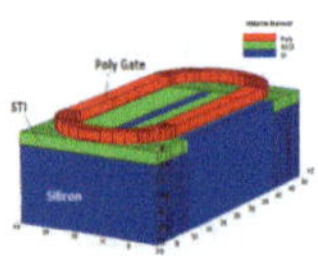	Step 5: Poly gate deposition
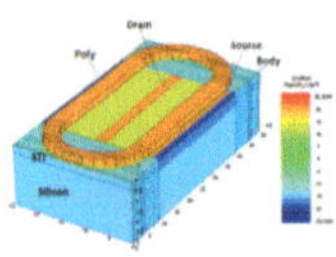	Step 6: Source/Drain n+ implant
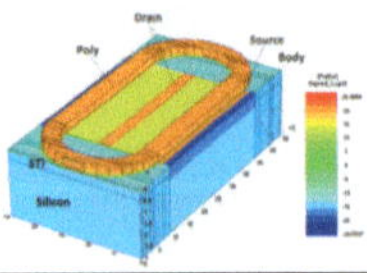	Step 7: P+ body contact implant

(continued)

Table 7.5 (continued)

Racetrack LDMOS	Process simulation steps
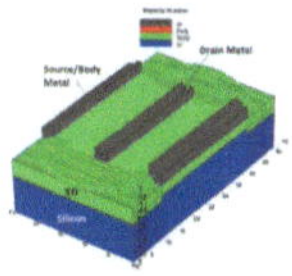	Step 8: Back-End of the Line (BEOL)
Racetrack LDMOS	Contact definitions for device simulation
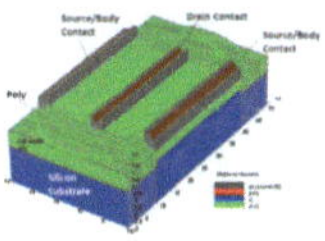	Step 9: Contact definitions for device simulation
Racetrack LDMOS	Device simulation
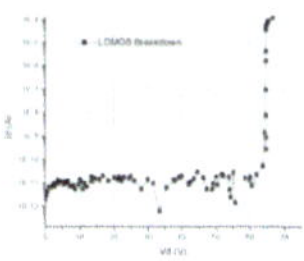	Step 10: Device simulation: breakdown voltage
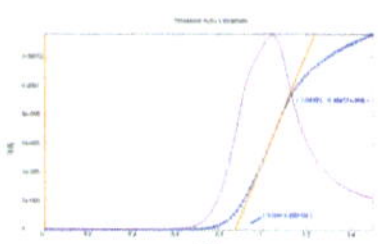	Step 11: Device simulation: threshold voltage
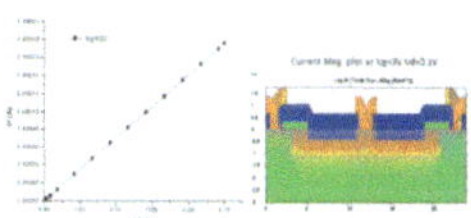	Step 12: Device simulation: on-state resistance
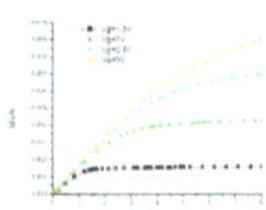	Step 13: Device simulation: I_D-V_D curves
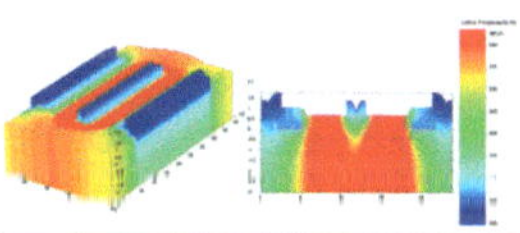	Step 14: Self-heating

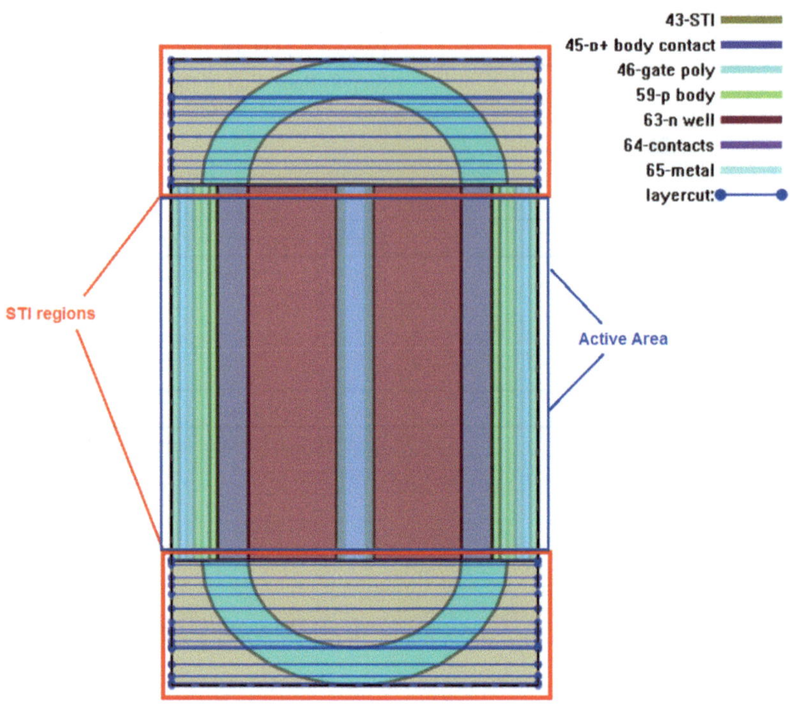

Fig. 7.6 STI and active regions in the racetrack LDMOS

```
line x loc=   0.00000 spacing=  0.5 tag=lft
line x loc=   2.00000 spacing=  0.5
line x loc=   2.37500 spacing=  1
line x loc=   2.75000 spacing=  0.5
line x loc=   7.37500 spacing=  1
line x loc=   12.0000 spacing=  0.5
line x loc=   16.6250 spacing=  1
line x loc=   21.2500 spacing=  0.5
line x loc=   21.6250 spacing=  1
line x loc=   22.0000 spacing=  0.5
line x loc=   24.0000 spacing=  0.5 tag=rht

line y loc=   0.00000 spacing=  0.5e-01 tag=top
line y loc=   0.10000 spacing=  0.5e-01
line y loc=   3.00000 spacing=  0.5 tag=bot

region silicon xlo=lft xhi=rht ylo=top yhi=bot
bound exposed xlo=lft xhi=rht ylo=top yhi=top
bound backside xlo=lft xhi=rht ylo=bot yhi=bot
```

STI Regions

```
line x loc=   0.00000    spacing=  0.277778      tag=lft
line x loc=   2.00000    spacing=  0.277778
line x loc=   2.37500    spacing=  0.694444
line x loc=   2.75000    spacing=  0.277778
line x loc=   7.37500    spacing=  0.694444
line x loc=   12.0000    spacing=  0.277778
line x loc=   16.6250    spacing=  0.694444
line x loc=   21.2500    spacing=  0.277778
line x loc=   21.6250    spacing=  0.694444
line x loc=   22.0000    spacing=  0.277778
line x loc=   24.0000    spacing=  0.277778      tag=rht

line y loc=   0.00000    spacing=  0.234375e-01 tag=top
line y loc=  0.100000    spacing=  0.234375e-01
line y loc=   3.00000    spacing=  0.234375      tag=bot

region silicon xlo=lft xhi=rht ylo=top yhi=bot
bound exposed xlo=lft xhi=rht ylo=top yhi=top
bound backside xlo=lft xhi=rht ylo=bot yhi=bot
```

Active Region

Fig. 7.7 Comparison of different cut plane mesh for STI regions and Active region

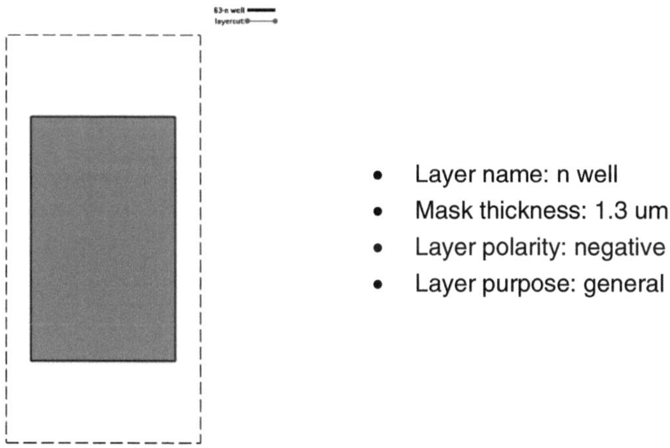

Fig. 7.8 N well implant mask layout for racetrack LDMOS

Process Simulation Code

```
# n well implant #
include file=racetrack.gds63.msk
struct outf=01_nwell_mask.str
implant phosphorus dose=1e12 energy=50 angle=0
implant phosphorus dose=2e12 energy=400 angle=0
etch photoresist all
diffuse time=5 temp=800  final_temp=1000
diffuse time=20 temp=1000
diffuse time=5 temp=1000  final_temp=800
struct outf=01_nwell.str
```

The `include` statement loads the mask generated by the MaskEditor GUI: Figure 7.8 shows the n well implant layout. The dashed line shows the outer simulation boundary (simulation area) and the drawn area is the negative mask where the photoresist will be removed and the implantation process will take place.

The phosphorous implantation is done by chaining together two implant commands: the lower energy implant and the higher energy implant. With the same thermal budget, an implant chain helps obtain a more uniform doping profile than what could be obtained with a single implantation. Figure 7.9 illustrates the implant chain.

The `etch photoresist all` command is used to remove all the leftover photoresist that was added when defining the implantation mask. This must be done before the diffusion can take place.

Three `diffuse` commands are used to define the thermal annealing process: the temperature is ramped up from 800°C to 1,000°C over 5 min, held constant for 20 min and brought back down to 800°C over 5 min. This chaining of `diffuse` commands is how a rapid thermal annealing process would be defined. Of course, the time scale used here is much slower than that of a RTA process. Figure 7.10 is the net doping chart after the n well implant anneal.

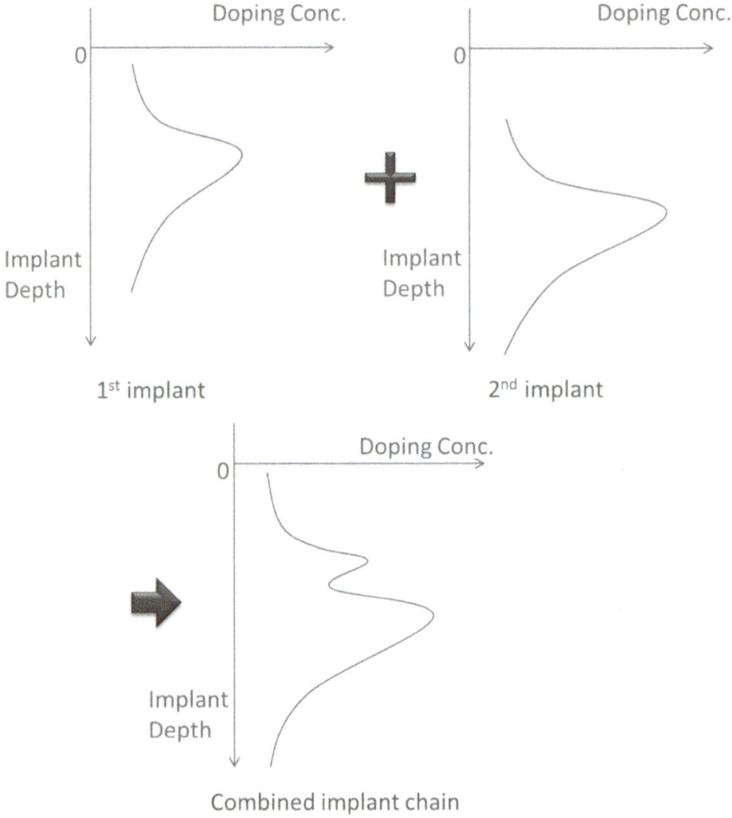

Fig. 7.9 Implant chain illustration

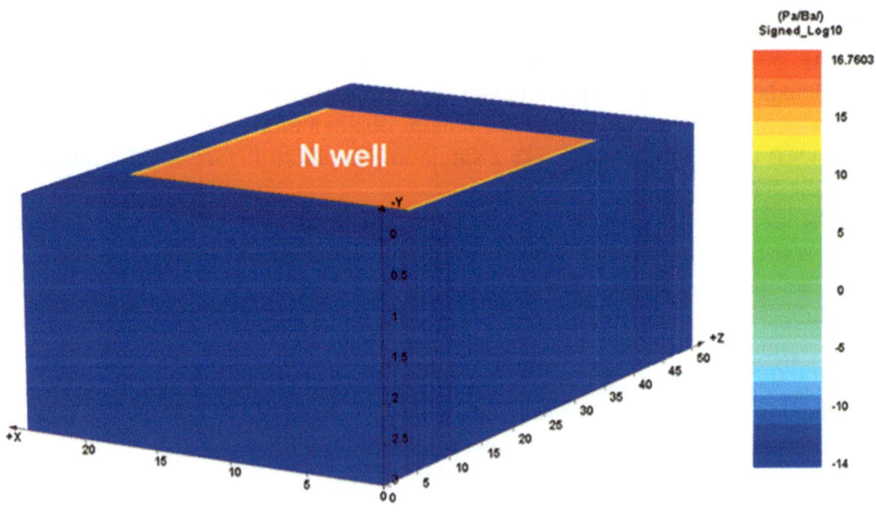

Fig. 7.10 N well implant of racetrack LDMOS showing net doping (01_nwell.str)

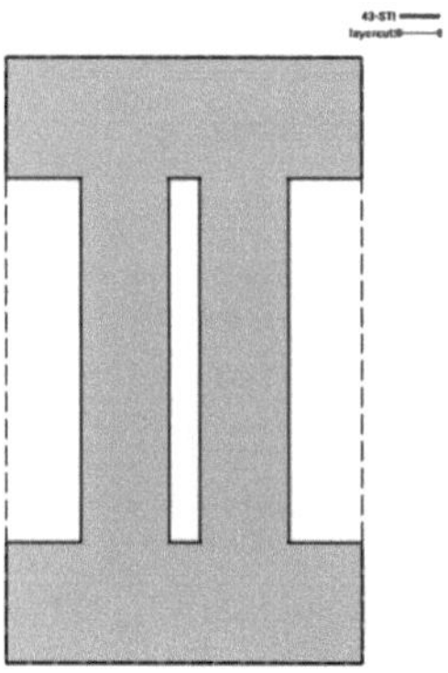

- Layer name: STI
- Mask thickness: 1.3 um
- Layer polarity: negative
- Layer purpose: change material
- Material to be changed: silicon
- Material change to: oxide
- Etch thickness: 0.4 um

Fig. 7.11 The mask layout for STI

7.3.4 STI Formation

The next step is to form the shallow trench isolation. As mentioned previously, the most important reason for using a STI process is to provide isolation between the devices. For this particular example, we only consider the standalone device so no device to device isolation is necessary. STI in this LDMOS is instead used to boost the breakdown voltage. As will be shown later, the peak electric field which triggers avalanche breakdown is designed to be at the edge of the gate and within the STI region. Since oxide has a higher critical electric field than silicon, the breakdown voltage of the LDMOS will be improved. If no STI or LOCOS had been used, the peak field would be within the silicon material which would degrade the breakdown voltage.

In the case of low voltage LDMOS (with a breakdown spec lower than 40 V), the STI is often neglected on purpose. This is because using STI within the LDMOS has some drawbacks such as increasing the on-state resistance (R_{on}). It also has long-term reliability issues as hot carrier injection into the STI region may degrade the device performance; this may be alleviated by using high quality STI liner oxide.

In this example, the STI mask is generated by the MaskEditor GUI. We use the simplified process step called "change material" to replace the silicon with oxide in the mask area rather than modeling all of the process steps. The angle of the etch process corresponds to an ideal vertical etch (angle = 0) which, of course, is not achievable in a real etch process. For more details on mask settings, refer to the earlier description of the MaskEditor GUI.

Process Simulation Code

```
# STI formation
include file=racetrack.gds43.msk
struct outf=02_STI.str
```

The `include` statement loads the STI mask layout from Fig. 7.11. The mask polarity is negative so the drawn area corresponds to the photoresist which would be removed in the full STI process. Here, it defines the material which is changed from

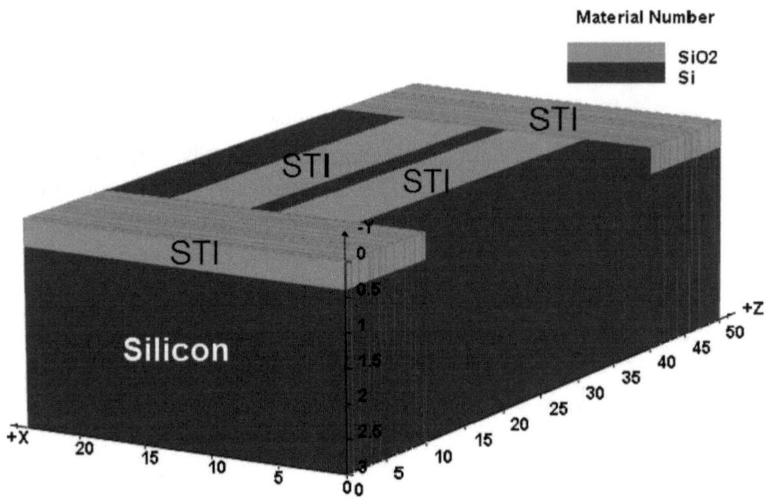

Fig. 7.12 STI formation (02_STI.str)

silicon to oxide because of the "change material" purpose of the mask layer. The resulting structure is shown in Fig. 7.12.

7.3.5 *P Body Implant*

P body implant is the next step. This step will prepare the p-type body region of the LDMOS.

Process Simulation Code

```
# p body implant #
include file=racetrack.gds59.msk
implant boron dose=3e12 energy=15 angle=0
implant boron dose=3e12 energy=40 angle=0
implant boron dose=3e12 energy=100 angle=0
implant boron dose=3e12 energy=180 angle=0
struct outf=03_body_mask.str
etch photoresist all
diffuse time=1 temp=800  final_temp=1000
diffuse time=10 temp=1000
diffuse time=1 temp=1000 final_temp=800
struct outf=03_body.str
```

The `include` statement will load the mask file for p body implant. Figure 7.13 is the mask layout view for the body implant. Negative mask is used; meaning the area drawn will be implanted.

An implant chain is used here to create p body region and achieve the appropriate channel doping for V_{th}. Since there is no separate V_{th} adjustment step in this

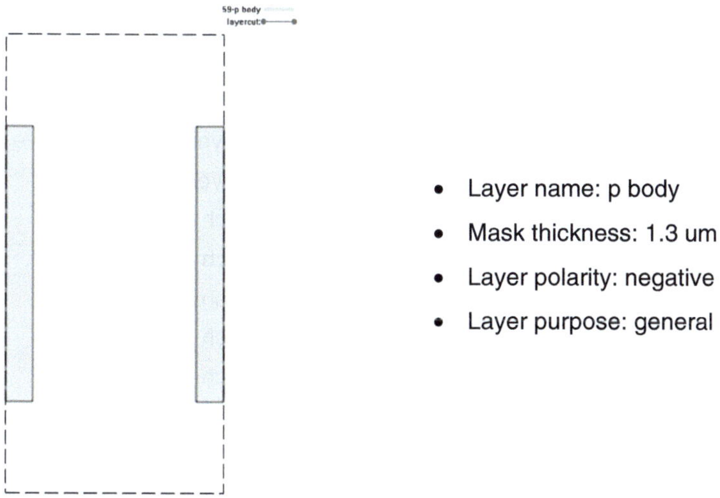

Fig. 7.13 The mask layout of p body implant

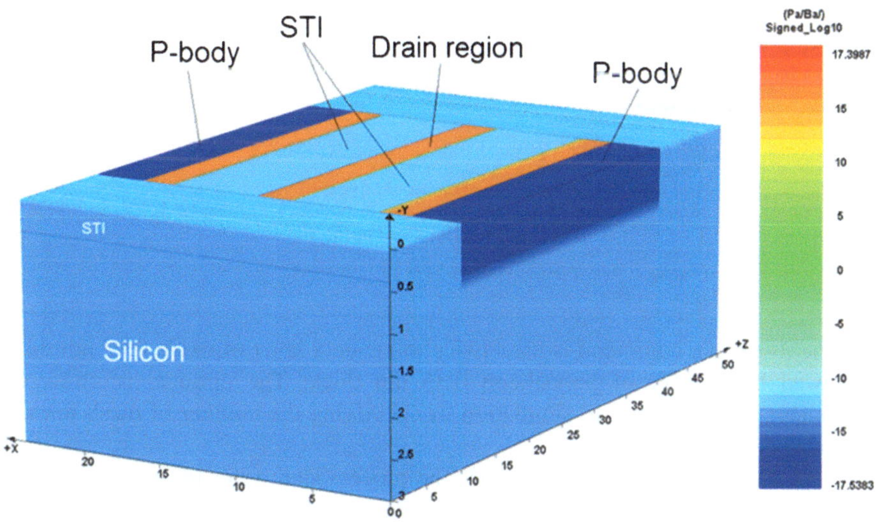

Fig. 7.14 After p body anneal showing net doping (03_body.str)

example, one should be careful in choosing the implantation parameters. Doping levels that are too high lead to high V_{th} while values that are too low yield a large leakage current from body punch through.

Figure 7.14 shows the net doping profile after p body implant anneal.

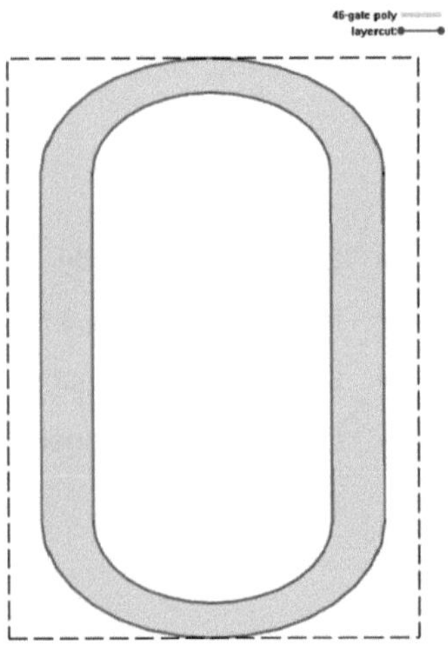

- Layer name: gate poly
- Mask thickness: 1.3 um
- Layer polarity: positive
- Layer purpose: etch
- Etch material: poly
- Etch depth: 0.35 um
- Etch angle: 0
- Etch material: oxide
- Etch depth: 0.02 um
- Etch angle: 0

Fig. 7.15 The mask layout for poly gate

7.3.6 Poly Gate Deposition

This step will form the racetrack shaped poly gate.

Process Simulation Code

```
# Gate oxide and Gate poly
deposit oxide thick=0.02 meshlayer=2
deposit poly thick=0.35 conc=1e20 phos meshlayer=2
include file=racetrack.gds46.msk
diffuse time=1 temp=800  final_temp=1000
diffuse time=10 temp=1000
diffuse time=1 temp=1000 final_temp=800
struct outf=04_poly.str
```

The `deposit` command is used twice to create a layer of oxide 0.02 um thick followed by a layer of polysilicon 0.35 um thick. The optional `meshlayer` parameter adds accuracy to simulation by specifying the number of mesh lines in the newly deposited material.

Deposited oxide is used to simplify the process step: refer to previous CMOS examples for the details on how to thermally grow gate oxide. Likewise, a single poly gate is used in this example; in a real smart IC process, a multiple gate stack might be used. Phosphorus is used for the in-situ doped poly gate.

The `include` statement loads the mask file shown in Fig.7.15. The "etch" purpose is set in the MaskEditor GUI which simplifies the process into a single command. The mask polarity is positive so the drawn area is protected while everything else is etched away.

After the deposition and etch, another `diffuse` sequence is used to anneal the poly gate. Figure 7.16 is the simulation result after poly gate is etched and subsequently annealed.

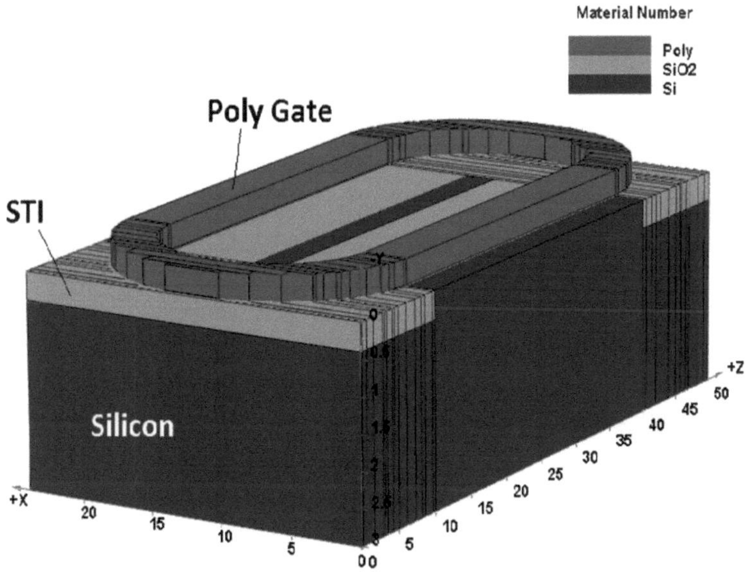

Fig. 7.16 Racetrack shaped poly gate (04_poly.str)

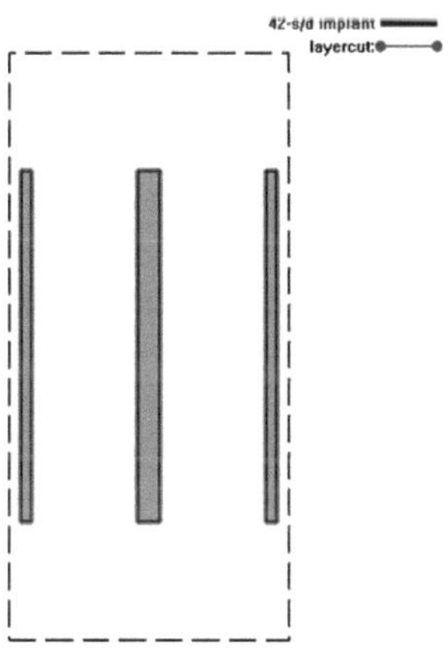

- Layer name: s/d implant
- Mask thickness: 1.3 um
- Layer polarity: negative
- Layer purpose: general

Fig. 7.17 Mask layout of source/drain n+ implant

7.3.7 Source/Drain N+ Implant

The implant for the source, drain and body contacts always comes in the last few steps of the front-end process flow. This is because the highly doped n+ and p+ regions need to be shallow enough and so cannot undergo a long thermal cycle. The mask layout from MaskEditor for this step is shown in Fig. 7.17.

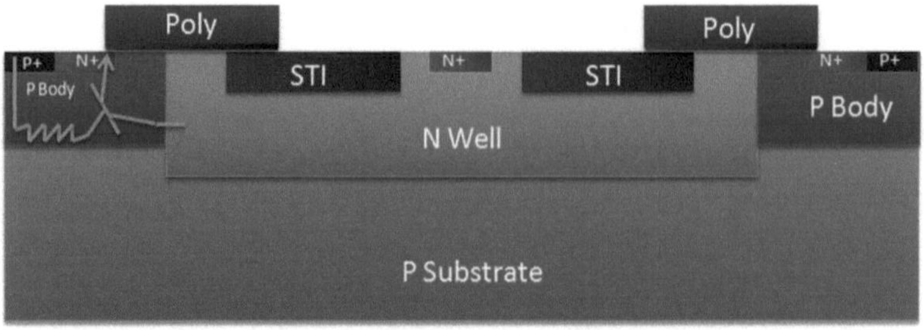

Fig. 7.18 Cross-section of a typical LDMOS showing parasitic BJT

Process Simulation Code

```
# source/drain n+ implant #
include file=racetrack.gds42.msk
implant phosphorus dose=1e15 energy=25
etch photoresist all
diffuse time=1 temp=950
struct outf=05_sd.str
```

7.3.8 P+ Body Contact Implant

One of the most notable differences between a MOSFET and LDMOS is the body region. LDMOS has a body region contact on top of the silicon surface which is connected with source region via a silicided contact and/or metal 1 layer. The body contact is fairly important in power device design because resistance in the body region may turn on the parasitic BJT shown in Fig. 7.18. The p+ body contact should not be left floating for breakdown measurement or simulation; otherwise the parasitic BJT may lower the BV by entering the so called BV_{CEO} mode (Collector to Emitter breakdown with base open).

Process Simulation Code

```
# p+ body contact implant #
include file=racetrack.gds45.msk
implant boron dose=1e15 energy=10
etch photoresist all
diffuse time=1 temp=950
regrid refine log10.change=12
struct outf=06_pplus.str
```

Most of the commands used here are similar to earlier process steps and will be skipped. The particular mask layout for this step is shown in Fig. 7.19.

The final command before outputting the structure is `regrid` which increases the mesh density near the p-n junction by imposing a limit on the change in doping

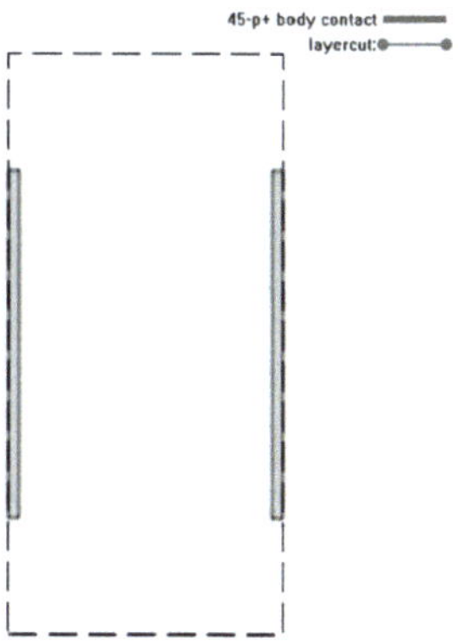

- Layer name: p+ body contact
- Mask thickness: 1.3 um
- Layer polarity: negative
- Layer purpose: general

Fig. 7.19 The mask layout of p+ body implant

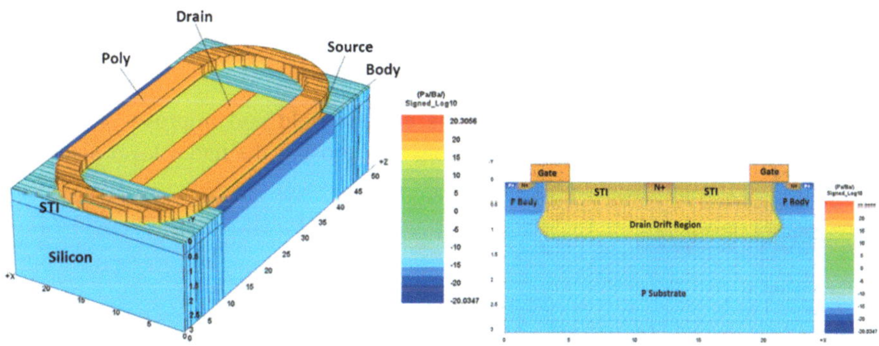

Fig. 7.20 After anneal of S/D n+ and body p+ regions (06_pplus.str)

concentration between neighboring mesh points. Figure 7.20 is the final result after p+ body implant with net doping shown. The cross-section view is taken at $z = 25$ um (the center of the device).

7.3.9 *Back-End of the Line (BEOL)*

The final steps are for Back End of the Line (BEOL). Oxide is used as the inter layer dielectric (ILD) in the simulation. Contact holes are opened and filled with aluminum. In simulation we just use aluminum for simplicity, but real process steps often use other materials such as tungsten for contact holes, and copper for metal layers. Copper has lower resistivity and is less susceptible to electron migration issues.

Note that the contacts in this simulation use large strips of metal which are consistent with the limited number of cut planes for our longitudinal mesh in the active area. A real process would use smaller metal squares.

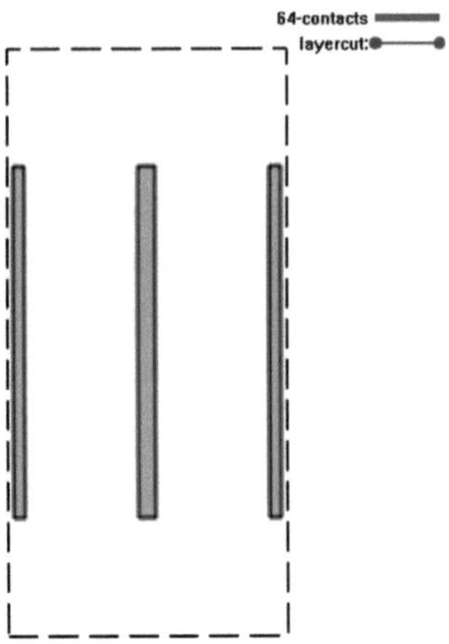

- Layer name: contacts
- Mask thickness: 1.3 um
- Layer polarity: negative
- Layer purpose: etch
- Etch material: oxide
- Etch depth: 1.0 um
- Etch angle: 0

Fig. 7.21 Mask layout for contacts

As the metal plate size may alter the electric field and improve or degrade breakdown voltage, one should be cautious in choosing the metal layer sizes and shapes when designing an LDMOS.

Process Simulation Code

```
# contacts
deposit oxide thick=0.7 meshlayer=2 space=0.1
include file=racetrack.gds64.msk
struct outf=07_contact_holes.str
deposit aluminum thick=1 meshlayer=3
etch aluminum dry thick=0.3
include file=racetrack.gds65.msk
struct outf=07_metal.str
export outf=racetrack.aps xpsize=0.005
```

The `deposit` command creates an ILD oxide layer 0.7 um thick. The `space` parameter is used to increase the mesh density and smooth out the curved areas of the deposit.

The contacts are created by etching holes in the oxide, refilling them with a blanket deposition of metal and etching away the unneeded material, leaving only the desired contacts. Figure 7.21 gives the mask layout for the contact holes while the final metal etch is done using the layout shown in Fig. 7.22. We note quickly the selective nature of the `etch` command which is set to only remove the deposited aluminum and leave the other materials intact.

The final structure is exported to the device simulator using the `export` command. The device simulator has a different convention than the process simulator regarding mesh points on material boundaries and requires that these points be split into two points on either side of the boundary. The parameter `xpsize` specifies the space of the gap. Figures 7.23 and 7.24 illustrates the final structure and cross section cut.

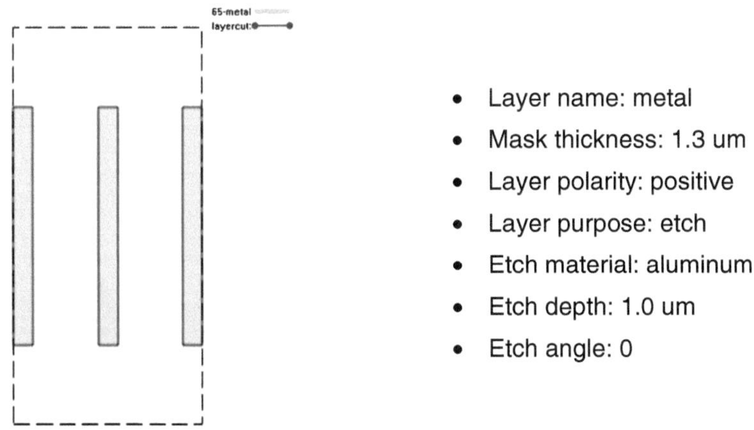

Fig. 7.22 Mask layout for metal 1 layer

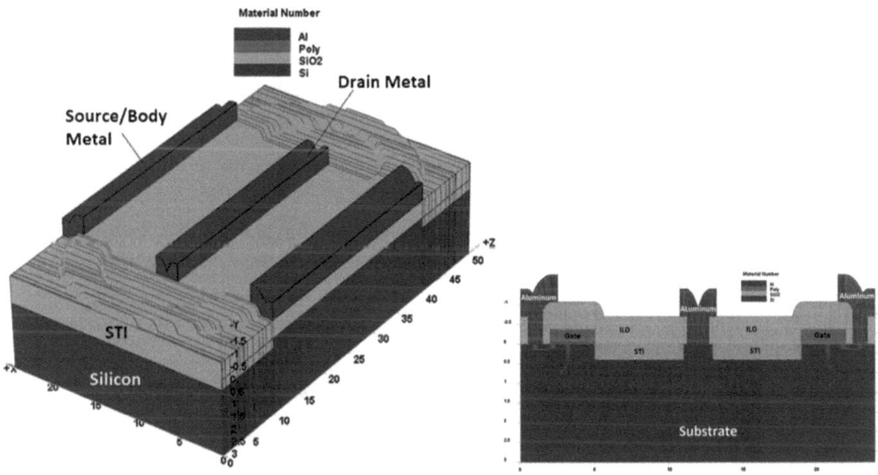

Fig. 7.23 Final structure after metal etching (07_metal.str)

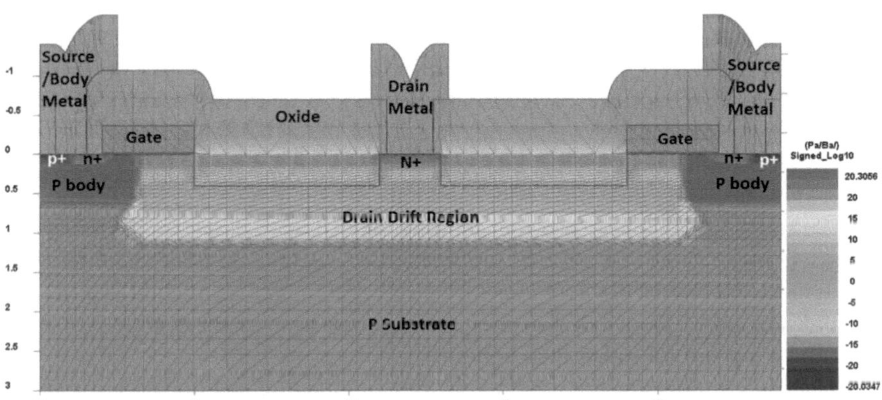

Fig. 7.24 Center cross-section view of the final net doping (07_metal.str)

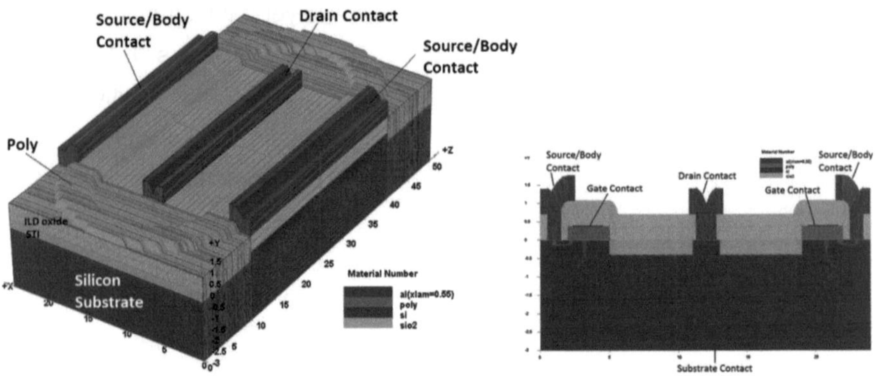

Fig. 7.25 Contact definitions for the device simulation

7.3.10 Contact Definitions for Device Simulation

Four contacts will be defined for the device simulation: source/body (sharing the same contact), gate, drain and substrate. Source/Body and Drain contacts are on the metal 1 layer and are shown in Fig. 7.25, gate contacts are placed on top of the polysilicon and the substrate contact is at the bottom of the silicon wafer.

7.3.11 Device Simulation: Breakdown Voltage

Now we have "fabricated" a racetrack LDMOS using our home-brewed process and have defined the necessary contact boundary conditions. The device simulation to simulate the electrical and thermal behavior of the device is the next step. As always, for a power LDMOS, we would like to model the breakdown voltage, on-state resistance, threshold voltage as well as the I_D-V_D curves.

For the racetrack LDMOS, the real active area is the middle part with silicon surface. The two ends are covered with STI, and no current should flow at the surface. However, for high voltage applied to the drain, the shape of the racetrack gate may alter the electric field and thus affect the breakdown voltage. This is a three-dimensional (z direction) effect and only 3D simulation can reveal this.

We can treat the device as a four terminal device, with drain, source, body and substrate controls. For the sake of simplicity, the substrate is connected directly at the bottom. Of course the thickness of the substrate in this simulation is arbitrary. Real device may have much deeper substrate or epi-layer, and substrate connect is usually at top surface rather than bottom.

The device simulator first imports the device structure and dopant information from the process simulation output file racetrack.aps; the device simulation output files will be named using bvds.out. Much of the syntax for the device simulator input file has already been discussed in previous chapters so we will concentrate only on the important parts.

Device Simulation Code

```
parallel_linear_solver

newton_par  damping_step=12.  print_flag=3  res_tol=1e-2  var_tol=1e-2

equilibrium

newton_par  damping_step=6. print_flag=3 res_tol=1.e-1 var_tol=1.e-1 &&
max_iter=50  opt_iter=25  stop_iter=4  mf_solver=3

$ scanline #2 breakdown voltage stops at current greater than 1e-6
scan  var=voltage_3 value_to= 100  print_step=10  &&
init_step= 0.5E-01 min_step= 0.5E-10  max_step=2.0 &&
auto_finish=current_4  auto_until=1.e-6  auto_condition=above
```

The `newton_par` statement controls many aspects of the non-linear Newton solver in the device simulator. The `damping_step` parameter used here limits the size of the solution update between iterations: smaller values yield slower but more reliable convergence. The values used here are larger than normal and should be reduced if the solution guess tends to oscillate.

The choice of the solver algorithm is controlled by the `mf_solver` option as well as the `parallel_linear_solver` statement. For 3D problems with many mesh planes and over 10^5 mesh points such as this one, it may be beneficial to use a parallelized GPU solver rather than multi-CPU algorithms: speed increases of factor of 2–3 are not unusual. This particular example uses default multi-CPU settings usable on a modern personal computer.

The `print_step` parameter turns on additional structural data output in the middle of a bias scan. This can be useful to recover a simulation that failed to converge all the way to the end.

The `scan` statement will ramp up the voltage on the drain to 100 V or until the current condition is met (i.e. breakdown), whichever comes first. Note that `auto_finish` conditions always imply an absolute value. The contacts were numbered during the definition in the previous step: contact 1 corresponds to the source/body; contact 2 is the gate; contact 3 is the drain contact; contact 4 is the substrate. So voltage_3 means the voltage of contact 3 (drain) and current_4 means current flowing through contact 4 (substrate).

Figure 7.26 is the 3D potential (equipotential) lines plot with a 2D cut in the center ($z = 25$ um). The potential in the STI (along z direction under the curved poly gate) are affected by the active region and gradually diminish towards each ends.

The electric field magnitude plot is shown in Fig. 7.27. From 3D field magnitude plot, an interesting butterfly shape is seen in the STI regions. A 2D cut in the device center reveals the peak field location in the STI region underneath the gate edge, which is expected. From surface plot of the field magnitude at the same 2D cut location, we can see more clearly where the electric field peaks are located.

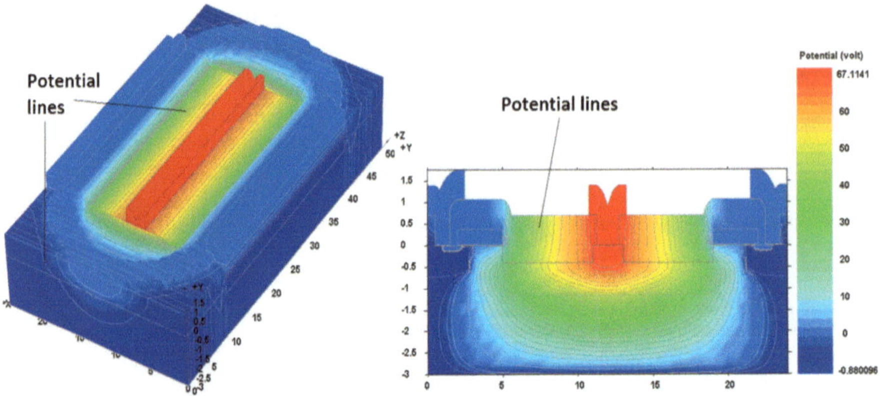

Fig. 7.26 Potential plot for racetrack LDMOS

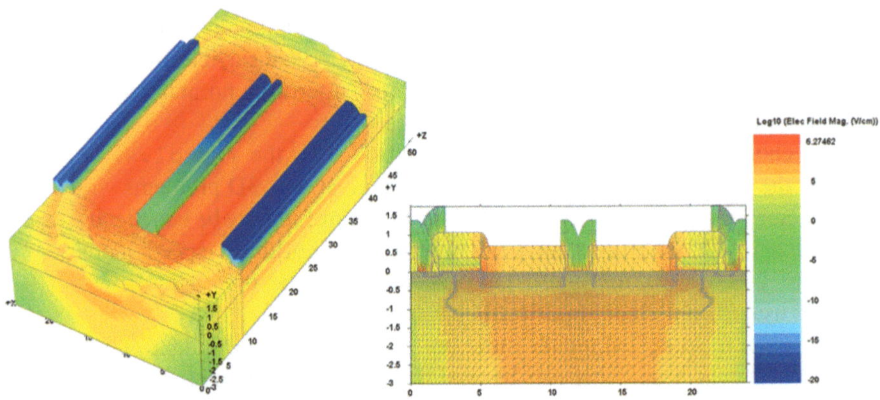

Fig. 7.27 3D plot of electric field magnitude and 2D cut view (x-y plane at z = 25 um)

The *I-V* curve is plotted in Fig. 7.28. Breakdown voltage for the LDMOS is about 64 V. The current increase is pretty sharp, indicating the current increase is due to avalanche breakdown rather than punch-through.

Power devices are built to withstand high breakdown voltages. For LDMOS, the L stands for lateral, which indicates the depletion region should progress laterally rather than vertically as VDMOS does. Naturally, in order to sustain a high breakdown voltage, the space required is quite large. This is especially true for high voltage IC devices, which can sustain up to 1,200 V. Traditionally high voltage (>120 V) power MOSFETs are reserved for vertical and discrete devices. These devices utilize the thick wafer substrate or epi-layer as drain drift region. The downside of these devices is that they are hard to integrate with CMOS and analog technologies.

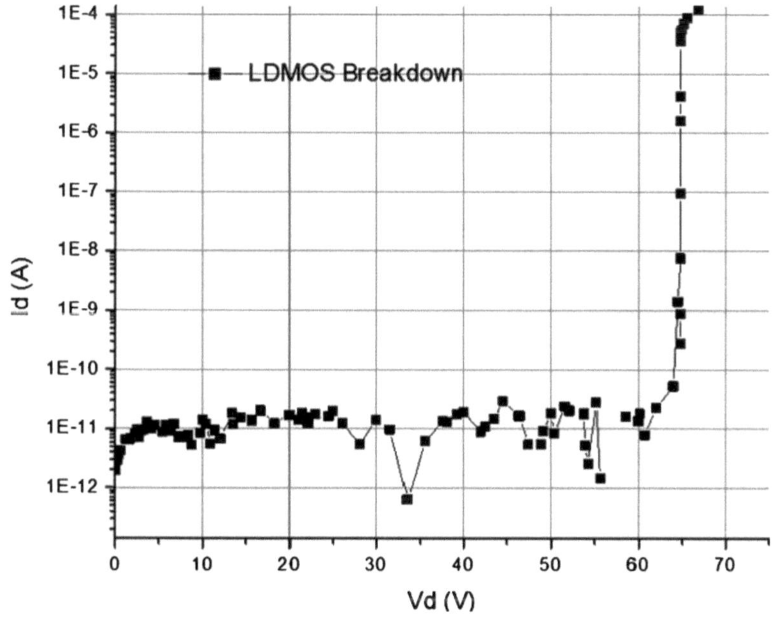

Fig. 7.28 Breakdown curve of the racetrack LDMOS

During the past decades, many research papers are published on how to improve the breakdown voltage without sacrificing the on-state resistance and cost. Among them are proposals to use single or multiple RESURF effect. We will talk about RESURF effect later in the book. But even with these novel ideas, silicon devices seem to reach a limit.

Recently, novel materials have been extensively researched for power devices, and some of them are quite promising. Silicon carbide (SiC) devices have been popular for quite a while. The specific on-state resistance of the drift region can be greatly reduced by replacing silicon with silicon carbide. Unfortunately, there are a few major drawbacks [72]:

1. The quality of the interface between the thermally grown oxide and silicon carbide surface has traditionally been poor.
2. The high electric field generated within the silicon carbide yield very high electric field within the gate oxide, which will lead to its rupture during operation.
3. Cost issues prevent SiC devices from being used in many commercial applications.

Another popular contender is GaN. GaN material has a wide band-gap and is already widely used for high-brightness light-emitting diodes. Recently, it has gained a lot of attention in the power device community.

7.3.12 Device Simulation: Threshold Voltage

For power devices, threshold voltage is usually higher than that of logic and analog devices. In this LDMOS example, the oxide thickness is chosen to be 200 Å, and the threshold voltage is expected to be around 1 V for n type devices. A threshold voltage that is too high introduces additional constraints in the design of the gate drive circuitry and should be avoided. However, if the threshold voltage is too low, it is possible for the V_{th} to be negative for n-channel power MOSFETs [73].

Device Simulation Code

```
parallel_linear_solver

newton_par damping_step=12. print_flag=3 res_tol=1e-2 var_tol=1e-2

equilibrium

newton_par damping_step=12. print_flag=3 res_tol=1e-1 var_tol=1e-1

scan var=voltage_3 value_to=0.1 init_step=1e-3 max_step=0.25 min_step=1e-6

scan var=voltage_2 value_to=2 init_step=1e-3 max_step=0.025 min_step=1e-6

stop
```

The drain contact (voltage_3) is biased at 0.1 V while 2 V is applied to the gate contact (voltage_2). The threshold voltage is extracted automatically by the plotting GUI. The extracting method is as follows:

1. Calculate the transconductance G_m using $\partial I_D / \partial V_G$.
2. Find the peak G_m and draw the Tangent of the I_D-V_G curve
3. V_{th} is defined as the crossing of the Tangent line and x coordinate.

The threshold voltage auto extraction from Fig. 7.29 is 0.87 V which is close to the 1 V value we expected.

7.3.13 Device Simulation: On-State Resistance

R_{on}, or the drain to source on-state resistance, is a very important parameter for power devices. Since power devices need to sustain high voltage and large current, the power dissipated in the power device is given by:

$$P = I_D V_D = I_D^2 R_{on} \tag{7.1}$$

For the demonstrated LDMOS, R_{on} is the combination of the following resistances [72]:

$$R_{on} = R_{cs} + R_{n+} + R_{ch} + R_d + R_{cd} + R_A \tag{7.2}$$

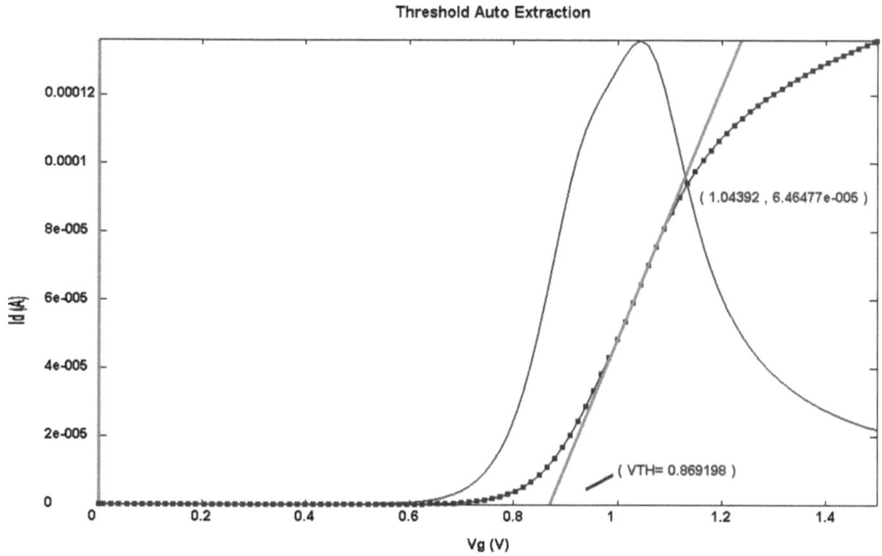

Fig. 7.29 Threshold voltage auto extraction

where R_{cs} is the source contact resistance, R_{n+} is the source n+ (including LDD) resistance, R_{ch} is the resistance from the channel region, R_d is the drain resistance, from both n drift region and drain n+ region. R_{cd} is the drain contact resistance. R_A, the accumulation resistance is unique to power devices, due to the n drift region underneath the gate when it is in accumulation mode (when positive bias is applied on the gate in n type LDMOS). Gate accumulation attracts electrons to the surface and modifies the resistance.

Among these resistance contributors, the most significant ones are R_{ch} and R_d. In a power LDMOS, the on-state current flow is established by the formation of n-channel region that connects n+ source with the n- drift region [72]. The channel resistance is given by:

$$R_{ch} = \frac{L_{ch}}{Z\mu_{ni}C_{ox}(V_G - V_{th})} \tag{7.3}$$

Where L_{ch} is the channel length, Z is the channel width. μ_{ni} is the inversion layer mobility, C_{ox} is the gate oxide capacitance. V_G is the gate voltage applied, and V_{th} is the threshold voltage. In this example, $V_G - 3V$ and $V_{th} = 0.87V$. Figure 7.30 illustrates the resistance components.

The resistance from drain region maybe larger or smaller than channel resistance: this depends on the doping concentration and the length of the drift region. This brings about a design dilemma: in order to increase the breakdown voltage, the most effective way is to lower the drift region doping concentration and increase

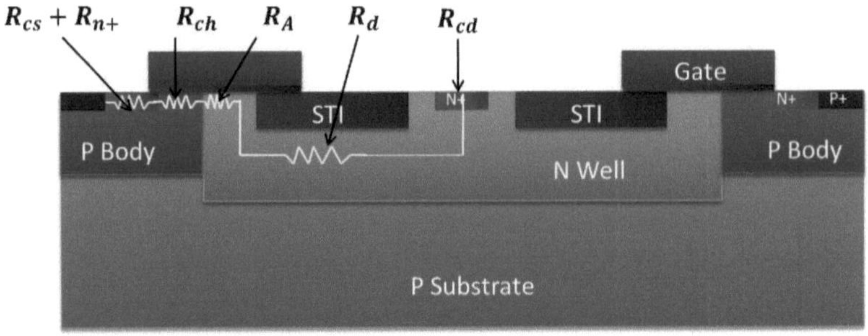

Fig. 7.30 On-state resistance components of an LDMOS

the drift region length. But this effectively increases the on-state resistance from the drift region since the resistance is proportional to the drift region length and inversely proportional to the doping concentration. There are some design guidelines for how long the drift region should be based on the desired breakdown voltage: for more information, please refer to Baliga [72] or other reference books on power devices.

In the power device industry, specific on-state resistance, $R_{on,sp}$ with unit such as mΩ mm^2 is often used. The specific on-state resistance takes device area into account so that the resistance comparison becomes meaningful for different devices and technologies.

Device Simulation Code

```
parallel_linear_solver

newton_par damping_step=12. print_flag=3 res_tol=1e-2 var_tol=1e-2

equilibrium

newton_par damping_step=12. print_flag=3 res_tol=1e-1 var_tol=1e-1

scan var=voltage_2 value_to=3 init_step=1e-3 max_step=0.25 min_step=1e-6 &&
var2=time value2_to=50

scan var=voltage_3 value_to=0.1 init_step=1e-4 max_step=0.01 min_step=1e-6

stop
```

The gate voltage (voltage_2) is first biased to 3 V in this case. Drain voltage (voltage_3) is ramped to 0.1 V. The *I-V* curve for R_{on} is linear so we can calculate the on-state resistance (R_{on}) from Fig. 7.31.

7.3.14 *Device Simulation:* I_D-V_D *Curves*

A family of I_D-V_D curves with different gate voltage applied. We can see a pronounced quasi-saturation at high gate voltages.

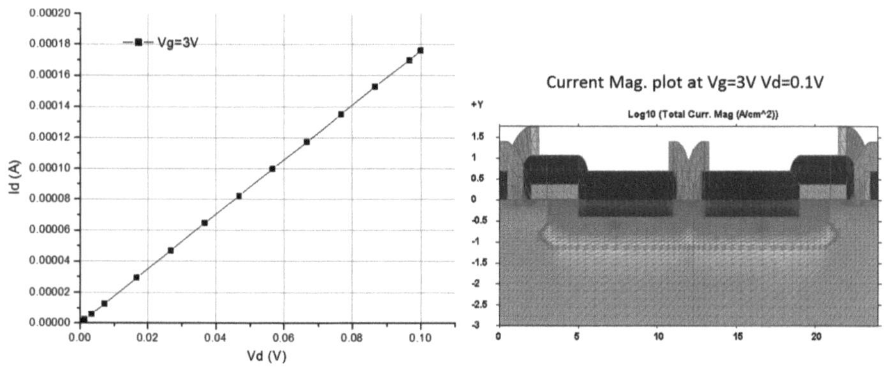

Fig. 7.31 *I-V* curve for R_{on} and 2D center cut (z = 25 um) showing total current magnitude

Device Simulation Code

```
parallel_linear_solver

$ Set equilibrium ---------------------------------------
newton_par damping_step=5. max_iter=100 print_flag=3
$ scan #1 equilibrium
equilibrium outfile_label=equi
$ Set equilibrium ---------------------------------------

$ Id-Vd curve family ---------------------------------------
newton_par damping_step=1. print_flag=3 var_tol=1.e-1 res_tol=1.e-1

$ scan #2 vg=1.5
scan var=Vg value_to=1.5 init_step=0.5000E-01

$ scan #3 vg=1.5 vd=8
scan var=Vd value_to=8.000E+00 init_step=0.5000E-01 max_step=0.4667E+00

$ scan #4 vg=2
scan var=Vg value_to=2 init_step=0.5000E-01 &&
  infile_label=equi
$ scan #5 vg=2 vd=8
scan var=Vd value_to=8.000E+00 init_step=0.5000E-01 max_step=0.4667E+00

$ scan #6 vg=2.5
scan var=Vg value_to=2.5 init_step=0.5000E-01 &&
  infile_label=equi
$ scan #7 vg=2.5 vd=8
scan var=Vd value_to=8.000E+00 init_step=0.5000E-01 max_step=0.4667E+00

$ scan #8 vg=3
scan var=Vg value_to=3 init_step=0.5000E-01 &&
  infile_label=equi
$ scan #9 vg=3 vd=8
scan var=Vd value_to=8.000E+00 init_step=0.5000E-01 max_step=0.4667E+00

end
```

Prior to the input statements shown above, `define_alias` is used to define user labels for interesting voltages rather than using the internal numbering of the device simulator. Please see the MOSFET and CMOS examples from the previous chapter for an example of the relevant syntax.

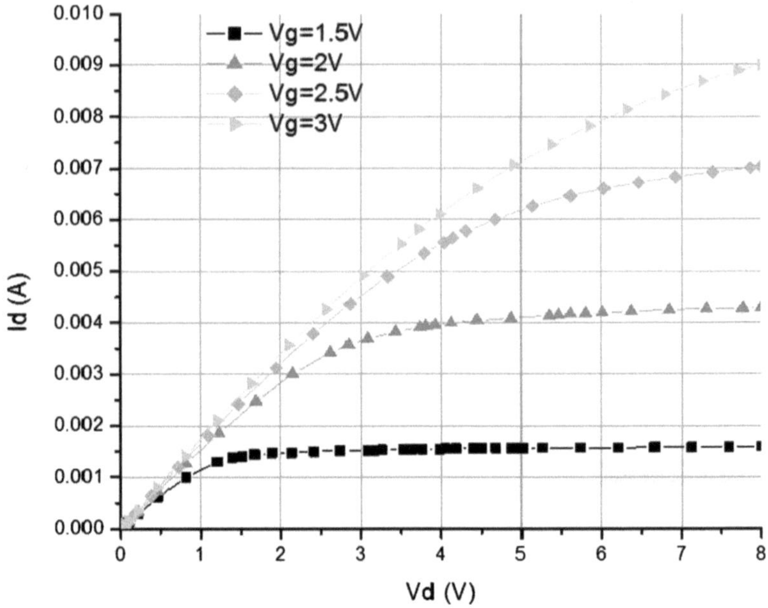

Fig. 7.32 I_D-V_D curves of racetrack LDMOS

Labels are also used to define output data sets and restart points for the various scan commands; this is useful when plotting multiple lines like in this case. The initial scan line (#1) is the equilibrium solution and each scan command afterwards is numbered accordingly. This is similar to the numbering of data sets that was discussed in earlier chapters with one crucial difference: printing intermediate structural data by adjusting the `print_step` parameter only increases the data set number, not the scan line.

We have 9 scan lines here and the equilibrium solution is saved using `outfile_label=equi`. This solution is then used to restart the various scans by using the `infile` parameter. For each I_D-V_D curve, a scan statement is issued to control the gate voltage ($V_G = 1.5, 2, 2.5$ and 3 V) before sweeping the drain voltage from 0 to 8 V.

Unlike the long channel MOSFET, presented in Chap. 6, the LDMOS has an n-type drift region which, as previously discussed, is a major contributor to R_{on}. It is well-known that in MOSFET, the drain current enters the saturation region when the pinch-off point is reached in the channel [61]. For LDMOS, this saturation phenomenon also exists but because of the drain drift region, there is another kind of saturation called quasi-saturation. This is essentially caused by velocity saturation of the carriers under high gate voltages [74]. The carrier velocity saturates because the relatively lightly doped drift region cannot sustain the required current with limited carrier mobility. This phenomenon can be viewed from the I_D-V_D curves in Fig. 7.32, at higher gate voltages (e.g. $V_G = 2.5$ V, $V_G = 3$ V), the curves become more "crowded" than at lower gate voltages.

7.3.15 *Self-heating*

This section illustrates how to set up a device simulation to include thermal effects. This model is usually neglected and must be explicitly turned on. Additional physical parameters are needed and it is important that the software introduce thermal dependence of many other physical parameters such as band-gap.

Device Simulation Code

```
heat_flow

contact num=1 thermal_type=3 thermal_cond=0.1
contact num=2 thermal_type=3 thermal_cond=0.1
contact num=3 thermal_type=3 thermal_cond=0.1
```

Here, the statement `heat_flow` turns on thermal self-heating model. With `thermal_type=3` statement, the contacts are assumed to be connected to a thermal conductor defined by the parameter `thermal_cond`. This thermal conductor is connected to a heat sink with a fixed temperature given by parameter `extern_temp`. The default value of `extern_temp` is 300 K.

The `thermal_cond` parameter corresponds to the thermal conductance of this particular thermal boundary model. The total heat flow (in Watt) to/from the device is given by $K(T - T_{ext})$ where K is the thermal conductance and T_{ext} is the external thermal contact. For a realistic 3D device, the unit is of the thermal conductance is Watt/K [58]. It should not be confused with the thermal conductivity which is defined in Watt/(m·K) [75].

Figure 7.33 gives the self-heating effect of LDMOS with Lattice temperature shown.

7.3.16 *Simulation Data*

Table 7.6 gives simulation data for the racetrack LDMOS.

7.4 Superjunction LDMOS

As mentioned before, the tradeoff between breakdown voltage and on-state resistance is a major issue for power semiconductor devices. In the past decade many ingenious designs have been proposed and implemented. One of the most promising design concepts is called the superjunction, which is similar to the RESURF (Reduced Surface Field) effect.

We will not explicitly show the process modeling for this device and only include a brief overview for interested readers. The device structure and net doping

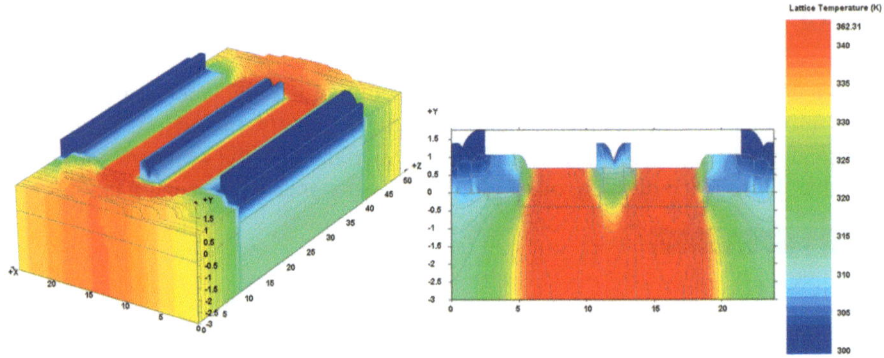

Fig. 7.33 Self-heating effect of racetrack LDMOS

Table 7.6 Simulation data for racetrack LDMOS

	Process simulation	Device simulation	Total mesh count	Total number of planes
Racetrack LDMOS	38 min	3 h (BV) 4 h (I_dV_d) 30 min (V_{th}) 30 min (R_{on})	42,061(regrid)	41
Computer configuration: HP desktop with Intel i7-860/6G/1T/NVIDIA GeForce GTX260/Win7				

of a simple superjunction LDMOS is shown in Fig. 7.34. The p well is adjacent to the n wells (n-type drain drift regions). The n drift region is first depleted by the p well, provided that the p well has a matching dose to that of n wells. The depleted n drift regions can sustain a higher breakdown voltage due to a flatter electric field distribution within the region.

This effect can be explained in a 2D rendering plot. Figure 7.35 illustrates the electric field (E field) in the n drift region with and without RESURF/superjunction corresponding to the top down view of the superjunction LDMOS structure. Figure 7.36 gives the potential and its contour plot of this device.

In comparison with LDMOS, superjunction LDMOS has a higher breakdown voltage with the same implant energy and dose for n well. On the other hand, with the same breakdown voltage, superjunction LDMOS can achieve higher dose and thus lower on-state resistance. Both approaches can greatly enhance the device figure of merit. As shown in Fig. 7.36 is a comparison of breakdown voltage between an LDMOS and superjunction LDMOS with the similar structure and doping (i.e. only difference is the center p well in superjunction LDMOS and n well in normal LDMOS).

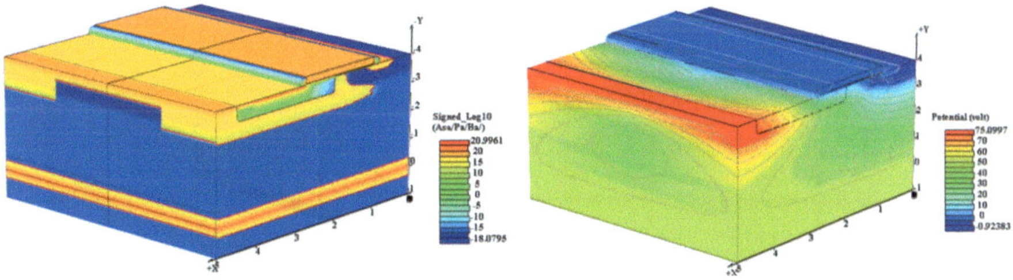

Fig. 7.34 Potential lines at breakdown of the superjunction LDMOS (right)

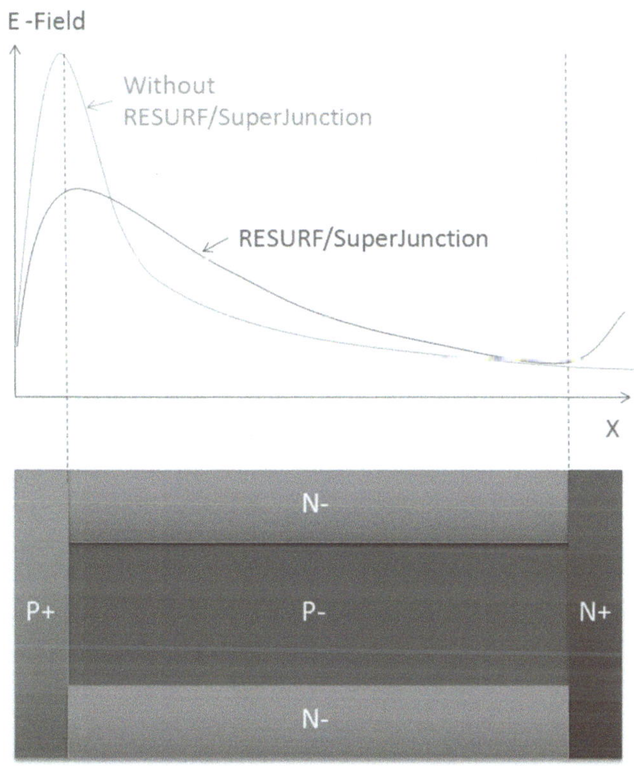

Fig. 7.35 Electric field in the n well and 2D top down view of superjunction concept

7.5 Hexagonal VDMOS

LDMOS is a lateral device, meaning the source/body gate and drain terminals are on top of device and current flows laterally. For high voltage devices, especially for discrete devices, a lateral solution may not be suitable. Vertical DMOS or VDMOS applies drain terminal to the back side of the wafer, and the thickness of the drain

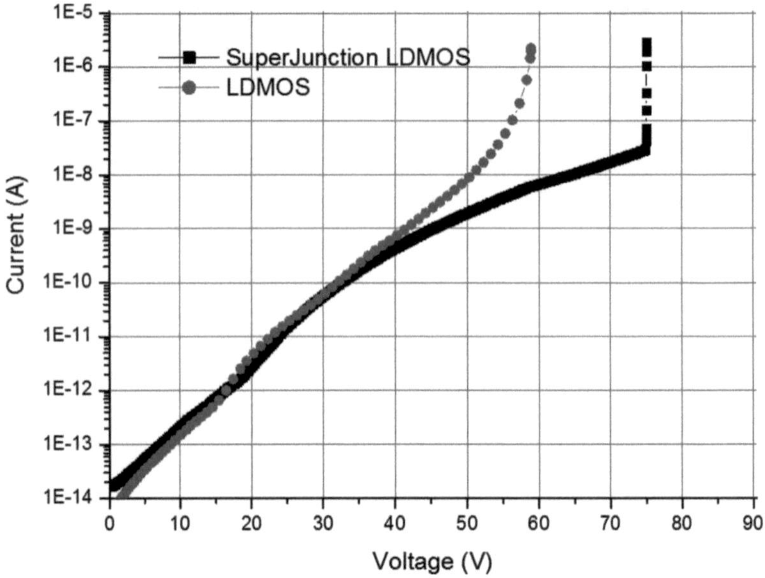

Fig. 7.36 Simulation result: breakdown voltage IV curve comparison

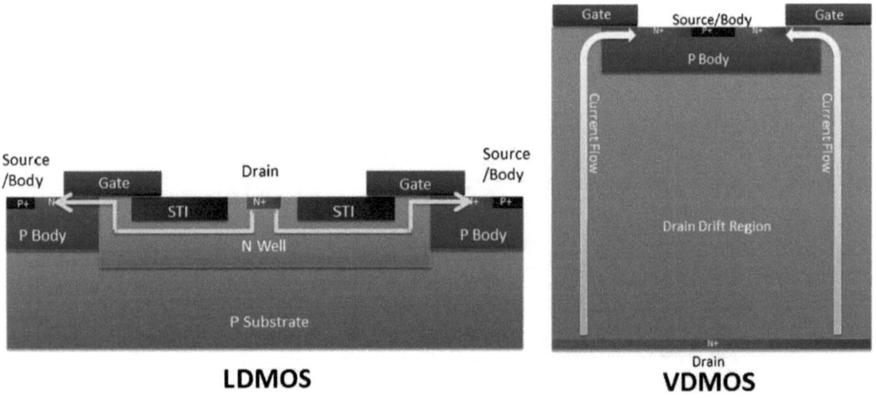

Fig. 7.37 Comparison of current flow between lateral DMOS and vertical DMOS

region is usually large. Current flows vertically from drain to source. Figure 7.37 compares the current flow direction between a lateral DMOS and vertical DMOS.

One of the most successful VDMOS structures is called a HEXFET [76] which has a hexagonal poly gate. The biggest advantage of HEXFET is that it offers very high packing density and the on-state resistance can be made very low. Seen from above, this unique structure looks like a honeycomb and the hexagonal cells give the HEXFET its name.

7.5.1 *Overview of Simulation Steps*

We will use the process simulator to create the structure of the Hexagonal VDMOS. This is followed by contacts definitions and device simulation. The breakdown voltage will be modeled for this device. An overview of the simulation steps is presented in Table 7.7.

7.5.2 *Process Simulation of the Hexagonal VDMOS*

A VDMOS with hexagonal shaped gate is built with imaginary process steps but follows the guidelines of references [16] and [76]. First a mask layout is created, as shown in Fig. 7.38. The cross-section view of the device is shown in Fig. 7.39 with normal current flow direction during on-state.

7.5.3 *Substrate*

Unlike many other simulations included in this book where the 3D simulation improves on a 2D model and shows new physical effects, this device absolutely requires a full 3D simulation due to its unique shape. The substrate is phosphorus doped with doping concentration of $1E + 15\ cm^{-3}$. Note that many real VDMOS uses highly doped substrate plus a lightly doped epi-layer as the drift region. Since there is no difference between the two for simulation purposes, the substrate acts as both n+ drain and n− drift regions in this book. Like the rest of the examples in this book, all the process step parameters are arbitrary and only used for demonstration purposes.

Process Simulation Code

```
mode three.dim
3d_mesh inf=geo
init phosphorus conc=1e15 orient=100
struct outf=00_sub.str
```

`mode three.dim` turns on the full 3D simulation which is used in this cxamplc. Thc `3d_mesh` command reads in mesh declaration statements generated by the MaskEditor GUI and the `init` command processes all the mesh commands and generates the initial substrate for the process simulation. Unlike previous examples, this device uses an n type substrate, because the substrate will act as the drain drift region. The `3d_mesh` statement loads the geo files. Universal geo files are used in this example, as shown in Text Box 7.1.

Table 7.7 Overview of simulation steps for the hexagonal VDMOS

Hexagonal VDMOS	Process simulation steps
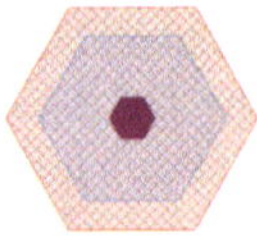	Step 1: Substrate
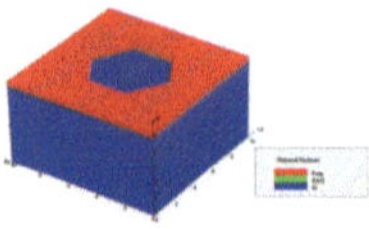	Step 2: Gate poly define
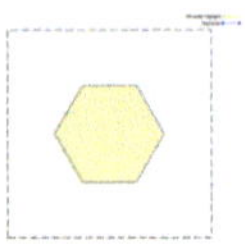	Step 3: P body, n+/p+ and drain implant
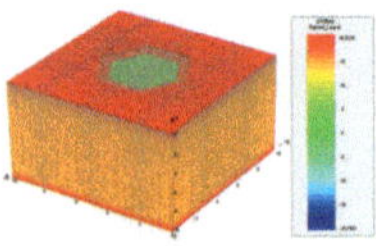	Step 4: Source/body contact metal deposit
Hexagonal VDMOS	**Contact definitions for device simulation**
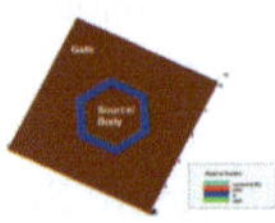	Step 5: Contact definitions for device simulation
Hexagonal VDMOS	**Device simulation**
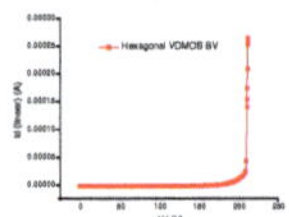	Step 6: Breakdown voltage simulation
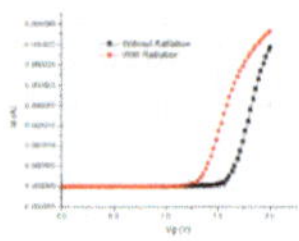	Step 7: Radiation hardening simulation

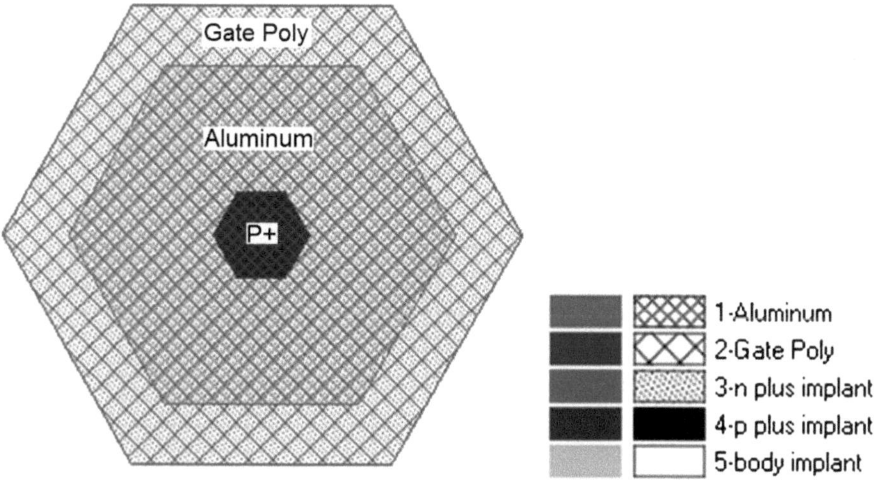

Fig. 7.38 Layout of the hexagonal VDMOS

Fig. 7.39 Cross section view of the hexagonal VDMOS

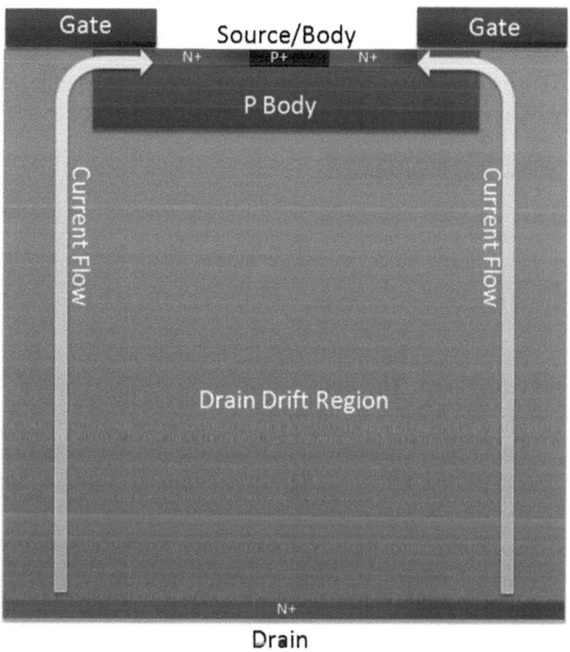

Text Box 7.1 Content of a geo File for the VDMOS

```
line x loc=   0.00000    spacing=    1.08642      tag=lft
line x loc=   2.00000    spacing=  0.162963
line x loc=   2.50000    spacing=    1.08642
line x loc=   3.00000    spacing=  0.162963
line x loc=   4.00000    spacing=    1.08642
line x loc=   5.00000    spacing=  0.162963
line x loc=   6.00000    spacing=    1.08642
line x loc=   7.00000    spacing=  0.162963
line x loc=   7.50000    spacing=    1.08642
line x loc=   8.00000    spacing=  0.162963
line x loc=   10.0000    spacing=    1.08642      tag=rht

line y loc=   0.00000    spacing=  0.892857e-01 tag=top
line y loc=   5.00000    spacing=  0.892857      tag=bot

region silicon xlo=lft xhi=rht ylo=top yhi=bot
bound exposed xlo=lft xhi=rht ylo=top yhi=top
bound backside xlo=lft xhi=rht ylo=bot yhi=bot
```

7.5.4 Gate Poly Define

The first process step is the deposition of the gate oxide and poly. An etch process will also be used to create the hexagonal shape.

Process Simulation Code

```
# Gate Poly Define #
deposit oxide thick=0.03 meshlayer=3
deposit poly phosphorus conc=1e19 thick=0.3 meshlayer=3
include file=hex.gds43.msk
struct outf=01_poly.str
```

As before, the `deposit` commands are used to deposit the oxide (30 nm) and poly (0.3 um). The poly is in-situ doped (N+) and as before, the oxide is deposited rather than grown to simplify the process simulation.

The `include` statement loads the mask layout file of poly gate generated by the MaskEditor GUI. Note that in previous examples, the poly masks were all positive so the drawn area would correspond to the leftover poly. In this case, we do the opposite because it is simpler to draw the hexagon hole rather than the leftover poly. We also simplify the process simulation by using the "etch" purpose of the mask to remove the oxide and polysilicon in the drawn area. Figure 7.40 shows the mask for gate poly and Fig. 7.41 gives the simulation result after etching.

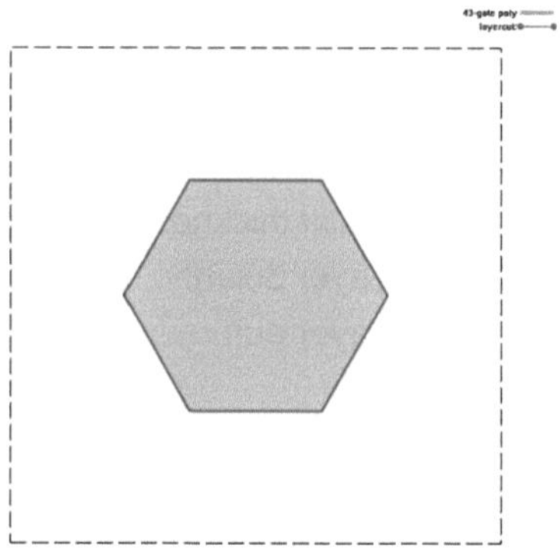

- Layer name: gate poly
- Mask thickness: 1.3 um
- Layer polarity: negative
- Layer purpose: etch
- Etch material: poly
- Etch depth: 0.3 um
- Etch angle: 0
- Etch material: oxide
- Etch depth: 0.03 um
- Etch angle: 0

Fig. 7.40 Mask layout of gate poly

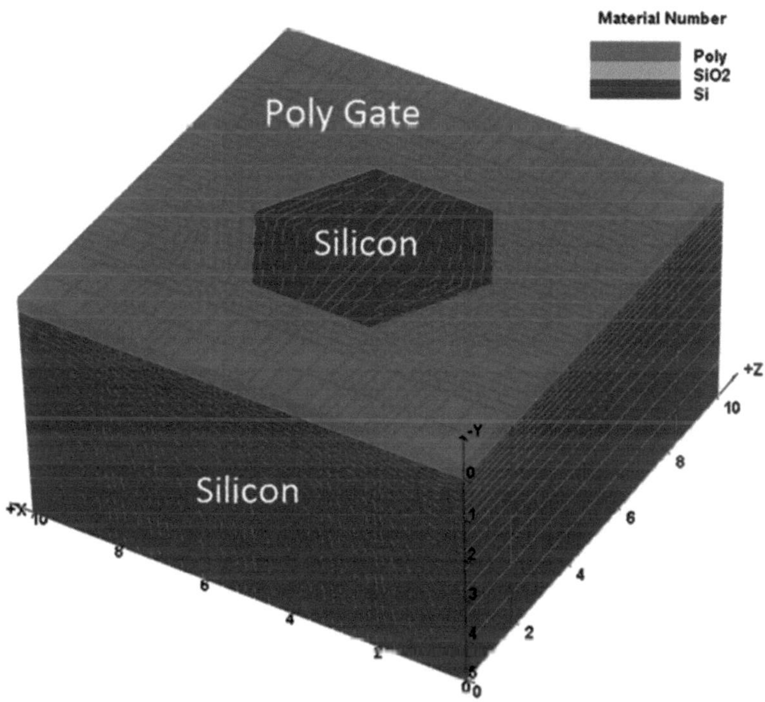

Fig. 7.41 After poly and gate oxide etch (01_poly.str)

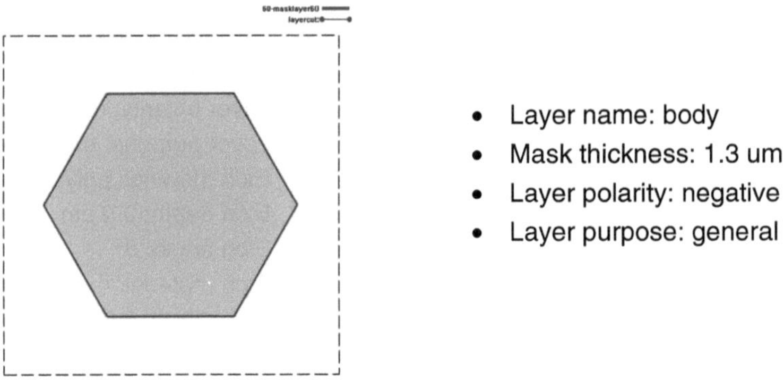

Fig. 7.42 Mask for p body implant

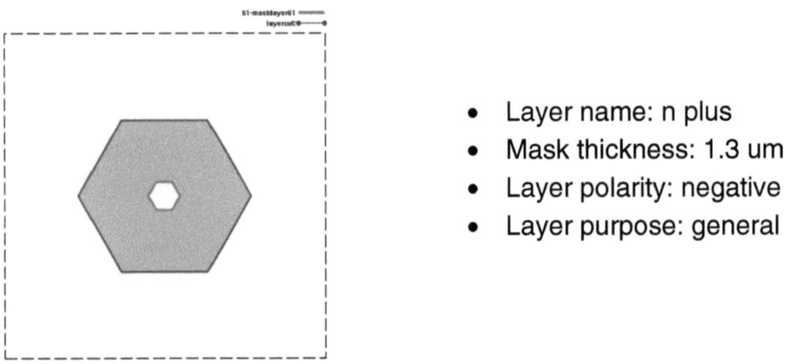

Fig. 7.43 Mask layout for n plus

7.5.5 P Body, N+/P+ and Drain Implant

The p-type body implant step comes next. An implant chain is used (Fig. 7.42).

Process Simulation Code

```
# Body implant #
include file=hex.gds60.msk
implant boron energy=100 dose=1e12
implant boron energy=160 dose=7e12
etch photoresist all
diffuse time=5 temp=1000
struct outf=02_body.str
```

The next step is the n+ and p+ implants to create the source and body contact regions. The relevant mask layouts are shown in Figs. 7.43 and 7.44. The `regrid` statement will increase the mesh density for all p-n junctions of the device.

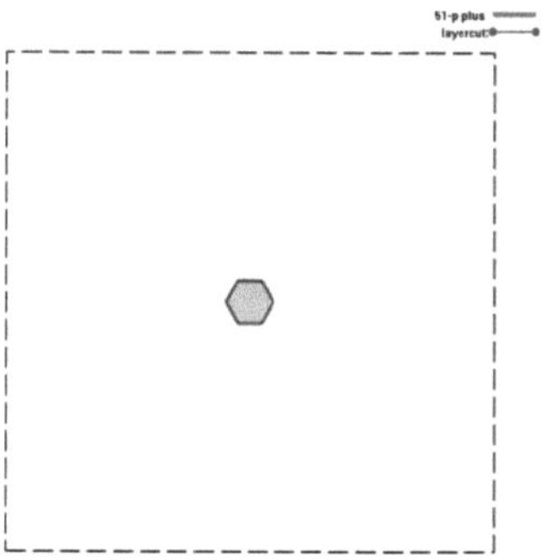

- Layer name: p plus
- Mask thickness: 1.3 um
- Layer polarity: negative
- Layer purpose: general

Fig. 7.44 Mask layout for p plus

Process Simulation Code

```
# nplus implant #
include file=hex.gds61.msk
struct outf=03_nplus_mask.str
implant phos energy=20 dose=1e15
etch photoresist all
diffuse time=10/60 temp=950
struct outf=03_nplus.str

# pplus implant #
include file=hex.gds51.msk
struct outf=04_pplus_mask.str
implant boron energy=7 dose=4e15
etch photoresist all
diffuse time=1 temp=1000
struct outf=04_pplus.str
regrid refine log10.change=7
```

Now, we need to flip the wafer to implant n+ drain onto the backside. This is because the device we are going to simulate is a vertical device, meaning the current flow is in the vertical direction rather than in the lateral direction that we are already familiar with. Please note that the drain n+ implant should not be at the beginning of the process, because the large thermal cycle of the later steps will drive the implanted n+ drain, increase doping in the drain drift region and degrade the breakdown voltage.

Process Simulation Code

```
# flip chip to define drain #
flip_y
implant phosphorus dose=1e15 energy=15
flip_y
diffuse time=1 temp=950
struct outf=05_drain.str
```

`flip_y...flip_y`. These statements are used to flip the device upside down, do the implant on the back side of the wafer and then flip it back to its original orientation.

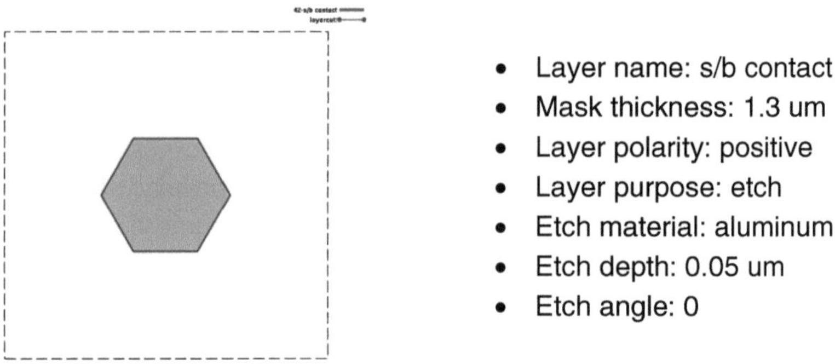

Fig. 7.45 Mask layout for source/body contact

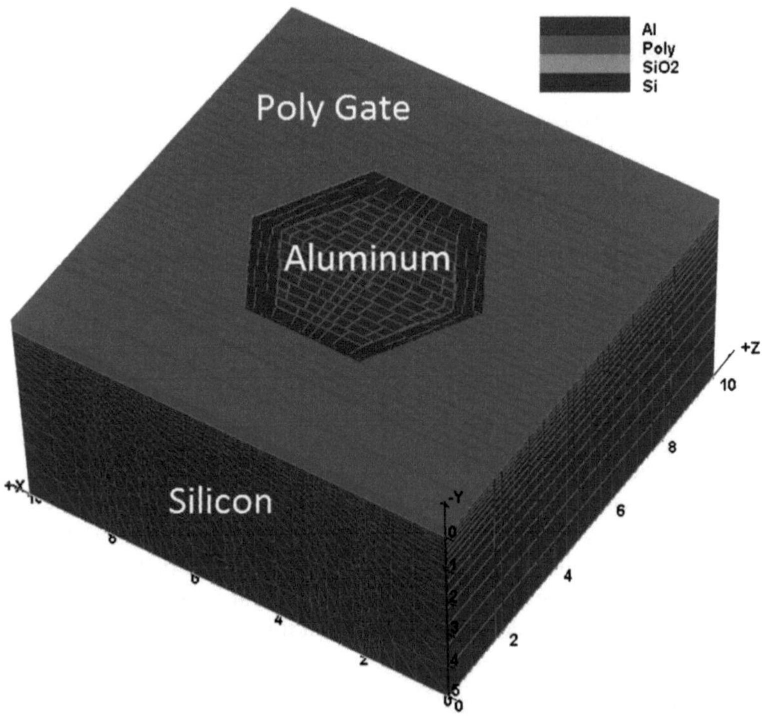

Fig. 7.46 Final structure of the hexagonal VDMOS

7.5.6 Source/Body Contact Metal Deposit

Finally, aluminum is deposited on top of source/body and etched away to make a contact region. Figure 7.45 is the mask layout for source/body contact and Fig. 7.46 illustrates the final hexagonal VDMOS structure. Net doping chart and 2D cross-section view are shown in Fig. 7.47.

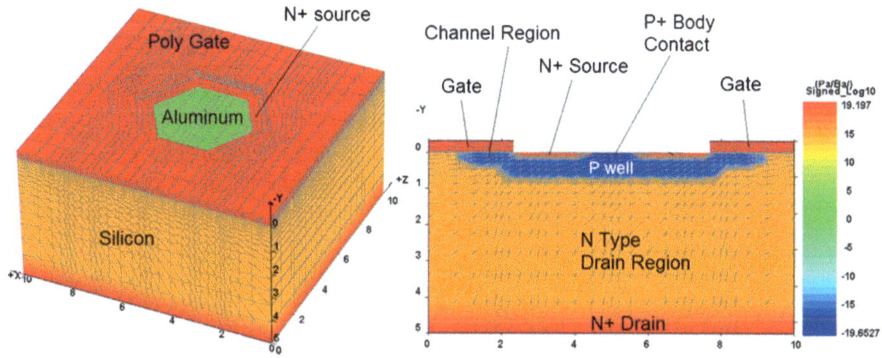

Fig. 7.47 3D and 2D cut ($z = 5$ um) net doping charts for the hexagonal VDMOS

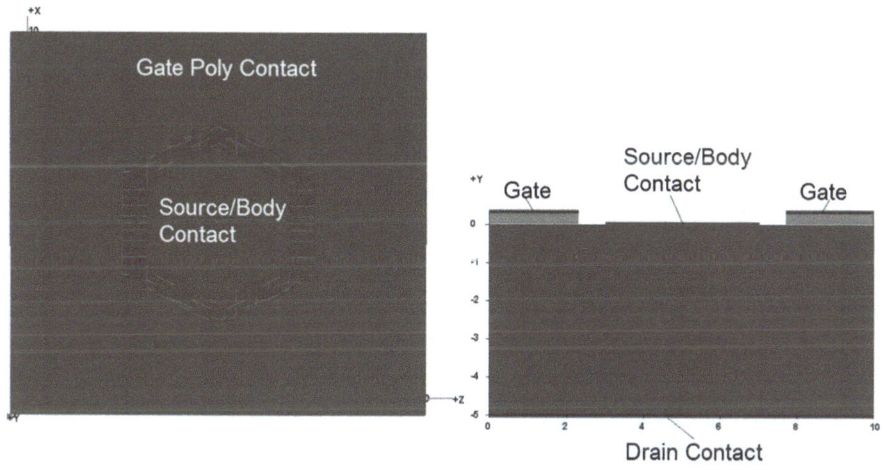

Fig. 7.48 Contact definitions for the hexagonal VDMOS

Process Simulation Code

```
# contact for source/body #
deposit aluminum thick=0.01
include file=hex.gds42.msk
struct outf=06_contact.str
export outf=hex.aps xpsize=0.001
```

7.5.7 *Contact Definitions*

Like most power MOSFETs, three terminals/contacts (source, gate and drain) are defined for this hexagonal VDMOS. The contact placement is shown in Fig. 7.48.

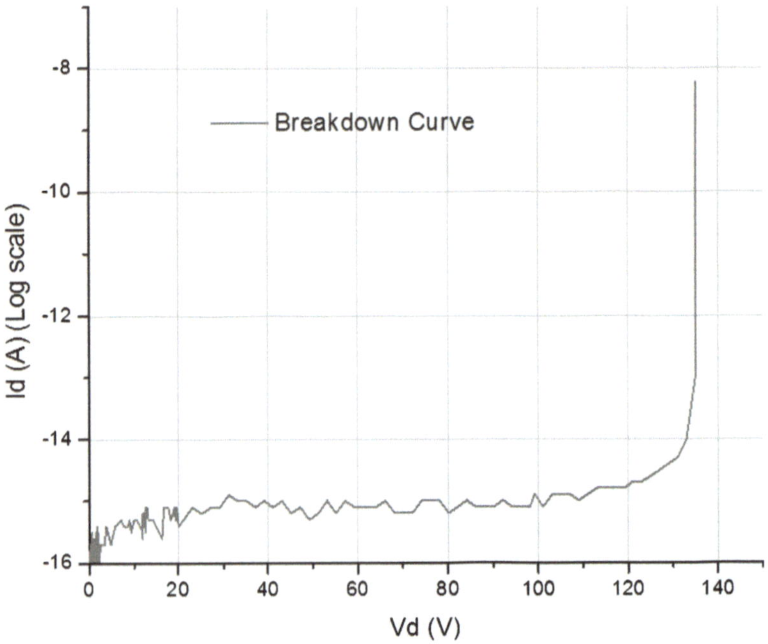

Fig. 7.49 Breakdown curve of Hexagonal VDMOS (linear)

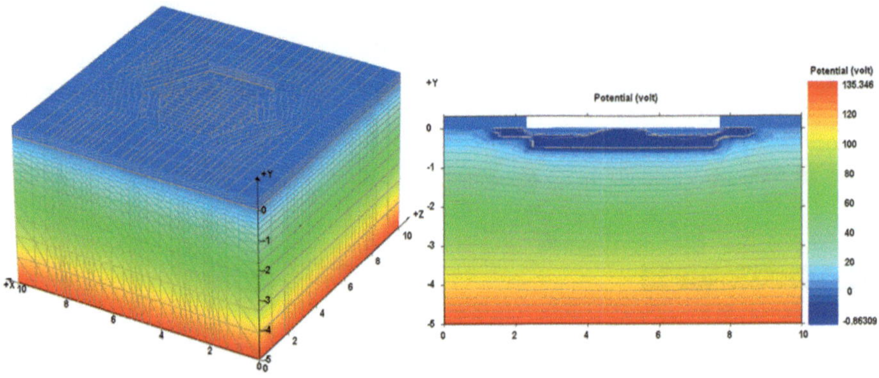

Fig. 7.50 Potential plot and 2D center cut ($z = 5$ um)

7.5.8 Breakdown Voltage Simulation

Vertical power devices are generally designed to have a higher breakdown voltage than lateral devices. Since vertical devices are quite often the discrete rather than integrated devices. The deep n type substrate acts as a long drain drift region. In this example the breakdown of the Hexagonal VDMOS exceeds 120 V. The breakdown voltage simulation result is shown in Fig. 7.49 and the potential plot with 2D cut are given in Fig. 7.50. The potential lines are fairly flat due to the vertical structure. Note that in the device simulation code, voltage_3 is the drain voltage while current_3 is the drain current.

Device Simulation Code

```
parallel_linear_solver

newton_par damping_step = 10 res_tol= 1.000000e-002 &&
   var_tol = 1.000000e-002 max_iter = 60 opt_iter = 30

equilibrium

newton_par damping_step = 12 res_tol= 1.000000e-001 &&
   var_tol = 1.000000e-001 max_iter = 60 opt_iter= 30

scan var=voltage_3 value_to=200 init_step=0.5e-2 &&
 min_step=0.5e-12 max_step=2 auto_finish=current_3 &&
 auto_condition=above auto_until=1e+1

end
```

7.5.9 Radiation Hardening Simulation

The performance of a MOSFET is known to degrade when subjected to ionizing radiation such as high energy protons, alpha particles, x-ray and gamma-rays. Device and circuit design methods to achieve resistance to radiation is called radiation hardening and devices designed to be resistant to radiation are known as radiation hardened device. This radiation hardening is necessary when designing MOSFETs for space applications but is also found in many military specifications.

Known effects of ionizing radiation include the generation of positive fixed sheet charge at oxide/silicon interface and the creation of acceptor-like deep level traps at oxide/silicon interface [77]. The effect of deep-level acceptors near the silicon channel is similar to an increase of channel p-doping which would shift the threshold voltage of an n-MOSFET upwards. On the other hand, adding a positive fixed charge to the channel interface has an effect similar to n-doping which shifts the threshold voltage downwards.

Figure 7.51 shows the downward shift of threshold voltage due to the introduction of positive fixed charges with a surface density of 8E + 11 cm^{-2} and deep acceptor traps of 2E + 11 cm^{-2} located at 0.4, 0.6 and 0.8 eV, respectively, below the conduction band of the silicon band-gap. The net effect of the radiation is a significant downward shift of the threshold. When calibrated with the total dose of

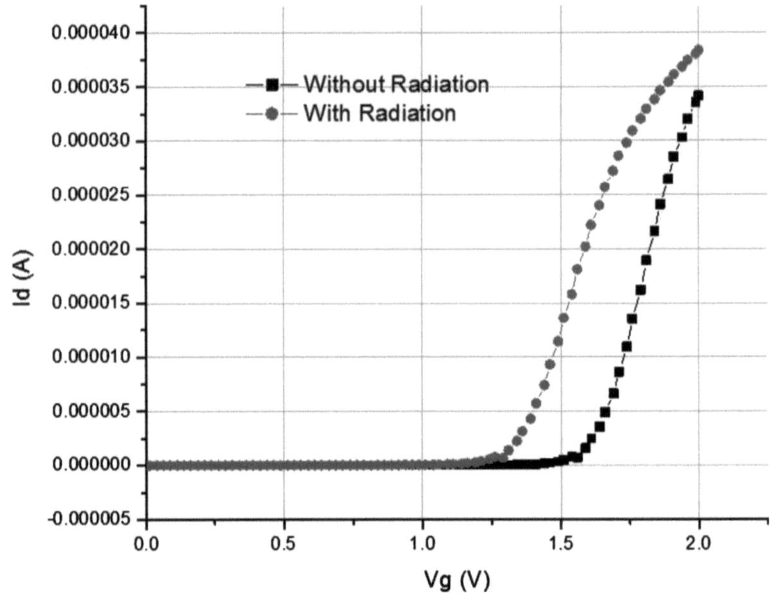

Fig. 7.51 Threshold voltage shift with and without radiation

ionizing radiation in an experimental setting, the simulation can help design a MOSFET which is hardened against the effects of radiation in the field.

The commands of the device simulator related to the radiation are listed as follows. Please note the surface charge density is in $1/m^2$.

Device Simulation Code

```
$ surface charge/trap density in unit of 1/m^2
interface model=charge within_y=(-0.001 0.001) fix_charge=8.e15


$ trap level measured below conduction band
interface model=trap trap_type=acceptor within_y=(-0.001 0.001) &&
trap_density=2.e15 trap_level=0.4  &&
trap_life_n=1.e-8 trap_life_p=1.e-8 surftrap_num=1

$ trap level measured below conduction band
interface model=trap trap_type=acceptor within_y=(-0.001 0.001) &&
trap_density=2.e15 trap_level=0.6  &&
trap_life_n=1.e-8 trap_life_p=1.e-8 surftrap_num=2

$ trap level measured below conduction band
interface model=trap trap_type=acceptor within_y=(-0.001 0.001) &&
trap_density=2.e15 trap_level=0.8  &&
trap_life_n=1.e-8 trap_life_p=1.e-8 surftrap_num=3

$ output select
more_output space_charge=yes impact_ionization=yes
```

Table 7.8 Simulation data for the hexagonal VDMOS

	Process simulation	Device simulation	Total mesh count	Total number of planes
Hexagonal VDMOS	4.5 h	4.3 h (GPU,BV) 30 min (GPU, V_{th})	29,636	27
Computer configuration: HP desktop with Intel i7-860/6G/1T/NVIDIA GeForce GTX260/Win7				

Please note that the three level positions are selected rather arbitrarily for demonstration. In real device application, it should be guided by Deep Level Trap Spectroscopy (DLTS) measurements. The trap life time above is a convenient quantity to be used to compute the electron capture cross section of the trap according to the following formula.

$$\sigma_{tn} = \frac{t_s}{v_{thm} N_s \tau_{tn}} \tag{7.4}$$

where τ_{tn} is the `trap_life_n` in the commands and v_{thm} is the thermal velocity:

$$v_{thm} = \sqrt{\frac{8k_B T}{\pi m_n^* m_0}} \tag{7.5}$$

where m_n^* is the effective mass ratio and other symbols have their usual meanings. N_s is the surface trap density and t_s is the thickness of the surface mesh element. The final trap recombination rate would be independent of t_s when integrated over the thickness of the surface element and the recombination would become surface recombination. The trap capture formula for holes is completely similar so that we omit its description here.

7.5.10 Simulation Data

Table 7.8 gives simulation data for the hexagonal VDMOS.

7.6 NPN Bipolar Junction Transistor

The last example of this chapter is 3D NPN Bipolar Junction Transistor (BJT). The power BJT is one of the most important building blocks for smart power IC technology. Most of the process steps are similar to those used previously so we will skip straight to the results. The mask layout and device cross-section are shown in Fig. 7.52.

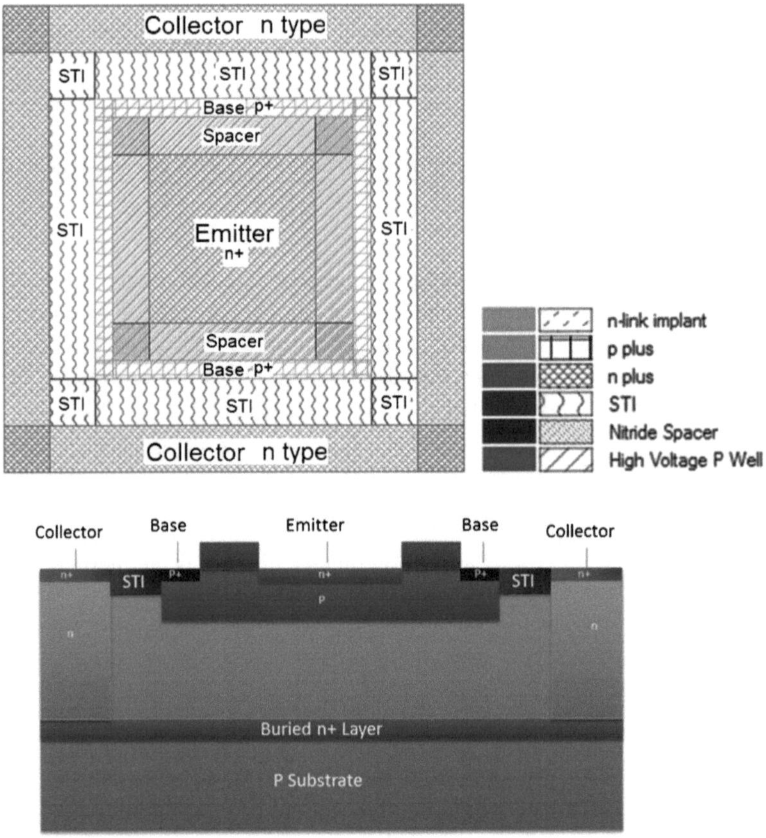

Fig. 7.52 Mask layout and cross-section (center cut) of NPN BJT

7.6.1 Process Simulation Result of NPN BJT

Unlike previous device examples, this BJT example has a buried n+ layer, which is used to connect n type collectors. Buried layer in smart power technology usually has multiple purposes, like isolation, provide RESURF (see Sect. 7.4) for power LDMOS, etc.

The n-type buried layer is usually created by high dose and low energy implant of arsenic or antimony (Sb) after substrate is initialized. Since this implant step is at the very beginning of the whole process flow, it has to go through a lot of thermal cycles later on. The implant energy has to be low to keep the buried layer within required thickness. Quite often, oxide is grown before the buried layer implant to help further reduce the implant depth. Antimony is a better choice over arsenic in terms of keeping buried layer thickness within spec for the n-type buried layer implant.

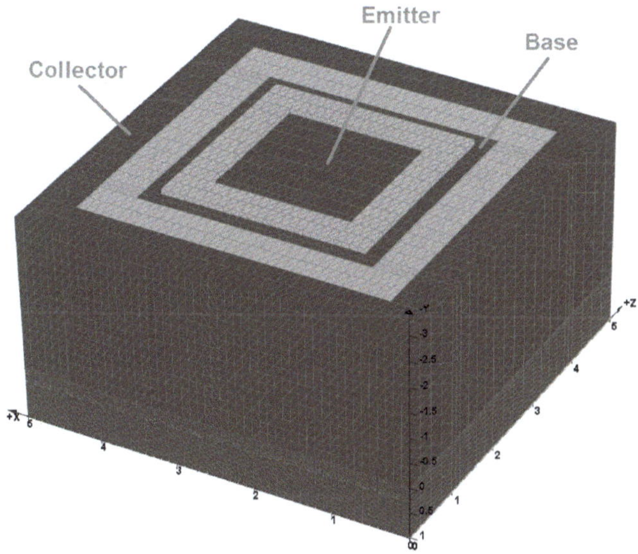

Fig. 7.53 Finalized structure of NPN BJT process simulation

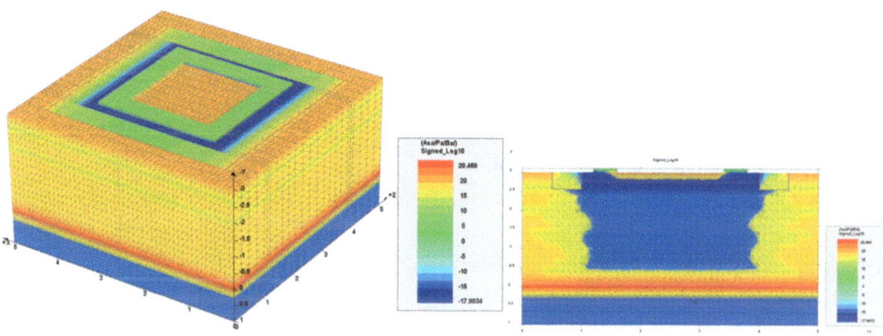

Fig. 7.54 3D and 2D center cut net doping charts of NPN BJT

Note that a thick epi-layer of silicon is grown right after the buried layer anneal. In fact, in the smart power IC process, the epi-layer (typically several um thick) is where most devices are built rather than in the substrate.

The process simulation result is shown in Fig. 7.53, which shows three terminals: collector, base and emitter. Figure 7.54 is the 3D and 2D center cut of the final net doping charts. The collectors are connected via the buried layer.

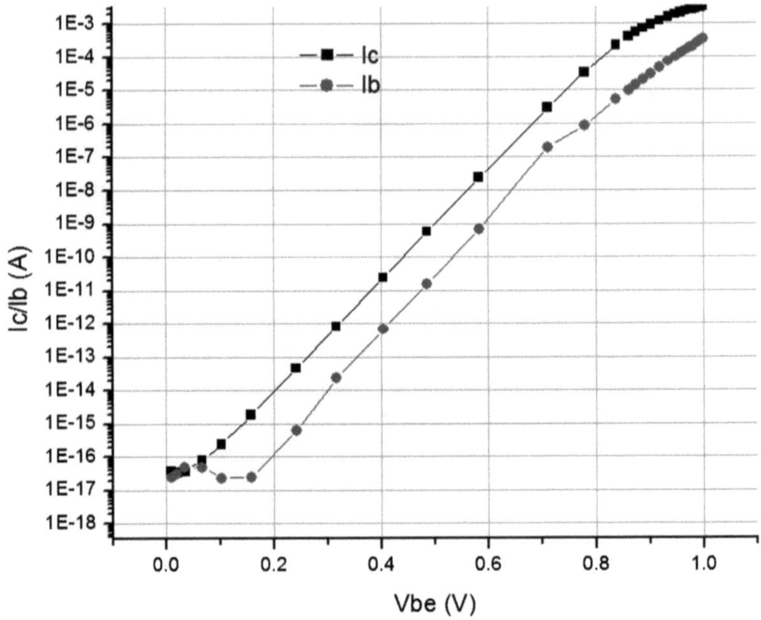

Fig. 7.55 Gummel plot of the NPN BJT

7.6.2 Device Simulation of NPN BJT

The Gummel plot of the NPN BJT is shown in Fig. 7.55. One can calculate the beta parameter of the BJT using this plot.

Chapter 8
3D Interconnect Simulation

In the past, device engineers have often neglected the contribution of metal interconnect to the on-state resistance (R_{on}). For larger sized MOSFETs and power devices, this is no longer true and extensive research has shown that interconnects are now one of the major contributors to R_{on} [78]. This chapter will build an example of a 3D metal interconnect, show how to calculate the interconnect resistance and demonstrate its influence on a large sized MOSFET.

8.1 Process Simulation of 3D Interconnect

In this section, a large MOSFET with a symmetrical source-drain MOSFET with large size has been simulated with an interdigitated source and drain. Double metal layer interconnect is built with metal layer #1 (metal 1) as stripes and metal layer #2 (metal 2) as blocks. The gates are connected with metal layers. To simplify the process, aluminum is used for all metal, contacts and vias Fig. 8.1 is the layout view.

8.1.1 Overview of Simulation Steps

We will use the process simulator to create the structure of the 3D interconnect on a large sized MOSFET. This is followed by the contact definitions and device simulation. The on-state resistance will be simulated for this device. An overview of simulation steps is presented in Table 8.1.

S. Li and Y. Fu, *3D TCAD Simulation for Semiconductor Processes, Devices and Optoelectronics*, DOI 10.1007/978-1-4614-0481-1_8,

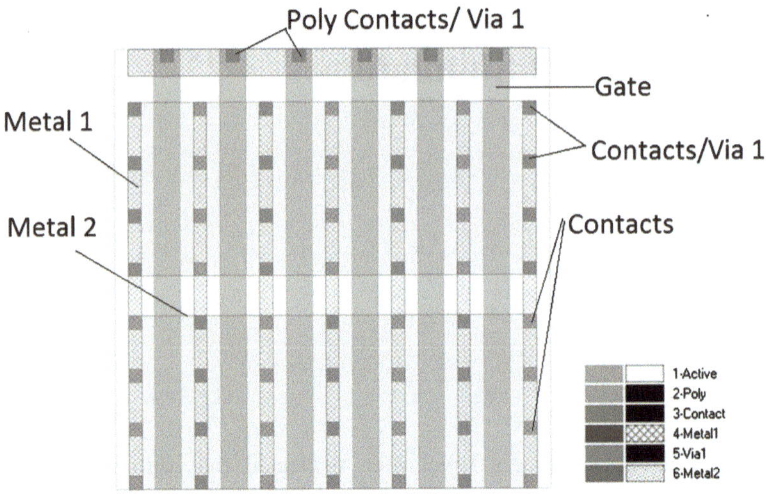

Fig. 8.1 GDS layout view of the symmetrical MOSFET

Table 8.1 Overview of simulation steps of the 3D interconnect

3D Interconnect	Process simulation steps
	Step 1: Substrate
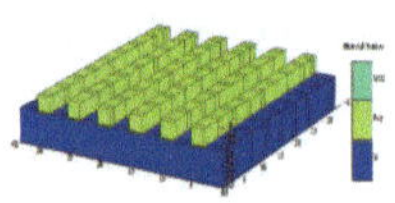	Step 2: Oxide and gate poly deposition
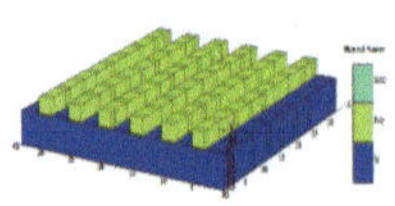	Step 3: Source/drain implant
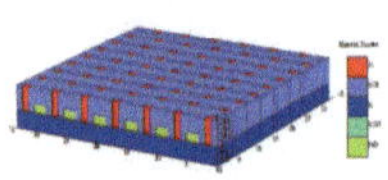	Step 4: Contacts
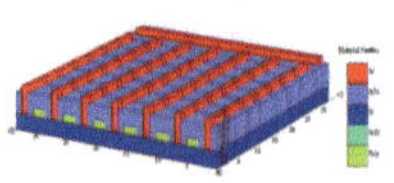	Step 5: Metal layer 1

(continued)

Table 8.1 (continued)

3D Interconnect	Process simulation steps
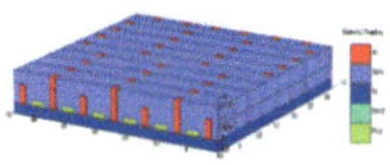	Step 6: Via1 placement
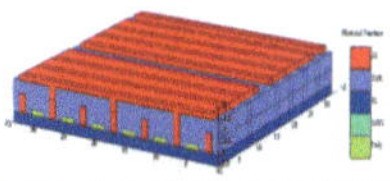	Step 7: Metal 2 formation
3D Interconnect	**Contact definitions for device simulation**
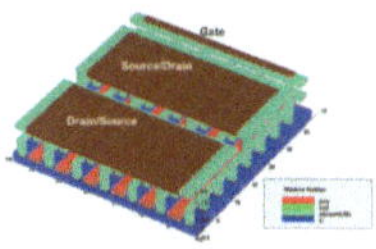	Step 5: Contact definitions for device simulation
3D Interconnect	**Device simulation**
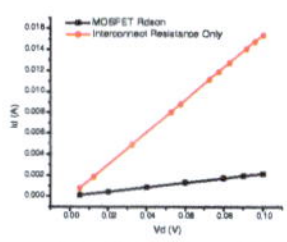	Step 6: Device simulation: on-state resistance

8.1.2 Substrate

The first step sets up the initial substrate. A quasi-3D approach is used for the process simulation to save time since no dopant diffusion takes place in the z direction.

Process Simulation Code

```
mode quasi3d
3d_mesh infile=geo
initial boron conc=1e16 orient=100
struct outf=01_sub.str
```

The content of the geo file is shown below in Text Box 8.1. Again, universal mesh define is used.

The `initial` statement processes all the mesh declaration statements and creates the initial structure for the process simulation. The substrate is boron doped with a concentration of 1E + 16 cm^{-3} and crystal orientation is [100]. Figure 8.2 shows the substrate simulation result.

Text Box 8.1 The Content of geo Files

```
line x loc=    0.00000     spacing=   0.412500       tag=lft
line x loc=    33.0000     spacing=   0.412500       tag=rht

line y loc=    0.00000     spacing=   0.195652e-01 tag=top
line y loc=   0.500000     spacing=   0.652174e-01 tag=bot

region silicon xlo=lft xhi=rht ylo=top yhi=bot
bound exposed xlo=lft xhi=rht ylo=top yhi=top
bound backside xlo=lft xhi=rht ylo=bot yhi=bot
```

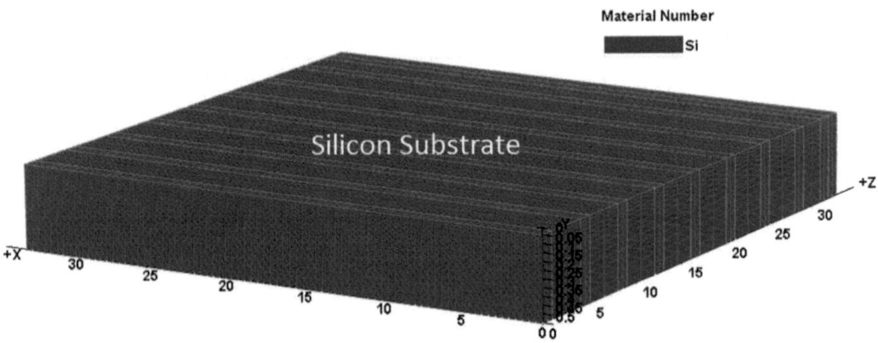

Fig. 8.2 Interconnect simulation result: substrate (01_sub.str)

8.1.3 *Oxide and Gate Poly Deposition*

Poly layer is deposited first. The poly gates are in stripes and will be used as the implant mask for source/drain implant as well. The nitride spacer deposition and etch steps is omitted in this example.

Process Simulation Code

```
# poly define #
deposit oxide thick=0.005
deposit poly thick=0.25 conc=1e20 phosphorus meshlayer=4
include file=interconnect.gds43.msk
struct outf = 02_poly_mask.str
etch oxide dry thick=0.015
struct outf=02_poly.str
```

A pair of `deposit` commands are used for the oxide (50 Å for low voltage NMOS) and polysilicon (0.25 um) layers. The oxide is deposited rather than

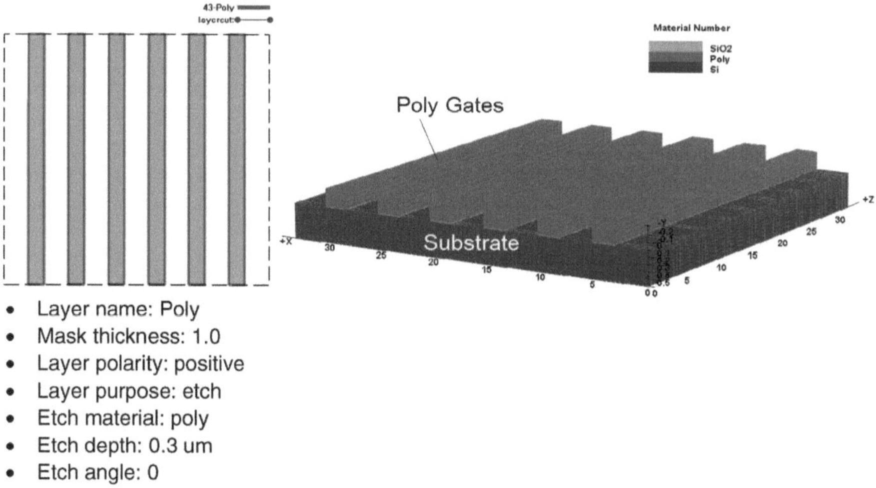

Fig. 8.3 Mask layout for poly layer and simulation result of poly gates step (02_poly.str)

thermally grown to simplify the process step. The poly gate layer is in situ doped with phosphorus with a concentration of 1E + 20 cm^{-3}. The `meshlayer` parameter is optional but helps increase the simulation accuracy by specifying additional mesh lines for the newly deposited material.

The `include` command loads the mask layout file generated by the MaskEditor GUI (Fig. 8.3). This mask is designed to etch the poly layer: a positive polarity is used so the drawn area will be protected and the rest will be etched away.

8.1.4 Source/Drain Implant

The source and drain implant step does not have specific mask layer: the already deposited poly is used as the mask for source and drain implant mask. This portion of process code is for interconnect with a large multi-fingered MOSFET device. We will show later a slightly modified code for interconnect without MOSFET device underneath. The difference is the latter has high substrate doping concentration, and the implant step for source/drain is neglected.

Process Simulation Code

```
# source/drainimplant #
implantphos dose=1e15 energy=10 angle=0
diffuse time=1/60 temp=950
structure outfile=03_sd.str
regrid refine log10.change=10
structure outfile=03_sd1.str
```

The `implant` command does a blanket implant of phosphorous which will also dope the poly. The `regrid` command refines the mesh around the p-n junctions.

8.1.5 Contacts

It's time to place contacts on top of source/drain as well as poly gates.

Process Simulation Code

```
# contacts definitions #
deposit oxide thick=1.2 meshlayer=4
etch start x=0 y=-3.5
etch continue x=0 y=-1.2
etch continue x=33 y=-1.2
etch done x=33 y=-3.5
include file = interconnect.gds44.msk
struct outf = 04_contact_mask.str
deposit aluminum thick=1.0
etch start x=0 y=-3.5
etch continue x=0 y=-0.7
etch continue x=33 y=-0.7
etch done x=33 y=-3.5
struct outf = 04_contact.str
```

The deposit command deposits a layer of oxide which will mimic the interlayer dielectric (ILD). Note that this is completely artificial and the ILD is an important part of a real device.

The `etch start...continue...done` commands does a preliminary chemical mechanical polish (CMP). Afterwards, the contact holes are etched using the mask layout shown in Fig. 8.4 and refilled by depositing aluminum. Another CMP step is then used to leave a flat surface for the next process step.

8.1.6 Metal Layer 1

In this design, the metal 1 layer consists of stripes that connect all the contacts. An aluminum layer 0.2 um thick is deposited and then removed using the etch mask of Fig. 8.5. As mentioned before, nowadays copper are used more often than aluminum due to lower resistivity and better electron migration properties.

Process Simulation Code

```
# metal 1 define #
deposit aluminum thick=0.2
include file = interconnect.gds45.msk
struct outf = 05_metal1.str
```

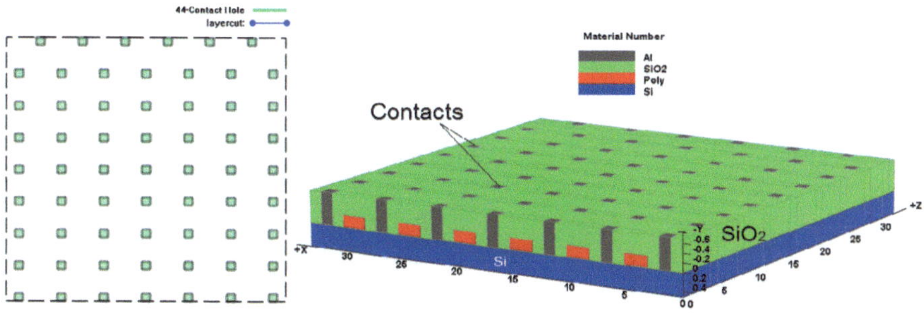

- Layer name: Contact Hole
- Mask thickness: 1.0
- Layer polarity: negative
- Layer purpose: etch
- Etch material: oxide
- Etch depth: 0.8 um
- Etch angle: 0

Fig. 8.4 Mask layout for contacts and simulation result of contacts step (04_contact.str)

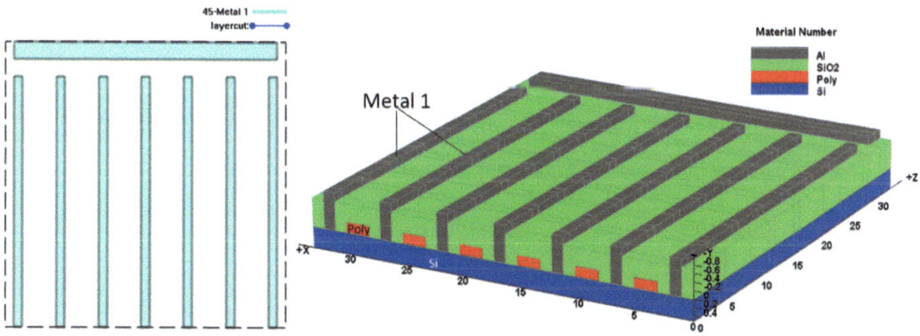

- Layer name: Metal 1
- Mask thickness: 1.0 um
- Layer polarity: positive
- Layer purpose: etch
- Etch material: aluminum
- Etch depth: 0.5 um
- Etch angle: 0

Fig. 8.5 Mask layout of metal1 and simulation result of the metal1 step (05_metal1.str)

8.1.7 Via 1 Placement

The next step is to place vias on top of the metal 1 layer. The vias are placed in such a way that the source and drain vias are interdigitated: this allows the metal 2 layer that will be deposited later to be used for both the source and drain.

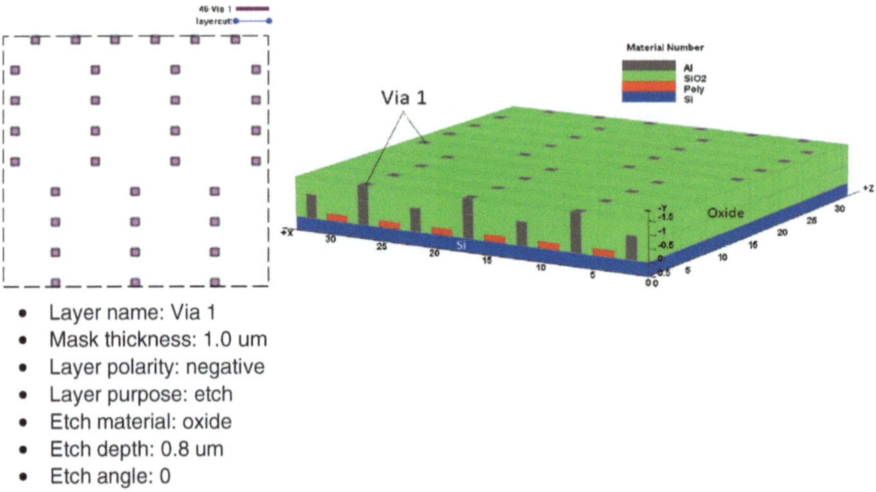

Fig. 8.6 Via 1 mask layout and simulation result of via1 step (06_via1.str)

Process Simulation Code

```
# via1 mask #
deposit oxide thick=0.8 meshlayer=3
etch start x=0 y=-5.5
etch continue x=0 y=-1.5
etch continue x=33 y=-1.5
etch done x=33 y=-5.5
includefile = interconnect.gds46.msk
struct outf = 06_via1_mask.str

# via1 placement #
deposit aluminum thick=0.6 meshlayer=3
etch alum start x=0 y=-5.5
etch alum continue x=0 y=-1.5
etch alum continue x=33 y=-1.5
etch alum done x=33 y=-5.5
struct outf=06_via1.str
```

As before, `deposit` is used to create an ILD layer. A CMP process is applied to flatten the structure afterwards.

Holes for the vias are etched into the oxide layer using the mask layout of Fig. 8.6. These holes are refilled aluminum and another CMP process is used to remove excess material and flatten the surface.

8.1.8 Metal 2 Formation

Finally, the metal 2 layer is deposited on top. Metal 2 layer is designed in metal blocks to reduce the metal resistance. A real process would likely have many more steps than in this simulation and would likely involve multiple metal layers.

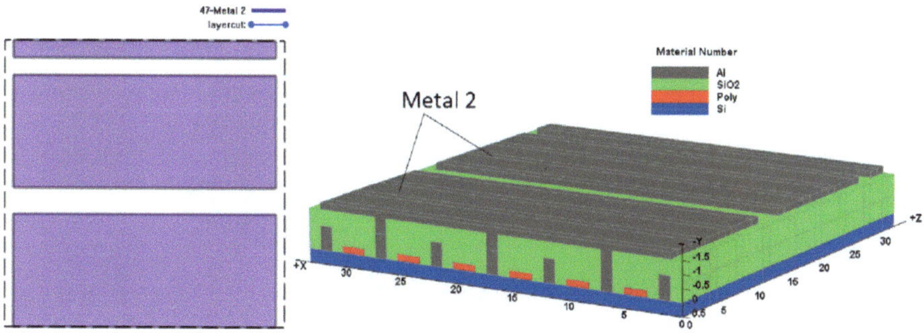

- Layer name: Metal 2
- Mask thickness: 1.0 um
- Layer polarity: positive
- Layer purpose: etch
- Etch material: aluminum
- Etch depth: 0.25 um
- Etch angle: 0

Fig. 8.7 Mask layout of metal layer 2 and the final simulation result (07_metal2.str)

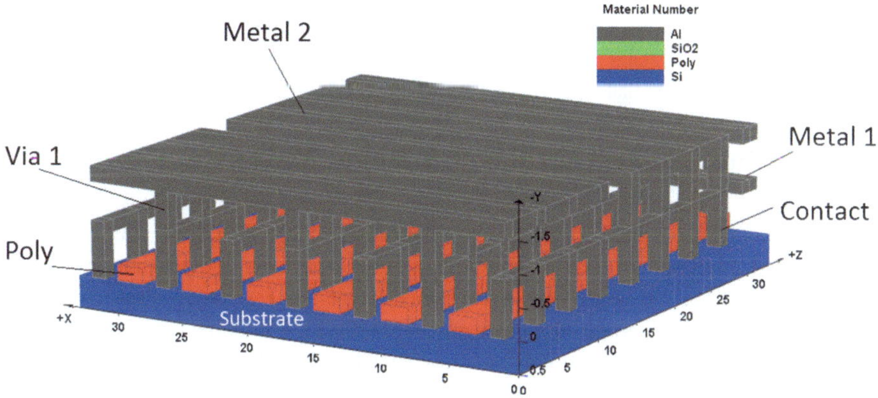

Fig. 8.8 Interconnect simulation result: metal layer2 (08_final.str) without ILD

Process Simulation Code

```
# metal 2 define #
deposit aluminum thick=0.25 meshlayer=3
include file = interconnect.gds6.msk
struct outf = 07_metal2.str
export outfile=int.aps xpsize=0.001
```

The process steps used here are similar to those of the metal 1 layer and will be omitted. The final result and the mask layout used for etching are shown in Fig. 8.7.

In order to have a better look at the 3D interconnect, the Inter dielectric layer (ILD) is intentionally removed from the view. The internal structure with the ILD removed is shown in Fig. 8.8.

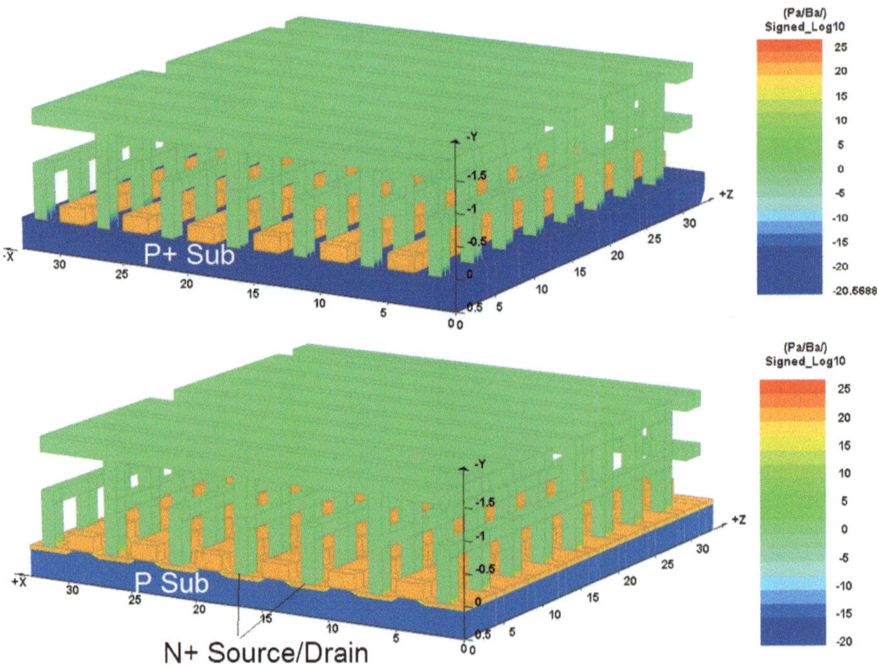

Fig. 8.9 Interconnect simulation result (net-doping): metal layer2 (08_final.str)

8.2 3D Interconnect Structure Without MOSFET

Up till now, we have finished process simulation of a large MOSFET with relatively complicated interconnect on top. The next step is to perform contact definitions and device simulation. However, we have noticed that the resistance simulation will inevitably include the contributions from both MOSFET channel and interconnect. In order to separate these two contributors, we need build a slightly modified structure. This structure has a highly doped substrate (boron 1E + 20 cm^{-3}) to rule out the influence of the substrate. The source/drain implantation step is eliminated so that no MOSFET structure exists. The substrate is basically used a highly conductive layer.

The final simulation results of both structures are shown in Fig. 8.9, with net doping chart shown. Please note the differences in substrate doping concentration and n+ source/drain implant.

8.3 Contact Definitions for Device Simulation

As always, we need to define the contacts before moving on to the device simulation. It is important not mix the term "contact" in the context of process and device simulation. In the former, it refers to metal electrodes that are deposited. In the latter case, it refers to equipotential boundary regions; sometimes, the metal is even

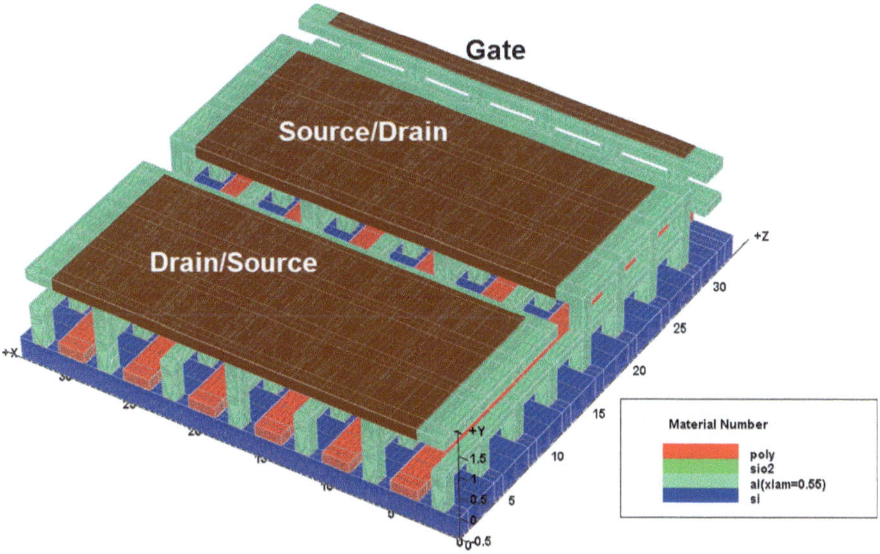

Fig. 8.10 Contact definitions for device simulation

ignored and the boundary is applied to the doped semiconductor itself. This will not be the case here since we are modeling the metal interconnect itself but it is important to apply the boundary region to the top of the metal.

Figure 8.10 shows the contact definitions for device simulation. Due to the symmetrical nature of MOSFET, the source and drain terminals are interchangeable.

8.4 Device Simulation: On-State Resistance

We are most interested to see how interconnect influences the total on-state resistance (R_{on}). Note that this device has a relatively large size (33 × 33 um) but in practice, many MOSFETS are much larger and the interconnect contribution to total resistance can be greater. Also, the FEOL (Front End of the Line) device can be an LDMOS or other devices.

Device Simulation Code

```
newton_par damping_step=12. print_flag=3 res_tol=1e-2 var_tol=1e-2

equilibrium

newton_par damping_step=12. print_flag=3 res_tol=1e-2 var_tol=1e-2

scan var=voltage_3 value_to=3 init_step=0.005 max_step=0.2 min_step=1e-6

scan var=voltage_1 value_to=0.1 init_step=0.005 max_step=0.02 min_step=1e-6

stop
```

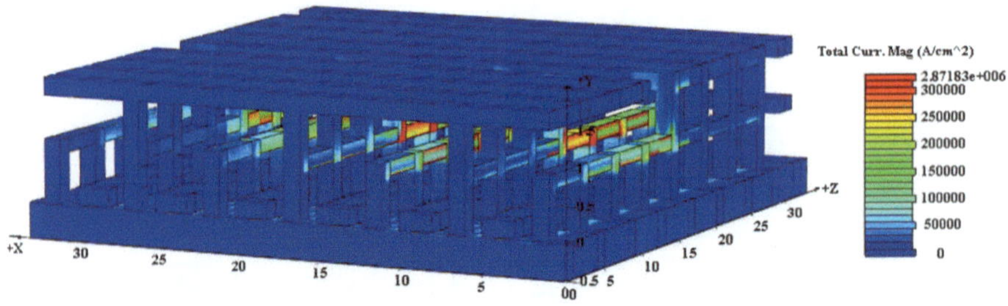

Fig. 8.11 Current magnitude within a large MOSFET device

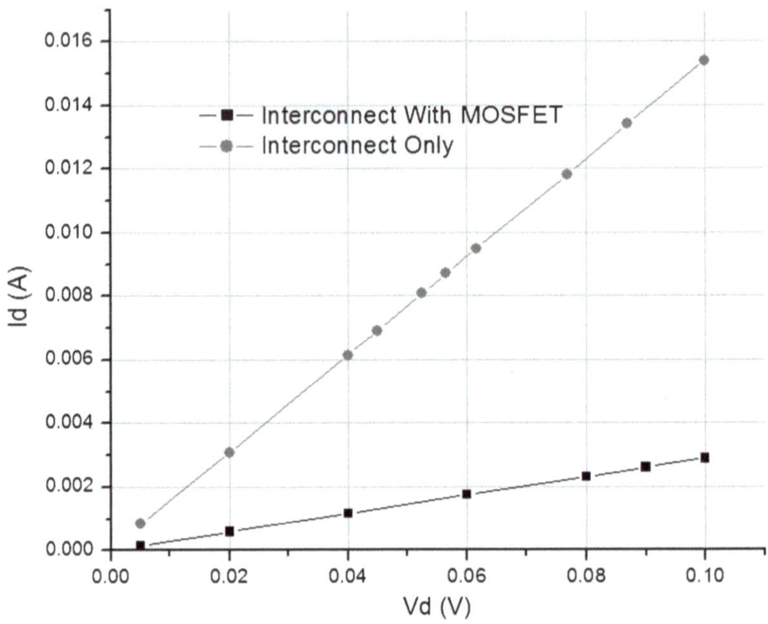

Fig. 8.12 Comparison of MOSFET R_{on} and interconnect resistance only

Here, contact 1 is the drain, contact 2 is the source and contact 3 is the gate. For this example, voltage_3 means the gate voltage and voltage_1 is the drain voltage. The on-state resistance simulation of the large size MOSFET starts with ramping up the gate voltage to 3 V to make sure the device is fully turned on. The drain is subsequently biased to 0.1 V to test the on-state resistance. Figure 8.11 shows the current magnitude within the device. As might be expected, the center of the interconnect has the highest current density.

A simple calculation points out that for this particular simulation structure, interconnect contributes approximately 15.8% of the total MOSFET R_{on} (with MOSFET, the resistance is 34.6 Ω, without MOSFET, the resistance is 6.5 Ω). Please note that many real fabricated devices have a much larger size than this example so the interconnect contribution can be considerably higher (Fig. 8.12).

Table 8.2 Simulation data for the 3D interconnect

	Process simulation	Device simulation	Total mesh count	Total number of planes
3D Interconnect	1 h	20 min (no MOS) 25 min (MOS)	67,566	38

Computer configuration: HP desktop with Intel i7-860/6G/1T/NVIDIA GeForce GTX260/Win7

8.5 Simulation Data

Table 8.2 gives simulation data for 3D interconnect simulation

Chapter 9
CMOS Image Sensor

Image sensors are widely used in today's consumer electronics and have been integrated with various hand held devices like cell phones. Growing interest has been placed on CMOS image sensor as technology scales and imager sizes approach that of CCDs [79]. This chapter will give a simplified mock 3D example of an active pixel CMOS image sensor. A transient simulation is explicitly used here.

9.1 Basics and Principle of Operation

In this chapter, a front-illuminated structure is used, this means the metal wiring is on top of photodiode and light goes through the metal layer [80]. Figure 9.1 illustrates this structure. Light is shone upon the lens and filtered by the color filters. It then goes through the metal wires and finally reaches the photodiode. The metal wiring may reduce the light collected by the photodiode. New method to prevent this kind of blocking has been developed using backside illumination [80].

While a real manufacturing CMOS image sensor is complicated and the detailed operation for that is out of scope for this book, a simplified structure based on an original research paper from Caltech [81] is presented here.

9.1.1 Operation Principle: Step 1

The basic structure of CMOS image sensor is shown in Fig. 9.2 [82]. Two poly gates are used: the first is called the transfer gate while the other is called the reset gate. The main light sensitive area is a p n junction photodiode.

S. Li and Y. Fu, *3D TCAD Simulation for Semiconductor Processes, Devices and Optoelectronics*, DOI 10.1007/978-1-4614-0481-1_9,

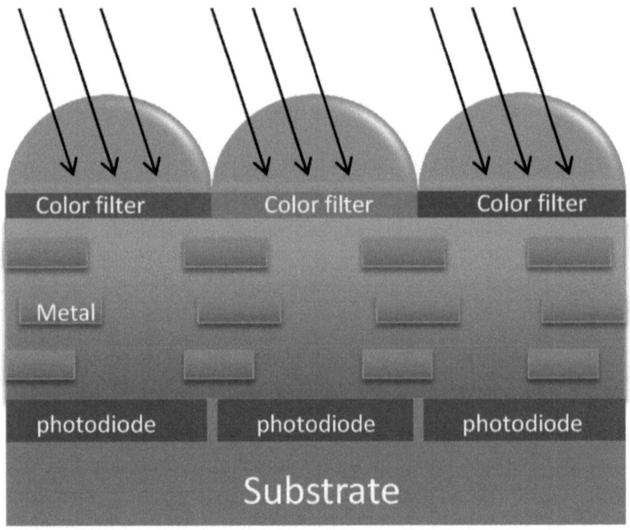

Fig. 9.1 CMOS image sensor with front-illuminated structure

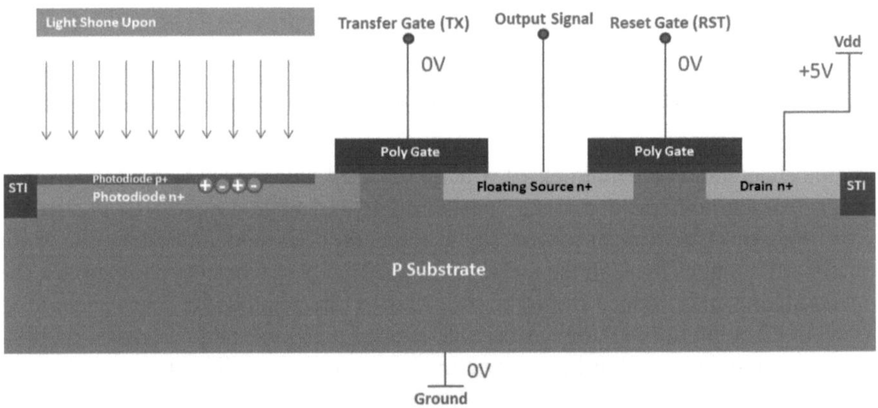

Fig. 9.2 2D structural view of CMOS image sensor operation: step 1

First, the reset FET drain (V_{dd}) is biased at 5 V while all other terminals (RST Gate, TX Gate and Substrate) remain 0 V. Light illuminates the surface of the photodiode and electron/hole pairs are generated. The potential at the output terminal is recorded under both dark and light conditions. The applied voltage on each contact and the potential at the output terminal as a function of time are shown in Fig. 9.3. Note that substrate contact terminal is always grounded.

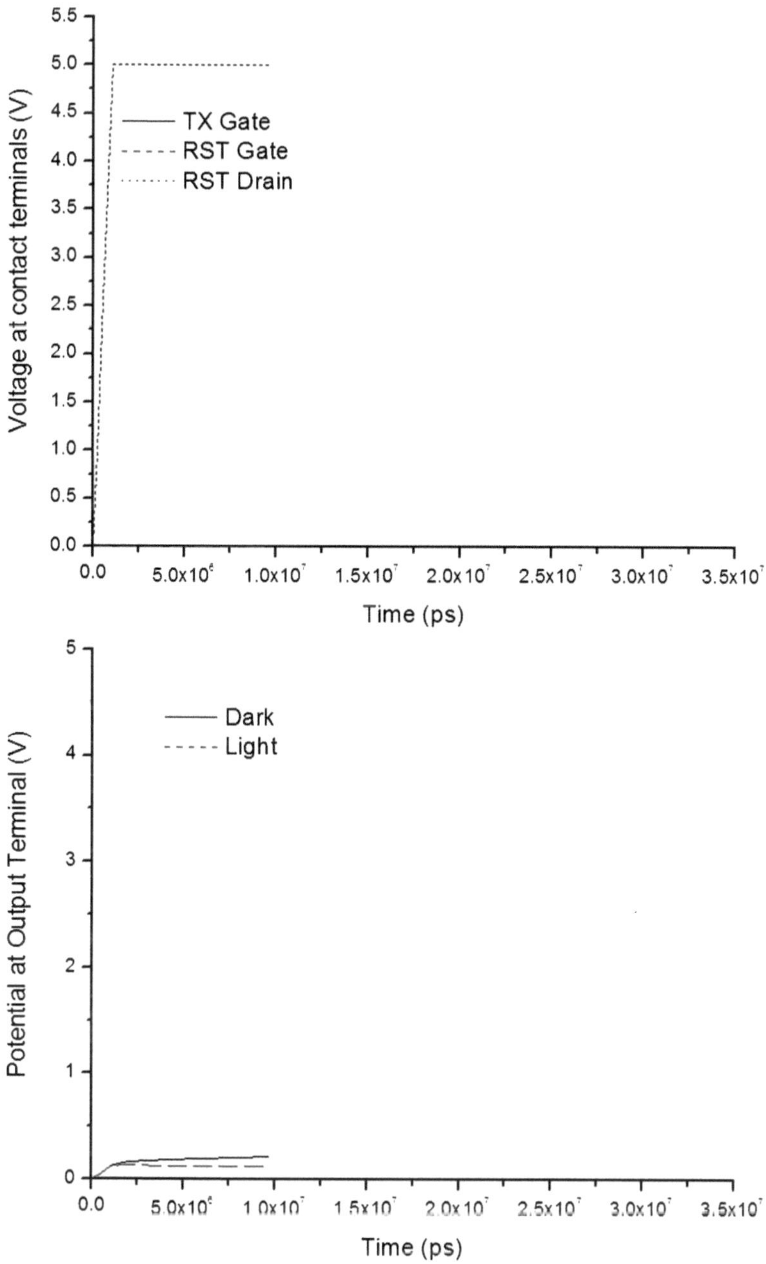

Fig. 9.3 Applied voltage on each contact (*left*) and potential at the output terminal: step 1

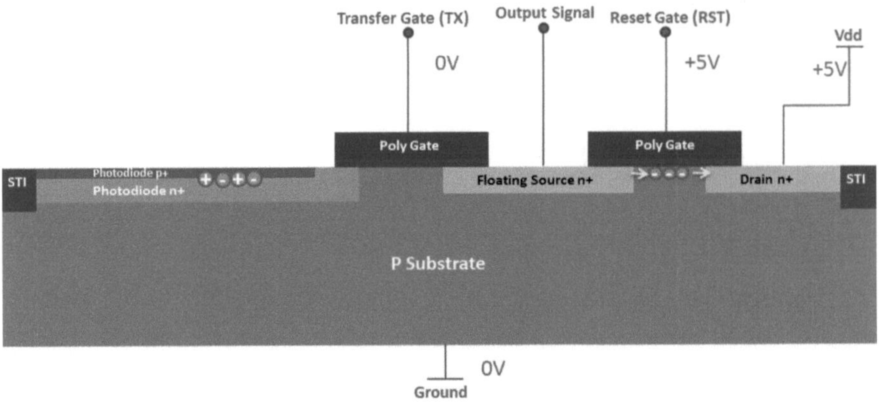

Fig. 9.4 Structural view of CMOS image sensor operation: step 2

9.1.2 Operation Principle: Step 2

The gate terminal of reset FET (Reset Gate) is biased to 5 V. This will allow excessive electrons being transported to the drain of reset FET to reset the potential of output terminal. The potential of output terminal will be pulled up by the drain of the reset FET.

In Fig. 9.4, electrons flow from the floating source to the drain. Figure 9.5 gives the applied voltage on each contact and the potential at the output terminal. Again, CMOS image sensor device simulation is a transient simulation, so time is used on the horizontal axis.

9.1.3 Operation Principle: Step 3

The Rest Gate (RST) is subsequently turned off for a little while (Fig. 9.6). From Fig. 9.7 we can see for both dark and light condition, potential drops at the output terminal.

9.1.4 Operation Principle: Step 4

Finally, the transfer gate is turned on. Excess carriers generated by the light are dumped to the floating source region and pull down the potential at the output terminal (Fig. 9.8). For the dark case, fewer carriers are available to be transferred to the floating source so a higher potential is observed at the output terminal, as shown in Fig. 9.9.

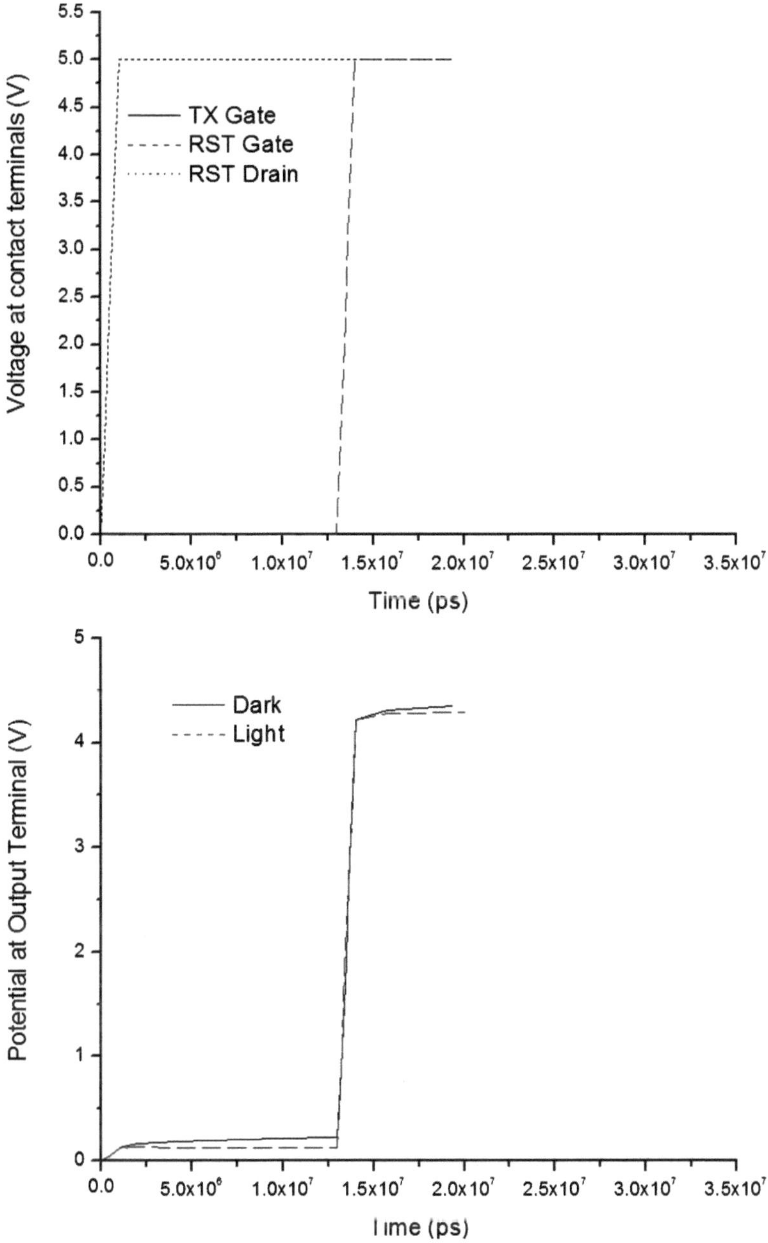

Fig. 9.5 Applied voltage on each contact (*left*) and potential at the output terminal: step 2

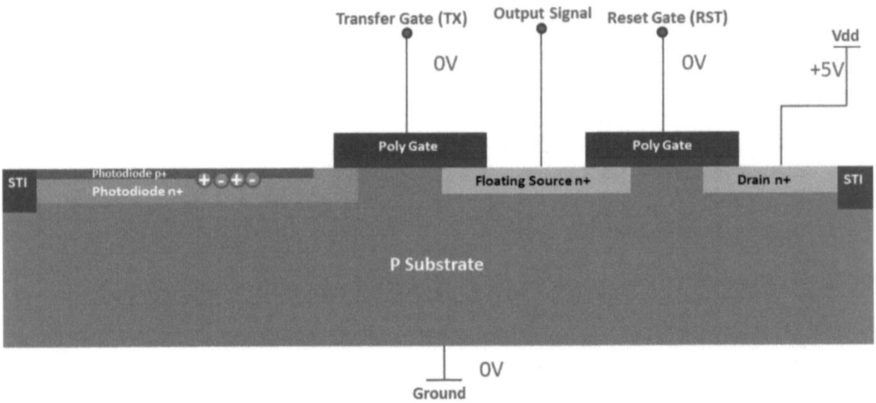

Fig. 9.6 Structural view of CMOS image sensor operation: step 3

9.2 Process Simulation of CMOS Image Sensor

A fake process is used to mimic the CMOS image sensor. The basic process steps are shown in Table 9.1. Lens and color filter step will be eliminated in this process simulation for the sake of simplicity. Figure 9.10 shows the layout and cut lines. Each mask will be shown one by one later.

9.2.1 Overview of Simulation Process Steps

We will use the process simulator to create the structure of the CMOS image sensor. This is followed by contact definitions and device modeling. A transient simulation will be used for this device. An overview of the simulation steps is presented in Table 9.1.

9.2.2 Substrate and STI Formation

The first step is to create a STI "fortress" to isolate the photodiode and reset FET regions.

Process Simulation Code

```
mater_define material_label=tungsten macro_name=tungsten
mater_define material_label=TEOS macro_name=TEOS

mode quasi3d
3d_mesh inf=geo

init boron conc=5.0e16
struct outf=01_sub.str

# STI formation
include file=cis.gds1.msk
struct outf=02_sti.str
```

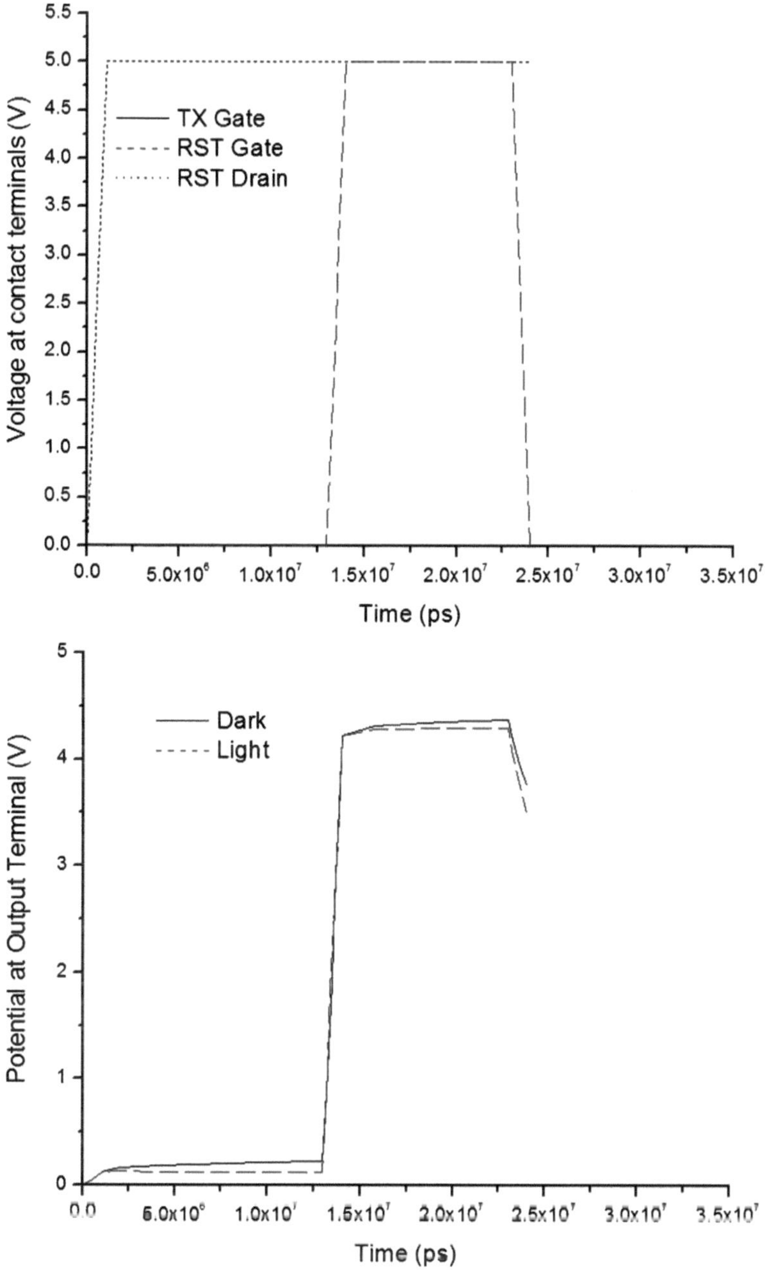

Fig. 9.7 Potential at the output terminal: step 3

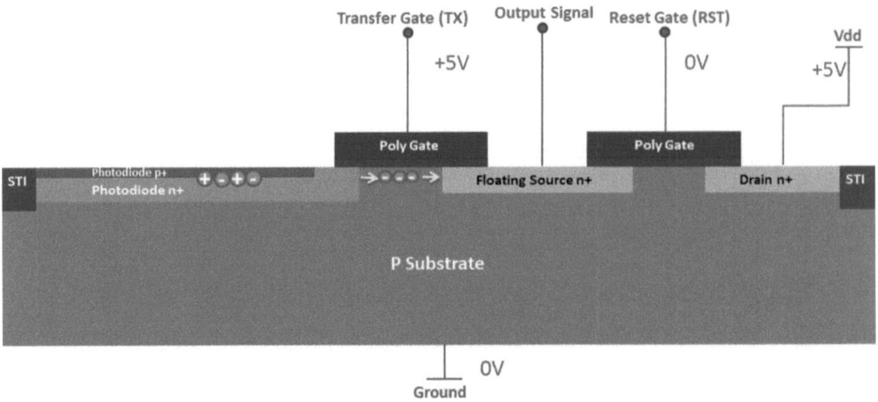

Fig. 9.8 Structural view of CMOS image sensor operation: step 4

The `mater_define` statement defines new user materials for the process simulation: tungsten for contacts in the back end process and TEOS for the ILD layer. It also creates a link with the appropriate macro defining the material parameters for the device simulation.

The `3d_mesh` command loads various statements generated by the MaskEditor GUI. Since this device is complicated, segmented mesh define is used and only the geo file for reset FET is shown in Text Box 9.1. The `init` statement sets up the initial substrate for the process simulation. The substrate is p-doped with a boron concentration of 5E + 16 cm^{-3}.

The `include` command loads the mask file for the STI. This file is created by the MaskEditor GUI and uses the "change material" purpose to replace the silicon material with oxide in a single process step. This is a simplification of the full process which would involve etching, re-growth and CMP: refer to earlier chapters for details. The mask layout from Fig. 9.11 has a negative polarity so the drawn area is etched away and replaced with the new material.

9.2.3 N+ Implant for Photodiode

The second step is n+ implant for photodiode.

Process Simulation Code

```
# photo diode n implant
deposit oxide thick=0.005
include file=cis.gds2.msk
struct outf=03_PDn_mask.str
implant phos energy=120 dose=5e13 angle=0 rot=0
implant phos energy=70 dose=5e13 angle=0 rot=0
etch photoresist all
etch start x=-1 y=-2
etch conti x=-1 y=0.
etch conti x=4. y=0.
etch done  x=4. y=-2
diffuse time=30 temp=1000
struct outf=03_PDn.str
```

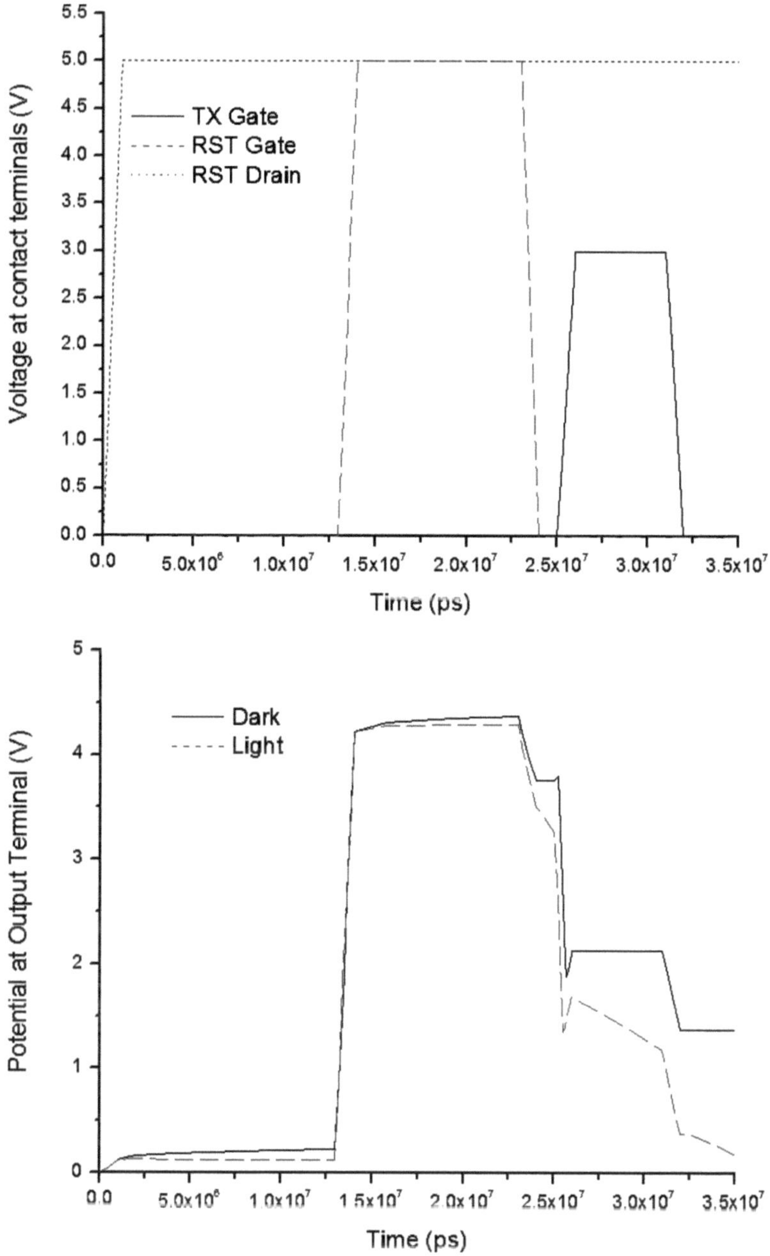

Fig. 9.9 Potential at the output terminal: step 4

Table 9.1 Overview of simulation steps of CMOS image sensor

CMOS image sensor	Process simulation steps
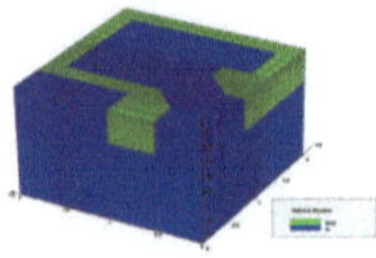	Step 1: Substrate and STI formation
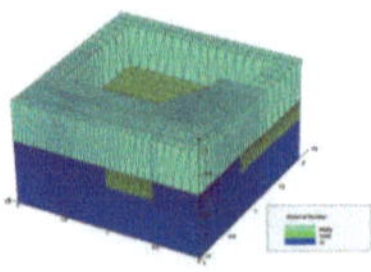	Step 2: N+ implant for photodiode
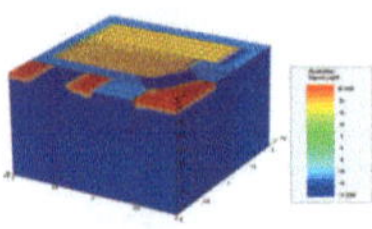	Step 3: Source/Drain n+ implant
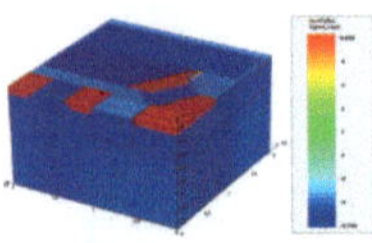	Step 4: Photodiode p-type implant
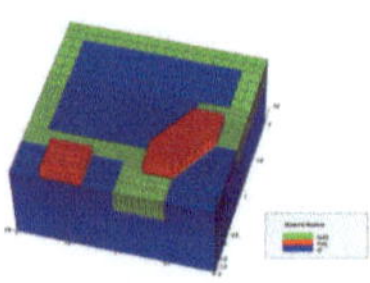	Step 5: Gate poly deposition
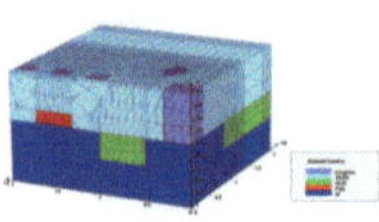	Step 6: ILD and contacts
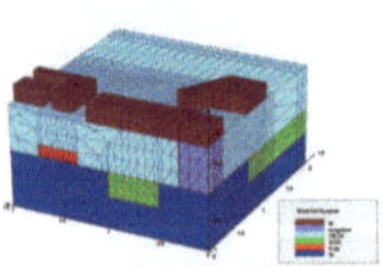	Step 7: Metal1 layer

(continued)

Table 9.1 (continued)

CMOS image sensor	Contact definitions for device simulation
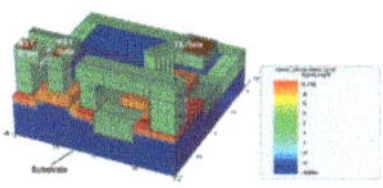	Step 5: Contact definitions for device simulation
CMOS image sensor	**Device simulation**
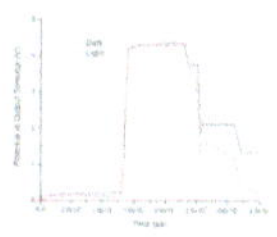	Step 6: Transient simulation steps

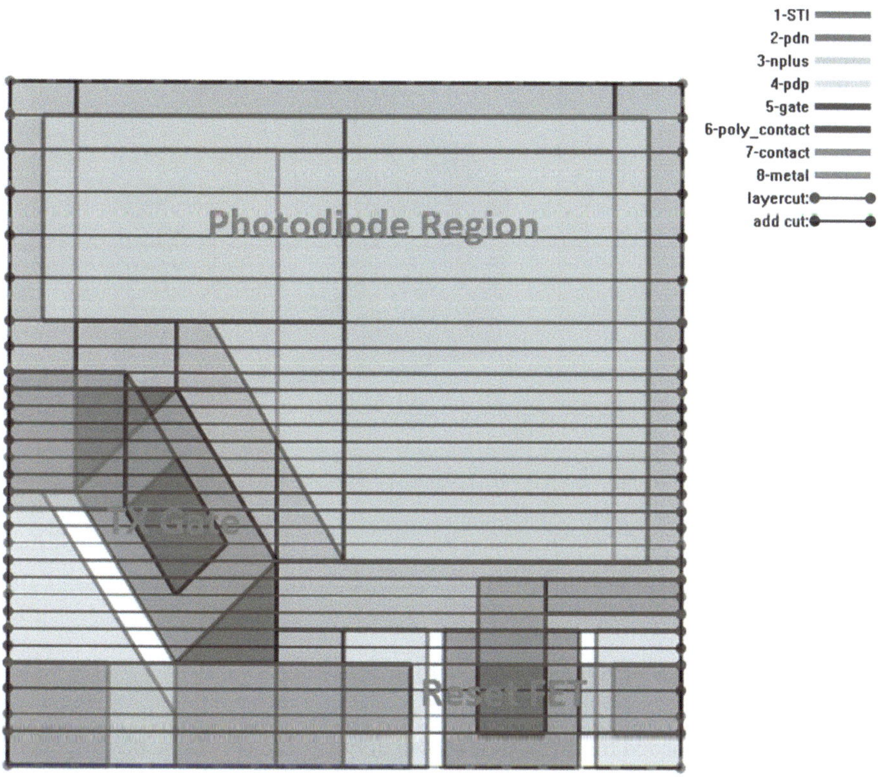

Fig. 9.10 Layout and MaskEditor cut lines for CMOS image sensor simulation

Text Box 9.1 Geo File for Reset FET of CMOS Image Sensor Simulation

```
line x loc=    0.00000 spacing=   0.100000 tag=lft
line x loc=    1.00000 spacing=   0.050000
line x loc=    1.50000 spacing=   0.025000
line x loc=    2.00000 spacing=   0.050000 tag=rht

line y loc=    0.00000 spacing=   0.214286e-01 tag=top
line y loc=    0.50000 spacing=   0.050000
line y loc=    1.50000 spacing=   0.214286 tag=bot

region silicon xlo=lft xhi=rht ylo=top yhi=bot
bound exposed xlo=lft xhi=rht ylo=top yhi=top
bound backside xlo=lft xhi=rht ylo=bot yhi=bot
```

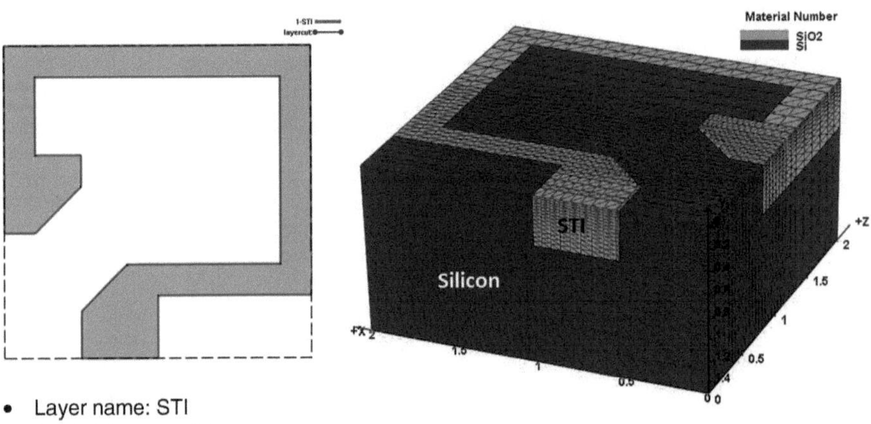

- Layer name: STI
- Mask thickness: 1.0 um
- Layer polarity: negative
- Layer purpose: change material
- Material to be changed: silicon
- Material change to: oxide
- Etch depth: 0.45 um
- Etch angle: 0

Fig. 9.11 The mask file for STI formation (02_sti.str)

The mask file loaded for this step is shown in Fig. 9.12. This mask has a "general purpose" setting which adds photoresist to protect unaffected device areas from implantation. The mask polarity is negative so the drawn area is where the photoresist is removed and the implantation takes place.

As in previous chapters, an implant chain is used to create a deep and more uniform doping profile. This is followed by a 30 min annealing step; note that leftover photoresist must be removed before this step.

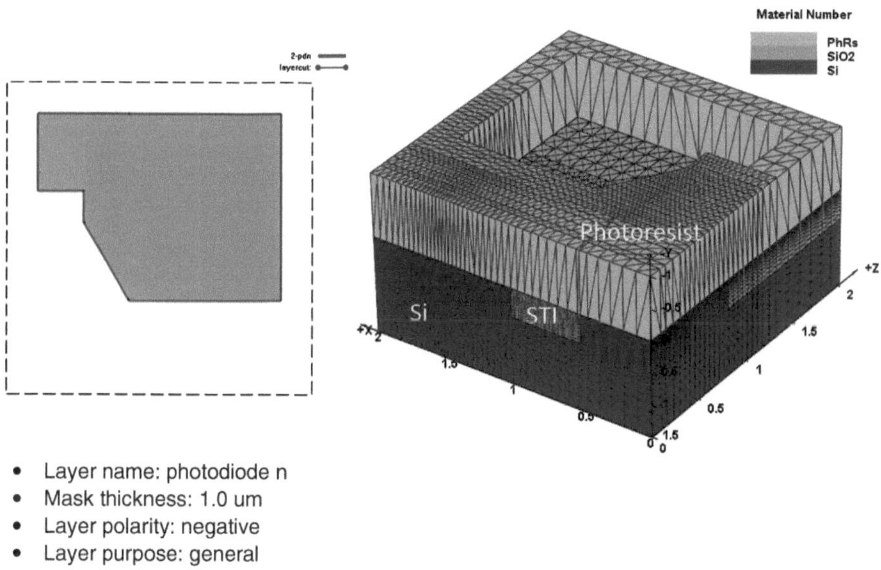

Fig. 9.12 The mask layout of n+ implant for photodiode and the photoresist of photodiode n-type implant (03_PDn.str)

9.2.4 Source/Drain N+ Implant

An n+ implant is used to create the floating source of the transfer FET and source/drain of the reset FET.

Process Simulation Code

```
# n+ source/drain implant
deposit oxide thick=0.009
include file=cis.gds3.msk
struct outf=04_Drain_mask.str
implant arsenic energy=10 dose=1e15 angle=0 rot=0
etch photoresist all
etch start x=-1 y=-2
etch conti x=-1 y=0.
etch conti x=4. y=0.
etch done  x=4. y=-2
diffuse time=1 temp=1000
struct outf=04_Drain.str
```

A `deposit` statement creates a thin layer (90 Å) of screening oxide for the implantation: this will be removed afterwards by using an `etch` command: note that the first call removes the leftover photoresist while the second call (`start` ... `done`) removes all material in the affected area and is what actually removes the oxide.

The implant mask is shown in Fig. 9.13. We note the low ion energy and `diffuse` time used here since the n+ regions is shallow.

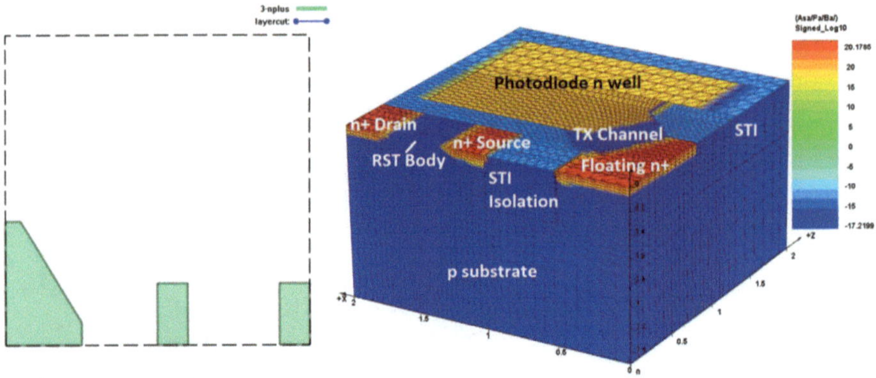

Fig. 9.13 Mask layout for source/drain implant (04_Drain.str)

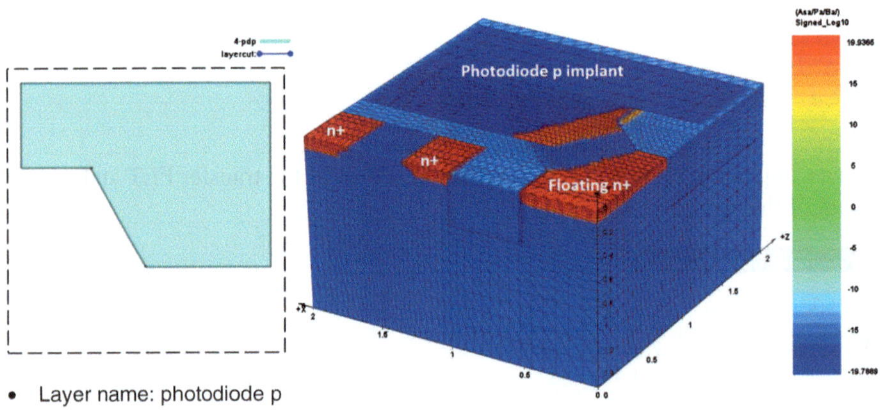

Fig. 9.14 Mask layout for photodiode p implant

9.2.5 *Photodiode P-Type Implant*

A second implant is needed to define the p region of the photodiode. Most of the process steps have been explained already and will be skipped. The relevant mask layout and simulation results are shown in Fig. 9.14.

Process Simulation Code

```
# photo diode p implant
deposit oxide thick=0.009
include file=cis.gds4.msk
struct outf=05_PDp_mask.str
implant boron energy=4 dose=1.0e14 angle=0 rot=0
etch photoresist all
etch start x=-1 y=-2
etch conti x=-1 y=0.
etch conti x=4. y=0.
etch done  x=4. y=-2
diffuse time=1 temp=1000
struct outf=05_PDp.str
```

9.2.6 Gate Poly Deposition

This step creates the gates for the transfer and reset FETs.

Process Simulation Code

```
# gate poly
deposit oxide thick=0.02 meshlayer=2
deposit poly thick=0.2 meshlayer=2 phos conc=1.0e19
include file=cis.gds5.msk
diff time=1 temp=1000
struct outf=06_poly.str
```

A pair of `deposit` statements is used for the gate oxide and polysilicon layers. The oxide is deposited rather than thermally grown to simplify the process simulation. The polysilicon is in-situ doped with phosphorous and the `meshlayer` parameter adds extra mesh lines to the newly deposited layers. After deposition, the mask file from Fig. 9.15 is used to etch the poly and oxide. A positive mask polarity is used so the drawn area represents the left-over area after etching.

A final diffusion step is also used to activate the dopant within the polysilicon. It is important that this be a quick step so that the previous n+ implantations remain shallow.

9.2.7 ILD and Contacts

The front end process steps arc now complete. The remaining steps deal with the back end: we start the ILD deposition and the creation of the contact holes.

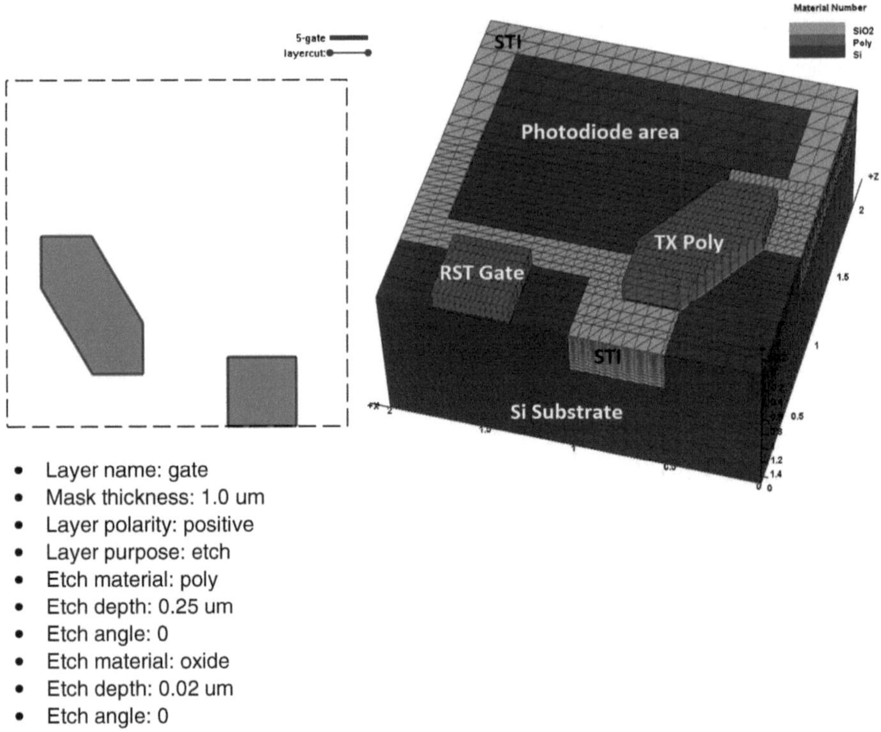

Fig. 9.15 The mask layout for poly gate and after poly formation (06_poly.str)

Process Simulation Code

```
# ILD
deposit TEOS thick=0.8 meshlayer=3
etch start x=-1 y=-2
etch conti x=-1 y=-0.8
etch conti x=4. y=-0.8
etch done  x=4. y=-2
struct outf=08_after_passivation.str

# contact
include file=cis.gds6.msk
struct outf=09_poly_contact.str
include file=cis.gds7.msk
struct outf=10_contact.str
```

The initial step involves the deposition of TEOS as an ILD layer followed by a CMP process to etch away excess material and leave a flat surface. Afterwards, two mask files (Fig. 9.16) are used to etch and refill contact holes for the poly and source/drain contacts. The "change material" simplification is used to replace the TEOS with tungsten in this step rather than doing the full etch/deposit/polish process.

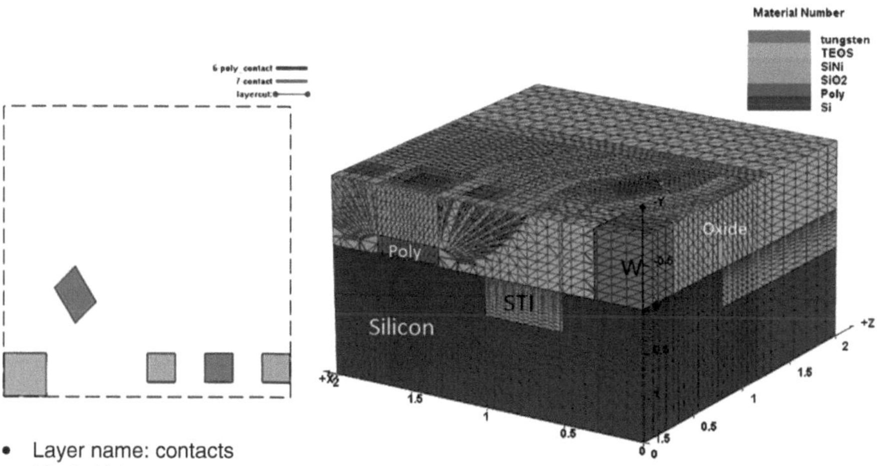

- Layer name: contacts
- Mask thickness: 1.0 um
- Layer polarity: negative
- Layer purpose: change material
- Material to be changed: TEOS
- Material change to: tungsten
- Etch depth: 0.8 um
- Etch angle: 0

Fig. 9.16 Mask layout for contacts

9.2.8 Metal 1 Layer

As usual, a metal layer will be placed on top to finalize the contacts. This is done using a blanket deposition of aluminum followed by etching using the mask file of Fig. 9.17. In Fig. 9.18, the ILD layer is hidden to help visualize the internal structure of the device: this is not a real process step and is used for visualization purposes only.

Process Simulation Code

```
# metal
deposit alum thick=0.3 meshlayer=2
include file=cis.gds8.msk
struct outf=11_metal.str
```

9.3 Contact Definitions for Device Simulation

Contacts need to be defined before moving forward to device simulation. Four terminals are defined, namely: TX gate (transfer gate, contact #1), RST gate (reset gate, contact #2), RST drain (contact #3) and substrate (contact #4). Figure 9.19 illustrates the contact locations with net doping chart. Note that the ILD is hidden to give the reader a better view.

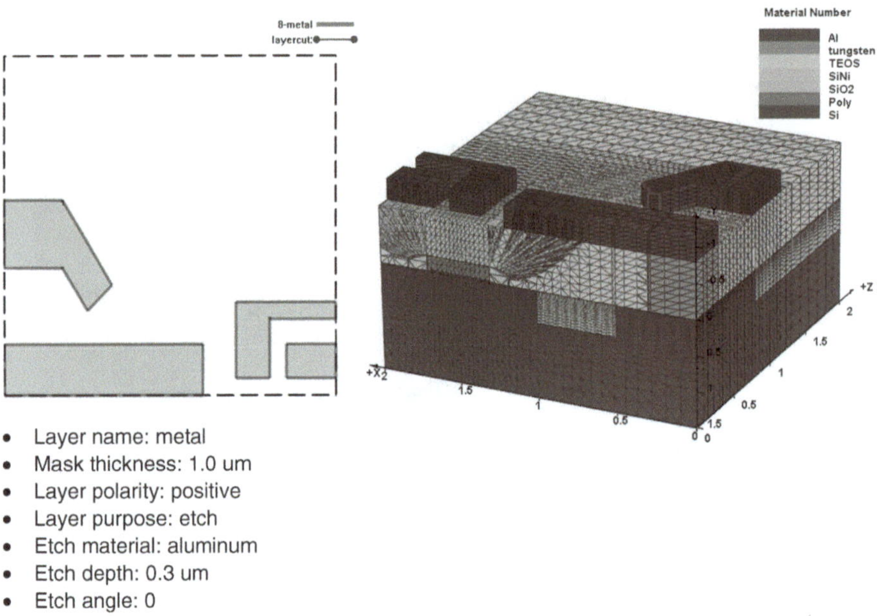

Fig. 9.17 Mask layout of metal layer and simulation result (11_metal.str)

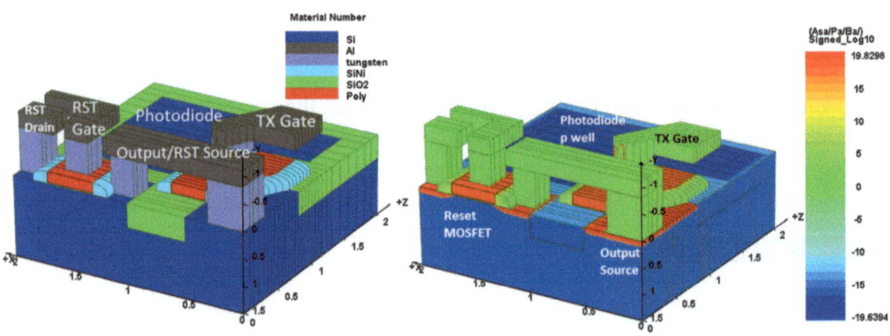

Fig. 9.18 The final view without ILD layer

9.4 Device Simulation

The CMOS image sensor is different than previous devices because the voltage input varies in time: this means the entire simulation must be done using transient analysis. It can be treated like a combination of devices or a simple circuit. As usual, only the final section of device simulation code is included.

9.4.1 Transient Simulation Steps

Two scenarios will be used here, one is the dark case (no light) and the other is the device under illumination. Both of these will undergo the same bias steps:

Step 1 (scanline=1):
Equilibrium. All contacts/terminals are set to 0. This also implies the light input is exactly 0.

Step 2 (scanline=2):
The voltage on the drain terminal of reset FET is biased to 5 V. Allow 1 μs to complete the process.

Step 3 (scanline=3):
The light is turned on, just like a voltage would be. For the light case, the illumination is set to 5 kW/m^2. For the dark case, a small non-zero value is used. The light value is ramped up over the course of 1 μs which brings the total time to 2 μs.

Step 4 (scanline=4):
Allow some time to collect the light-generated signal charge, total time is now 13 μs.

Step 5 (scanline=5):
Reset gate is turned on by ramping up the voltage on RST gate to 5 V. Allow 1 μs to complete this process. Total time is now 14 μs.

Step 6 (scanline=6):
Wait 9 μs for the charge to clean up. Total time is now 23 μs.

Step 7 (scanline=7):
Turn reset FET back to off by scanning the voltage on the RST Gate to 0 V. This process takes 1 μs and total time is now 24 μs.

Step 8 (scanline=8):
Transfer gate is now turned on by ramping up the voltage on the TX Gate to 5 V. Allow 1 μs to finish this process. Total time is now 26 μs.

Step 9 (scanline=9):
Wait to let device stabilize: total time is now 25 μs.

Step 10 (scanline=10):
Allow 5 μs for the transfer to complete. Total time is now 31 μs.

Step 11 (scanline=11):
Turn off transfer gate by lowering the voltage on the TX gate to 0 V. Allow 1 μs to finish the process. Total time is now 32 μs.

Step 12 (scanline=12):
Allow 3 μs of reading time which brings the total to 35 μs. At the output terminal, dark and light scenarios will show different potential curves due to the difference in optically-generated electron-hole pairs.

Device Simulation Code

```
light_power incident_power=5.e3 wavelength=0.55 profile=(0. 4.75 0.01 0.01)
start_loop symbol=%ii value_from=1 value_to=2
real_func symbol=%lit value_from=1.e-9 value_to=1

$ Solve for equilibrium condition
newton_par damping_step=5. var_tol=1.e-3 res_tol=1.e-3 &&
  max_iter=100 opt_iter=15 stop_iter=50 print_flag=3

$scanline=1
equilibrium

$  Gate_TX-1; Gate_Reset-2; Drain_Reset-3
newton_par damping_step=3. res_tol=1.e0 var_tol=1.e0 &&
  max_iter=30 opt_iter=15 stop_iter=15 print_flag=3 &&
  change_variable=yes

$ scanline=2
scan var=voltage_3 value_to=5.0 &&
var2=time value2_to=1.e-6

$ scanline=3
scan var=light value_to=%lit &&
var2=time value2_to=2.e-6

$ scanline=4
$ use 10 us to collect signal charge
scan var=time value_to=13.e-6

$ scanline=5
$ reset briefly to clear up charge at FD
scan var=voltage_2 value_to=5.0 &&
var2=time value2_to=14.e-6

$ scanline=6
$ takes 9 us to clean up
scan var=time value_to=23.e-6

$ scanline=7
$ turn reset MOSFET back to off
scan var=voltage_2 value_to=  0.0 &&
var2=time value2_to=24.e-6 min_step=0.5e-12

$ scanline=8
$ wait a little until other transistors are ready
scan var=time value_to=25.e-6

$ scanline=9
$ transfer the charge by increasing voltage
scan var=voltage_1 value_to=3.0 &&
var2=time value2_to=26.e-6

$ scanline=10
$ allow 5 us for the transfer to complete
scan var=time value_to=31.e-6

$ scan11
$ set TX gate voltage back to zero
scan var=voltage_1 value_to=0. &&
var2=time value2_to=32.e-6

$ scan12
$ allow reading time
scan var=time value_to=35.e-6

end_loop
```

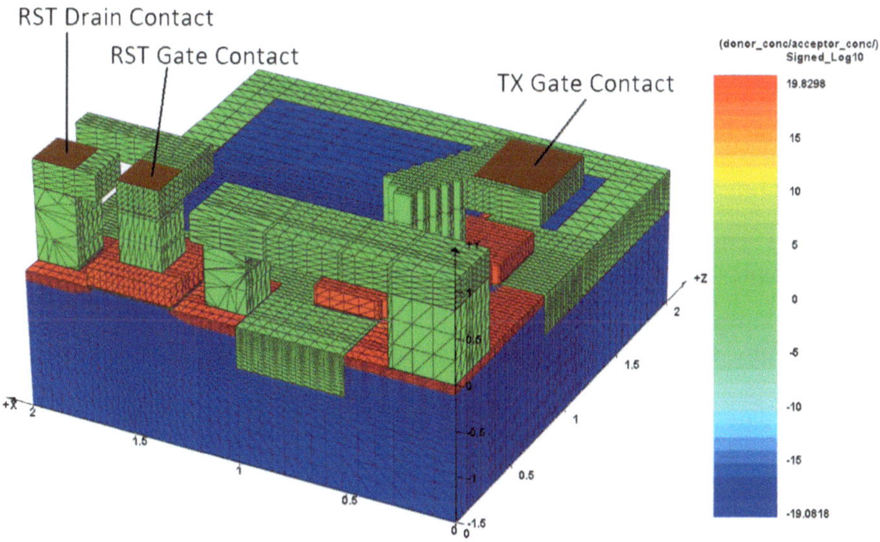

Fig. 9.19 Contacts defined for device simulation

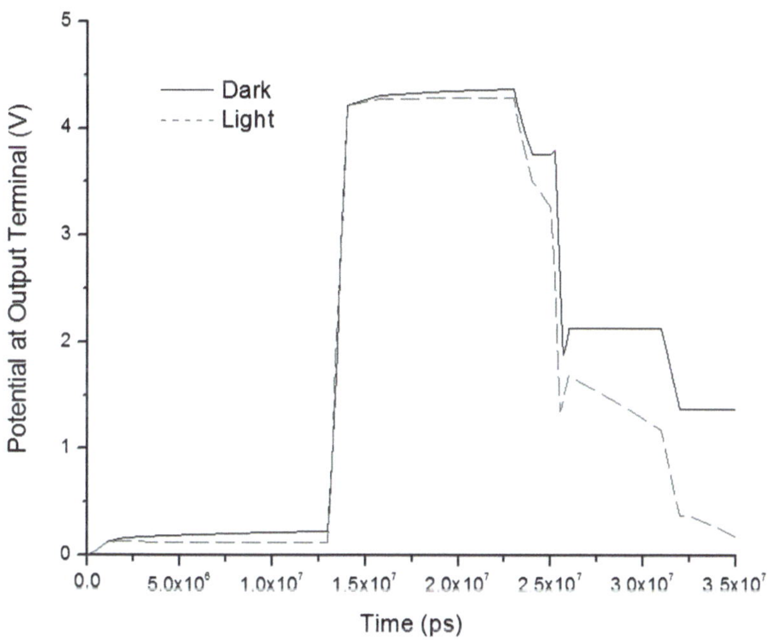

Fig. 9.20 Transient device simulation result of CMOS image sensor

`light_power incident_power=5.e3 wavelength= 0.55 profile=(0. 4.75 0.01 0.01)`. This statement sets the incident light power to be 5 kW/m2, with a wavelength of 0.55 um. `profile` is the cross section profile of the incident

Table 9.2 Simulation data for the CMOS image sensor

	Process simulation	Device simulation	Total mesh count	Total number of planes
CMOS image sensor	30 min	15 min	27,846	45
Computer configuration: HP desktop with Intel i7-860/6G/1T/NVIDIA GeForce GTX260/Win7				

light. The actual light value at any given time is this value multiplied by the light scan variable (which starts at zero at equilibrium).

```
start_loop symbol=%ii value_from=1 value_to=2
real_func symbol=%lit value_from=1.e-9 value_to=1
......
$ scanline=3
scan var=light value_to=%lit &&
var2=time value2_to=2.e-6
......
end_loop
```

The `start_loop...end_loop` instructs the simulator to repeat the simulation within the loop and model the dark and light scenarios in the same input file. The `real_func` statement defines a real-valued loop variable which can be used elsewhere in the input. Here, it defines the maximum value of the `light` scan variable: the dark case corresponds to a total light input of 1E-9 $\times$ 5 kW/m^2 which essentially means 5E-6 W/m^2.

Figure 9.20 shows the simulation result of the potential at the output terminal for both dark and light conditions.

9.5 Simulation Data

Table 9.2 gives simulation data for the CMOS image sensor simulation.

Chapter 10
Hybrid Silicon Laser

10.1 Introduction

As silicon transistors get smaller and faster, the bottleneck of the IC boils down to interconnects which traditionally use copper or aluminum wires with limited speed and bandwidth. Silicon-based optoelectronic integration offers the promise of low-cost solutions for optical communications and interconnects. A major advance in this field is the use of III–V lasers bonded on silicon and coupled to silicon waveguides. Such an approach is called hybrid integration and lasers built using such this approach are called hybrid silicon lasers.

The hybrid silicon laser is a key enabler for silicon photonics which will be integrated into silicon photonic chips that could enable the creation of optical data pipes carrying terabits of information. These terabit optical connections will be needed to meet the bandwidth and distance requirements of future servers and data centers powered by hundreds of processors.

We set up a laser diode structure in 3D similar to that reported in Ref. [83] with device schematic shown in Fig. 10.1. We assume a simple design such that one side of the laser cavity is coated with an anti-reflection (AR) layer and the output side uses a distributed Bragg reflector (DBR) to provide wavelength selection and feedback. A taper structure is used to couple the optical power of the cavity into the silicon waveguide as confined by air gaps on either side.

In this chapter, we describe the setup steps using MaskEditor. The actual process steps are much more complicated since the III–V and silicon parts are grown separately and bonded afterwards. We then describe the basic processing steps used to generate the device mesh using the process simulation program.

For the device simulation, we will point out the unique physical model that must be enabled to perform the optoelectronic simulation. We note that the device simulation of lasers is a complex subject in and of itself. This chapter is only a short introduction to the subject matter.

S. Li and Y. Fu, *3D TCAD Simulation for Semiconductor Processes, Devices and Optoelectronics*, DOI 10.1007/978-1-4614-0481-1_10,

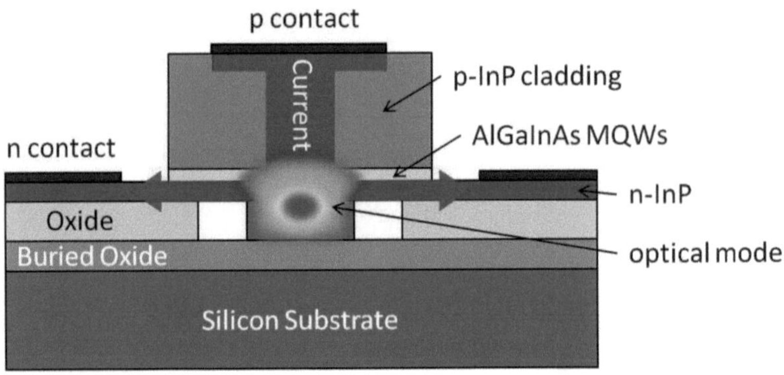

Fig. 10.1 Schematic for laser diode

We will briefly point out that laser simulation requires several advanced physical models such as beam propagation methods (BPM), photon rate equations, transfer matrices/coupled wave equations, thermal/self-heating modeling, etc. Very often, the laser modeling tool will be a separate component of a device/process modeling suite.

10.2 Device Setup Using Masks

We use a simple four layer mask structure to set up this LD. As displayed by MaskEditor in Fig. 10.2 (device top view), layer 1 is a narrow stripe in the center of the two Hybrid Silicon Laser structure to be used to pattern the silicon waveguide which forms the lower part of the LD. Layer 2 is used to pattern the silicon outside of the waveguide this is used to form the air bridge. The air gap would be between layer 1 and layer 2.

Layer 3 is used to form the main active waveguide of the laser diode. Please note the taper with the tip pointing to the silicon waveguide. The active MQW are contained within such an active waveguide. Layer 4 is used to pattern the Cu contact on both sides of the active waveguide.

Please note that since the device is symmetric, we only simulate the right half to save on the total mesh.

10.3 Processing Steps

The actual fabrication process steps may be complicated but the mesh set up simplified using a few deposition and etching steps. We shall only discuss the steps that are unique for this device and special for processing compound materials here.

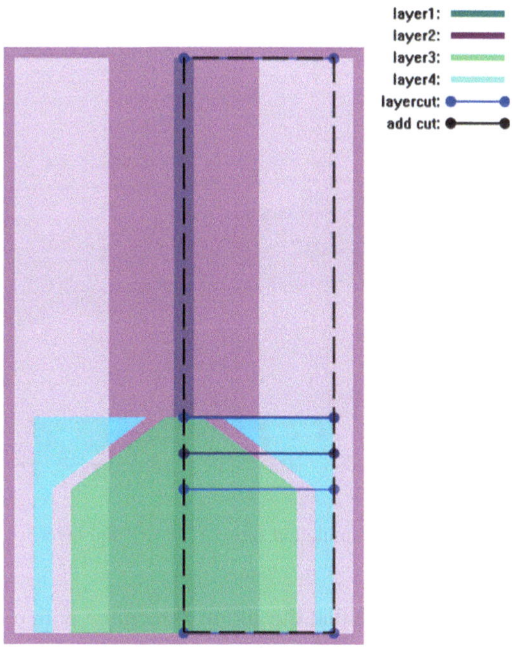

Fig. 10.2 The four layer mask layout for the LD

Please note the following lines in the process simulation input file:

Process Simulation Code

```
mater_define material_label=air macro_name=air
mater_define material_label=InP macro_name=inp
mater_define material_label=Cu macro_name=cu
mater_define material_label=InGaAsP_bar macro_name=ingaasp var_symbol1=y
var1=0.3
mater_define material_label=InGaAsP_qw macro_name=ingaasp
active_macro=InGaAsP/InP var_mode quasi3d
#mode three.dim
3d_mesh inf=geo
#restart file=06_mask4.str
```

Compared with earlier device processing files, this device consists of compound material which the process simulator may not recognize. The command `mater_define` allows the association of new material name with material macros in the device simulator so that material database can easily be located after process simulation. The material database of the device simulator contains hundreds of materials and more can be added by the end user.

The following input commands forms the silicon waveguide using layer1 and layer2 masks.

Process Simulation Code

```
init
struct outf=01_sub.str
1.3 Processing steps 3
deposit silicon thick=0.52 meshlayer=4
include file=taper_mask1.msk
struct outf=02_mask1.str
include file=taper_mask2.msk
struct outf=02_mask2.str
```

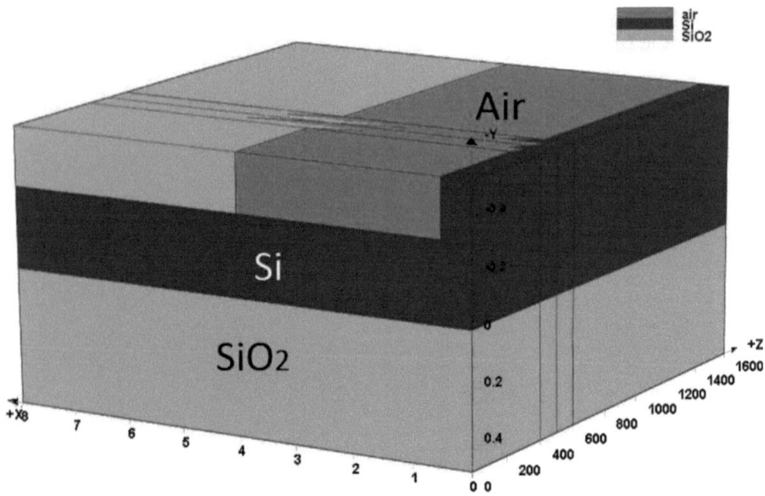

Fig. 10.3 The simulation result of layer1 and layer2

Please note that the action associated with mask 1 (layer 1) is to cover up part of the device, change the exposed area into air and finally remove the photoresist. For mask2, we cover up the middle part where the waveguide was already formed and change exposed outside area into oxide to make the air bridge. The results of these two steps are shown in Fig. 10.3.

Next, we deposit all the material layers to form the active MQW:

Process Simulation Code

```
deposit oxide thick=0.01
deposit InP thick=0.1 meshlayer=4 n_doping conc=2.e18
# this would define the whole InP base
deposit InGaAsP_bar thick=0.1 meshlayer=4 n_doping conc=5.e17
deposit InGaAsP_qw thick=0.008 meshlayer=3
deposit InGaAsP_bar thick=0.01 meshlayer=3
deposit InGaAsP_qw thick=0.008 meshlayer=3
deposit InGaAsP_bar thick=0.01 meshlayer=3
deposit InGaAsP_qw thick=0.008 meshlayer=3
deposit InGaAsP_bar thick=0.1 meshlayer=3
deposit InP thick=1. meshlayer=7 ratio=1.3 p_doping conc=2.e18
# add up the above to etch to InP, allow 0.2 um for lateral injection
# tot=1.2+0.044
include file=taper_mask3.msk
struct outf=03_mask3.str
etch dry thick=1.244
etch photores all
struct outf=04_wavegd.str
```

After we have all the MQW layers in place, we deposit and pattern the photoresist to cover up the middle part of the active waveguide using mask3 above. Then we use dry etch to form the active waveguide before removing the photoresist. The results at this step are shown in Fig. 10.4 which clearly indicates the shape of a tapered waveguide.

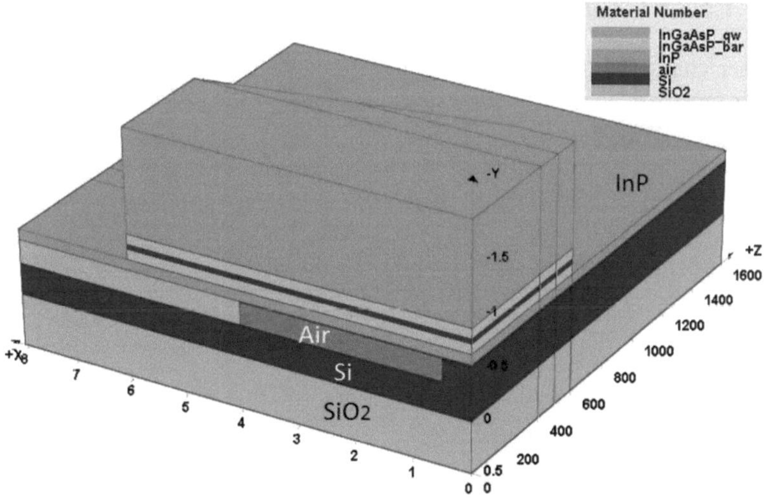

Fig. 10.4 The simulation result of layer3

Finally, we make the Cu side contacts using mask4.

Process Simulation Code

```
deposit Cu thick=0.3 meshlayer=4
include file=taper_mask4.msk
struct outf=05_mask4.str
etch InP all segm=5
etch InP all segm=6
struct outf=06_inp_etch.str
deposit air thick=1.444 meshlayer=5
struct outf=07_dep_air.str
etch air start x=-1 y=-1.5
etch air cont x=10 y=-1.5
etch air cont x=10 y=-5.5
etch air done x=-1 y=-5.5
struct outf=08_air_etch.str
activation.model n_doping fraction=1.0 force
activation.model p_doping fraction=1.0 force
struct outf=09_final.str
export outf=taper.aps xpsize=0.001
quit
```

Please note that air is a special insulating material which we use to fill up the side and some top part of the waveguide. Air is necessary to provide the material refractive index for optical calculation although it does not affect the electrical model. Figure 10.5 shows the final structure (minus the air) with 3D mesh allocation.

10.4 Laser Characteristics

The laser characteristics can be obtained from the following solution input file which is mostly generated by the export command in the process simulation:

Device Simulation Code

```
begin
$ modified from template.sol by adding suprem_contact and contact commands
$ then, add equlibrium and scan action commands.

include file=zmesh.zst &&
ignore1=load_mesh ignore2=output ignore3=export_3dgeo
load_mesh mesh_inf=taper.aps suprem_import=yes
...
...
...
$ Ajudst refractive index which affects the optical confinement
real_index value=3.9 mater=7
real_index value=3.45 mater=5
affinity value=4.03 mater=6
$ Adjust work function to make sure Cu has no
$ injection barrier, or tunneling would have to be used.
affinity value=4.03 mater=6
$ Wave boundary is used only when optical modes are solved
wave_boundary point_ll=[ 0.000000000000E+000 -0.5] &&
point_ur=[ 3. 1.5]
init_wave backg_loss=500. &&
boundary_type=(2 1 1 1 ) init_wavel=1.3 &&
wavel_range=(1.25, 1.35)
direct_eigen
newton_par max_iter=300
equilibrium
start_loop symbol=%zk value_from=1 value_to=6
rtgain_phase density=2.0e24 zseg_num=%zk
end_loop
$ Set Newton parameters for solution with bias
newton_par var_tol=1.e-3 res_tol=1.e-3 change_variable=yes &&
max_iter=40 opt_iter=20 damping_step=5
$ use rtgain checking here just to initialize data for the next scan
scan var=voltage_1 value_to=-1.2 max_step=0.1 &&
auto_finish=current_1 auto_until=1.e-4 auto_condition=above
scan var=current_1 value_to=100.e-3 max_step=0.5e-3 &&
auto_finish=rtgain auto_until=0.9 init_step=0.1e-3

scan var=current_1 value_to=100.e-3 max_step=2.e-3 &&
solve_rtg=yes init_step=0.1e-3
end
$ *************************************************************
begin_zsol
$
$ AR coated on the right side
$
longitudinal left_f_refl=0.9 right_f_refl=0.0 ref_wavel=1.30e-6
$ loop is not yet ready for this session
section kappa_real=0. &&
sec_num=1 mesh_points=8 zseg_num=1
section kappa_real=0. &&
sec_num=2 mesh_points=8 zseg_num=2
section kappa_real=0. &&
sec_num=3 mesh_points=8 zseg_num=3
section kappa_real=0. &&
sec_num=4 mesh_points=4 zseg_num=4
section kappa_real=2000. &&
sec_num=5 mesh_points=8 zseg_num=5
section kappa_real=2000. &&
sec_num=6 mesh_points=8 zseg_num=6
mode_srch omega_xrange=25. omega_yrange=5.0 adjust_range=yes
end_zsol
```

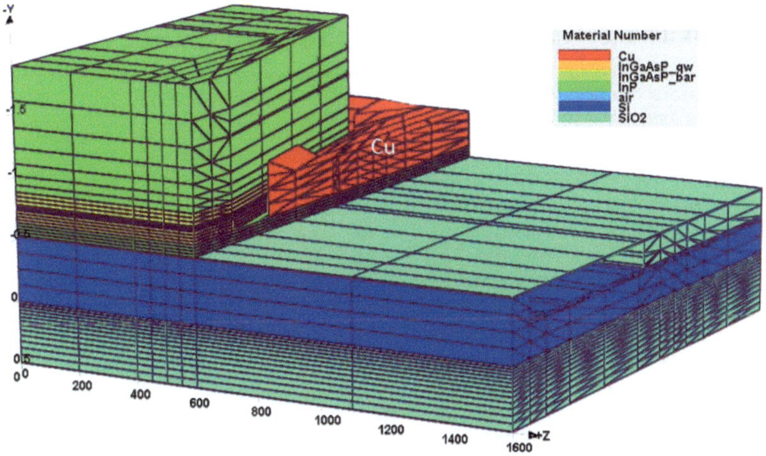

Fig. 10.5 Final structure with 3D mesh allocation

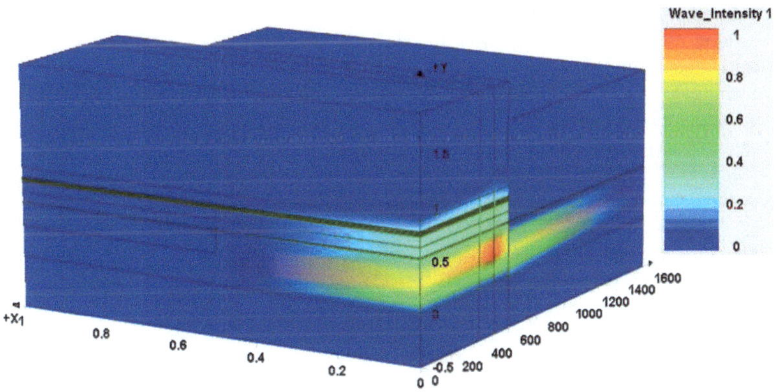

Fig. 10.6 Modal power in both MQW and lower WG

Please pay attention the following points which are unique to the simulation of MQW laser diodes.

- `real_index` command is used to adjust the refractive index which determines the optical confinement within the waveguide (WG) and couples the optical power from active InP waveguide to passive silicon waveguide underneath. Careful design of the index profile is critical for the hybrid laser to work properly, to achieve large optical confinement within the MQW while also providing sufficient power transfer to the silicon passive WG. The optical modes can be seen in Fig. 10.6 which shows modal power in both MQW and lower WG.

Text Box 10.1 A Fragment of Run Time Message

```
Cmplx lateral modal index for z-segment:   6
     1 (3.28816473918325,5.412059994762079E-005)
photon#              lambda          long.mode       lat.mode
0.244272E+03         0.130196E+01        1              1
0.294662E+03         0.130187E+01        2              1
0.370348E+03         0.130153E+01        3              1
0.375758E+03         0.130119E+01        4              1
0.496542E+03         0.130075E+01        5              1
0.819769E+03         0.130041E+01        6              1
0.865770E+06         0.130007E+01        7              1
0.230580E+04         0.129978E+01        8              1
0.477961E+03         0.129937E+01        9              1
0.451737E+03         0.129902E+01        10             1
0.336417E+03         0.129869E+01        11             1
0.332875E+03         0.129825E+01        12             1
0.312083E+03         0.129791E+01        13             1
Use long. mode wavelength (um)= 1.30006873367826
Seg: 1 Modal gain (1/m): 0.167495E+04
At lambda= 1.30006873367826
Use long. mode wavelength (um)= 1.30006873367826
Seg: 2 Modal gain (1/m): 0.149310E+04
At lambda= 1.30006873367826
Use long. mode wavelength (um)= 1.30006873367826
Seg: 3 Modal gain (1/m): 0.170720E+04
At lambda= 1.30006873367826
Use long. mode wavelength (um)= 1.30006873367826
Seg: 4 Modal gain (1/m): -0.114713E+05
At lambda= 1.30006873367826
=========WARNING=========
Optical confinement factor for
active region number 4
is very small: 0.000000000000000E+000
=====end of WARNING======
Use long. mode wavelength (um)= 1.30006873367826
Seg: 5 Modal gain (1/m): -0.500000E+03
At lambda= 1.30006873367826
=========WARNING=========
Optical confinement factor for
active region number 4
is very small: 0.000000000000000E+000
=====end of WARNING======
Use long. mode wavelength (um)= 1.30006873367826
Seg: 6 Modal gain (1/m): -0.500000E+03
At lambda= 1.30006873367826
MQW Reg.# 1 Average Conc. (n&p)= 0.2260E+25 0.2805E+25
Reg.# 1 Levels: Gamma= 1 L= 3 HH= 3 LH= 2
MQW Reg.# 2 Average Conc. (n&p)= 0.2239E+25 0.2775E+25
Reg.# 2 Levels: Gamma= 1 L= 3 HH= 3 LH= 2
MQW Reg.# 3 Average Conc. (n&p)= 0.2715E+25 0.3451E+25
Reg.# 3 Levels: Gamma= 1 L= 3 HH= 3 LH= 2
MQW Reg.# 4 Average Conc. (n&p)= 0.8907E+23 0.1305E+22
Reg.# 4 Levels: Gamma= 1 L= 3 HH= 3 LH= 2
```

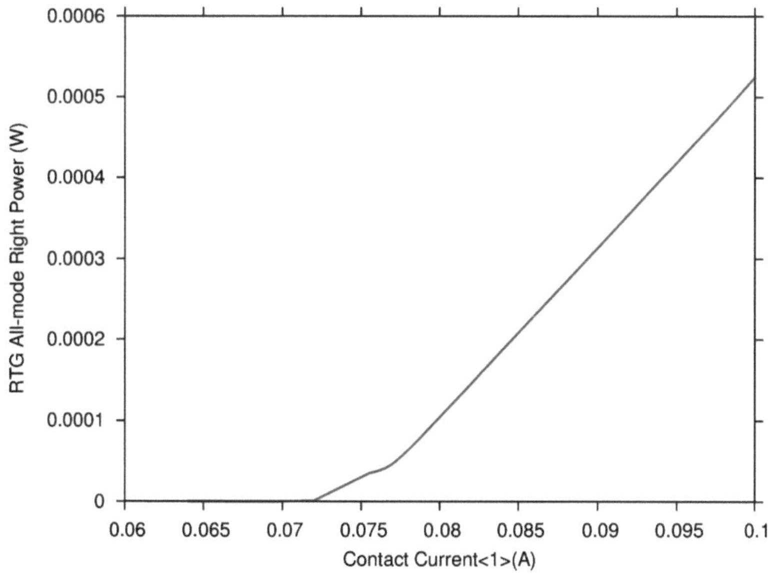

Fig. 10.7 The lasing power from the DBR side

- The work function of Cu is adjusted to provide easy injection of carriers into p-InP. Without such adjustment, quantum tunneling would have to be used to inject hole across a thick barrier, which we avoid here. It would also be possible to use an ohmic boundary condition provided the semiconductor was sufficiently doped.
- The commands `wave_boundary,` `init_wave` `direct_eigen` and `rtgain_phase` are needed for laser diode simulation to define the model mode boundaries, setting the wavelength range for the mode solver and initializing the longitudinal rate equations.
- An extra section starting with `begin_zsol` is needed to define parameters relating to the longitudinal modes. The Bragg gratings are not explicitly shown in the electrical or process simulation but are merely defined by the optical coupling coefficient kappa. This approach is warranted because in reality, gratings are often created by etching the top surface to perturb the effective index of the mode which does not perturb the electrical solution very much.
- When turning on the current/voltage using the scan command, special parameters are used to monitor the round-trip-gain of the laser cavity until it is close enough to lasing. Then finally, the flag `solve_rtg` is turned on to allow the solution of rate equation for each longitudinal mode along with the usual drift-diffusion equations related to electrical current.

When running the simulation, the run-time message will contain information about the MQW quantum states, optical modal gain and cavity photon number. Text Box 10.1 is a fragment of the run-time message. Some of the warnings below

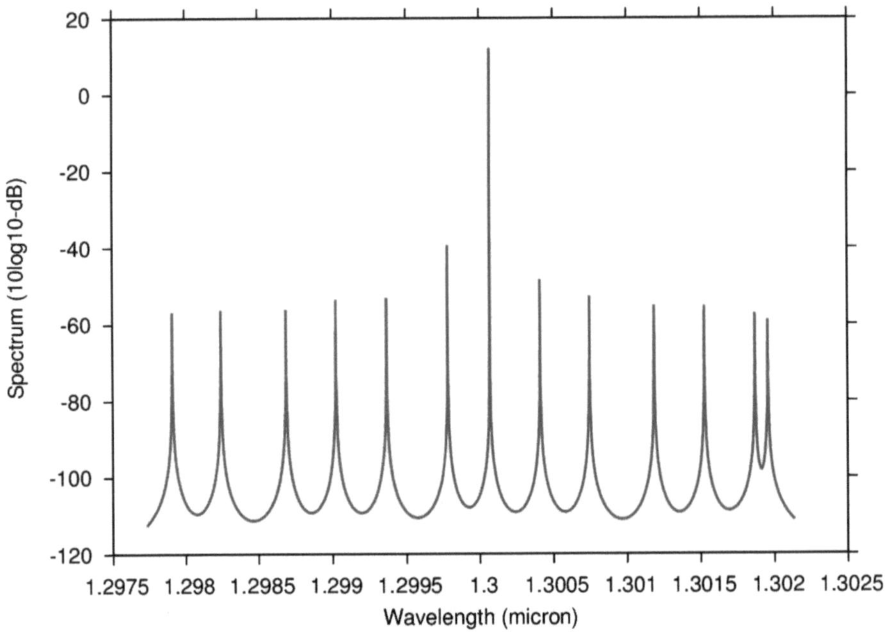

Fig. 10.8 The output spectrum

Table 10.1 Simulation data for the hybrid laser

	Process simulation	Device simulation	Total mesh count	Total number of planes
Hybrid laser	3 min	15 min	3,674	6
Computer configuration: Toshiba Laptop with Intel P8700-2.53G/4G/Win7				

are issued because the waveguide design is unusual for a laser (because of the transfer of power to the silicon).

The results of the simulation are briefly presented here. The lasing power from the DBR side (adjustable using DBR kappa and length of DBR) is indicated in Fig. 10.7 while the output spectrum is shown in Fig. 10.8.

In summary, simulation of hybrid silicon laser, or any other type of laser can be set up using masks in the same way as other conventional silicon based devices. The only difference is in the material definition in process simulation and use of optical models which are made accurate by quantum mechanical calculation of MQW.

10.5 Simulation Data (Table 10.1)

Bibliography

1. Semiconductor Industry Association. [Online] http://www.sia-online.org/.
2. EEtimes. [Online] www.eetimes.com.
3. Scanning electron microscope. *Wikipedzia.* [Online] http://en.wikipedia.org/wiki/Scanning_electron_microscope.
4. Moore's Law. *Intel.* [Online] http://www.intel.com/technology/mooreslaw/.
5. Tape out. *Wikipedia.* [Online] http://en.wikipedia.org/wiki/Tape out#cite_note 1.
6. Follow Heuristic Guidelines To Make Surface-Mount PC-Board Footprints. *Electronic Design.* [Online] 2006. http://electronicdesign.com/article/digital/follow-heuristic-guidelines-to-make-surface-mount-.aspx.
7. The Spice Page. *UC Berkeley.* [Online] http://bwrc.eecs.berkeley.edu/classes/icbook/spice/.
8. Stanford TCAD. *Stanford University.* [Online] http://www-tcad.stanford.edu/.
9. **Crosslight Software.** 2009. *CSuprem Manual.* Burnaby,BC,Canada
10. Crosslight. *Crosslight.com.* [Online] www.crosslight.com.
11. *Advanced 3D simulation of semiconductor processes, devices and optics.* **Yue Fu, Yegao Xiao, Jiajan Guo and Simon Li.** Vancouver,BC,Canada : CMOS Emerging Technologies, 2009
12. *Si–Ge interdiffusion in strained Si/strained SiGe heterostructures and implications for enhanced mobility metal-oxide-semiconductor field-effect transistors.* **Guangrui Maggie Xia, Judy L. Hoyt and Michael Canonico.** 101, 044901, s.l. : JOURNAL OF APPLIED PHYSICS, 2007.
13. *General relationship for the thermal oxidation of silicon.* **Deal, B.E. Grove, A.S.** pp:3770–3778, s.l. : J.App.Phys., 1965, Vol. 36.
14. *On inerstitial and vacancy concentrations in presence of injection.* **Hu, S.M.** pp:106, s.l. : J. Appl. Phys., 1985, Vol. 57.
15. *Process physics determinig 2D- impurity profiles in VLSI devices.* **Griffin, P.B., Plummer, J.D.** pp:522, Los Angeles : International Electron Devices Meeting, 1986.
16. **S.M.Sze.** *Physics of semiconductor devices, 2nd edition.* John Wiley & Sons, 1981.
17. *calculations of heterojunction discontinuities in the Si/Ge system.* **Martin, C.G. Van de Walle and R.M.** 8, p.5621, s.l. : Phys.Rev. B, 1986, Vol. 34.
18. **J.R. King, SIAM.** 1, s.l. : J. Appl. Math., 1989, Vol. 49. pp. 264–280.
19. *Stokes' Hypothesis for a Newtonian, isotropic fluid.* **Gad-el-Hak, Mohamed.** 1, s.l. : Journal of Fluids Engineering, 1995, Vol. 117. pp. 35.
20. *Energy transport numerical simulation of graded AlGaAs/GaAs heterojunction bipolar transistors.* **Azoff, E.M.** pp.609-616, s.l. : IEEE Trans. ED, 1989, Vol. 36.
21. *Close-form method for solving the steady-state generalised energy momentum conservation equations.* **Azoff, E.M.** 25–30, s.l. : COMPEL, 1987, Vol. 6.

S. Li and Y. Fu, *3D TCAD Simulation for Semiconductor Processes, Devices and Optoelectronics*, DOI 10.1007/978-1-4614-0481-1,

22. **Smith, R. A.** *Semiconductors 2nd edn.* Cambridge, England : Cambridge University Press, 1978.
23. **D. Bednarczyk and J. Bednarczyk.** 64A, 409, s.l. : Phys. Letters, 1978.
24. *Use of Fermi statistics in two dimensional numerical simulation of heterojunction devices.* **Z.-M. Li, S.P. McAlister and C.M. Hurd,.** 408, s.l. : Semicond. Sci.Techn., 1990, Vol. 5.
25. *Physical modeling of degenerately doped compound semiconductors for high-performance HBT design.* **James C. Li, Marko Sokolich, Tahir Hussain, Peter M. Asbeck.** 1440C1449, s.l. : Solid-State Electronics, 2006, Vol. 50.
26. *A physically based mobility model for numerical simulation of non-planar devices.* **C. Lombardi, S. Manzini, A. Saporito, and M. Vanzi.** no.11 pp.1164-1171, s.l. : IEEE CAD, Nov.1988, Vol. 7.
27. **Selberherr, S.** *Analysis and simulation of semiconductor devices.* New York : Springer-Verlag, 1984.
28. *Ionization rates for electrons and holes in silicon.* **Chynoweth, A.G.** p. 1537, s.l. : Phys., Rev., 1958, Phys. Rev, Vol. 109.
29. **Selberherr, S.** *Analysis and simulation of semiconductor devices.* Wien-New York : Springer-Verlag, 1984.
30. **Sze, C.R. Crowell and S.M.** no.6 pp. 242–244, s.l. : Appl. Phys. Lett.,, 1966, Vol. 9.
31. **Blokhintsev, D.I.** *Quantum mechanics.* Dordrecht-Holland : D. Reidel Publishing Company, 1964.
32. *Efficient band-structure calculation of strained quantum-wells.* **Chuang, S.L.** B, 43, 9649–9661, s.l. : Phys. Rev., 1991.
33. *A model for GRIN-SCH-SQW diode lasers.* **S.R. Chinn, P.S. Zory and A.R. Reisinger.** QE-24, 2191–2214, s.l. : IEEE J. Quantum Electron., 1988.
34. *Threshold current of single quantum well lasers: The role of the confining layers.* **J. Nagle, S. Hersee, M. Krakowski, T. Well, and C. Welsbuch.** No.20, 1325, s.l. : Appl. . Lett., 1986, Vol. 49.
35. *Auger recombination in strained and unstrained InGaAs/InGaAsP multiple quantum well lasers.* **G. Fuchs, C. Schiedel, A. Hangleiter, V. Harle, and F. Scholz,.** 396–398, s.l. : Appl. Phys. Lett., 1993, Vol. 62.
36. *Band mixing effects on quantum well gain.* **S. Colak, R. Eppenga, and M.F.H. Schuurmans.** QE-23, 960–967, s.l. : IEEE J. PilkuhnQuantum Well Electron, 1987.
37. *New k.p thoery for GaAs/Ga1−xAlxAs-type quantum wells.* **R. Eppenga, M.F.H. Schuurmans, and S. Colak.** 1554–1564, s.l. : Phys. Rev.,, 1987, Vol. 36.
38. *Superlattice band structure in the envelope-function approximation.* **Bastard, G.** 5693–5697, s.l. : Phys. Rev.,, 1981, Vol. 24.
39. **L.C. Andreani, A. Pasquarello, and F. Bassani.** Phys. Rev., s.l. : Phys. Rev., 1987.
40. **Chuang, S. L.** *Physics of Optoelectronic Devices.* New York : Wiley,, 1995.
41. **F.Wooten.** *Optical properties of solids.* New York and London : Academic Press, 1972.
42. *Anisotropy and broadening of optical gain in a GaAs/AlGaAs multiquantum-well laser.* **M. Yamada, S. Ogita, M. Yamagishi, and K. Tabata.** QE-21, 640–645, s.l. : IEEE. J. Quantum, 1985.
43. *Systematics of laser Ishigurooperation in GaAs/AlGaAs multiquantum well heterostructures.* **E. Zielinski, H. Schweizer, S. Hausser, R. Stuber, M. H. Pilkuhn, and G.Weimann.** QE-23, 969–976, s.l. : J. Quantum Electron., 1987.
44. *Optical gain and loss processes in GaInAs/InP MQW laser structure.* **E. Zielinski, F. Keppler, S. Hausser, M. H. Pilkuhn, and R. Sauer and W.T.Tsang.** QE-25, 1407–1416, s.l. : J.Quantum Electron., 1989.
45. *Corrections to the expression of gain in GaAs.* **R.H. Yan, S.W. Corzine, L.A. Coldren, and I. Suemune.** QE-26, 213–216, s.l. : IEEE J. Quantum Electron., 1990.
46. *Optical Gain of Strained Wurtzite GaN QuantumWell Lasers.* **Chuang, Shun Lien.** No.10 p1791, s.l. : IEEE JOURNAL OF QUANTUM ELECTRONICS, OCTOBER 1996, Vol. 32.

47. *k.p method for strained wurtzite semiconductors.* **Chang, S. L. Chuang and C. S.** pp. 2491–2504, s.l. : Phys. Rev. B, July, 1996., Vol. 54.
48. *Rigorous thermodynamic treatment of heat generation and conduction in semiconductor Transmodeling.* **Wachutka, G.K.** CAD-9, 1141–1149, s.l. : IEEE Trans., 1990.
49. *Electrical current in solids with position dependent band structure.* **Vilet, A. Marshak and K. van.** 21, 417–427, s.l. : Solid State Electron.,, 1978.
50. *Thermal effects on the characteristics of AlGaAs/GaAs heterojunction bipolar transistors using two dimensional numerical simulation.* **L. Liou, J. Ebel, and C Huang.** ED-40, 35–43, s.l. : IEEE Trans. ED, 1993.
51. *Semiconductor current flow equations (diffusion and degeneracy).* **R. Stratton.** ED-19, 1288–1292, s.l. : IEEE ., ED, 1972.
52. GDSII. *Wikipedia.* [Online] http://en.wikipedia.org/wiki/GDSII.
53. **James D.Plummer, Michael D.Deal, Peter B.Griffin.** *Silicon VLSI Technology, Fundamentals,Practice and Modeling.* s.l. : Prentice Hall, 2000. 0-13-085037-3.
54. Photoresist. *Wikipedia.* [Online] [Cited: March 25, 2011.] http://en.wikipedia.org/wiki/Photoresist.
55. MEMS and Nanotechnology Exchange. [Online] http://www.mems-exchange.org/users/masks/guidelines.html.
56. Photoresists. *www.drlitho.com.* [Online] http://www.drlitho.com/cms/website.php?id=/en/research/resists.html.
57. **Jaeger, Richard C.** *Introduction to Microelectronic Fabrication.* Upper Saddle River, New Jersey : Prentice Hall, 2002. 0-201-44494-1.
58. **Crosslight Software.** *APSYS Manual.* Burnaby,BC,Canada : Crosslight Software, 2006.
59. CUDA. *Nvidia CUDA.* [Online] http://www.nvidia.com/object/cuda_home_new.html.
60. Acceleware Delivers 100X Speed Up for Solar Cell Simulations. *Acceleware.* [Online] http://www.acceleware.com/acceleware-delivers-100x-speed-solar-cell-simulations.
61. **S.M.SZE.** *Semiconductor Devices Physics and Technology.* s.l. : John Wiley and Sons, 1985. 0-471-87424-8.
62. **G.Streetman, Ben.** *Solid State Electronic Devices.* s.l. : Prentice Hall, 1995. 0-13-168767-6.
63. **Zeghbroeck, B. Van.** *Principles of Semiconductor Devices.* s.l. : http://ecee.colorado.edu/~bart/book/book/title.htm, 2007.
64. *Comparison of MOSFET-threshold-voltage extraction methods .* **Kazuo Terada, , Katsuhiko Nishiyama and Kei-Ichi Hatanaka.** 1, s.l. : Solid-State Electronics, 1 January 2001, Vol. 45.
65. **Robert W. Dutton, Zhiping Yu.** *Technology CAD computer simulation of IC processes and devices.* s.l. : Kluwer Academic Publishers, 1993. 0-7923-9379-1.
66. CMOS. *Wikipedia.* [Online] http://en.wikipedia.org/wiki/CMOS.
67. **Bhushan, Bharat.** *Handbook of nanotechnology, 2nd edition.* s.l. : Springer, 2007.
68. **S.M.SZE.** *VLSI Technology.* s.l. : McGraw-Hill, 1983. 0-07-062686-3.
69. Chemical Mechanical Polishing. *Wikipedia.* [Online] http://en.wikipedia.org/wiki/Chemical-mechanical_planarization.
70. *Deep Trench Isolation for a 50V 0.35 μm Based Smart Power Technology.* **F. De Pestel, P. Coppens, H. De Vleeschouwer, P. Colson, S. Boonen, T. Colpaert,P. Moens, D. Bolognesi, G. Coudenys, M. Tack.** Estoril,Portugal : 33rd European Solid-state Device Research Conference, ESSDERC, 2003.
71. *Improvement of Breakdown Characteristics of LDMOSFETs with Uneven Racetrack Sources for PDP Driver Applications.* **Tae Moon Roh, Dae Woo Lee, Jongdae Kim, Jin Gun Koo, and Kyoung-Ik Cho.** Osaka : Proceedings of the 13th International Symposium on Power Semiconductor Devices and ICs, 2001. ISPSD '01. , 2001.
72. **Baliga, B.Jayant.** *Fundamentals of Power Semiconductor Devices.* s.l. : Springer, 2010. 978-0-387-47313-0.
73. **Baliga, B.Jayant.** *Modern Power Devices.* s.l. : Krieger Publishing Company, 1992.
74. **Grabinski, Wladyslaw.** *Power/HVMOS Devices Compact Modeling.* s.l. : Springer, 2010.
75. Wikipedia. *Thermal conductivity.* [Online] http://en.wikipedia.org/wiki/Thermal_conductivity.

76. HEXFET. [Online] International Rectifier. http://www.irf.com/technical-info/guide/device.html.
77. *Influence of total-dose radiation on the electrical characteristics of SOI MOSFETs.* **J.A. Felix, J.R. Schwank, C.R. Cirba,R.D. Schrimpf, M.R. Shancyfelt, D.M.Fleetwood, P.E. Dodd,.** 332–341, s.l. : Microelectronic Engineering, 2004, Vol. 72.
78. *Modeling and Analysis of Metal Interconnect Resistance of Power IC's.* **Chen, Y. Fu, Y. Cheng, X. Wu, T.X. Shen, Z.J.** s.l. : 19th International Symposium on Power Semiconductor Devices and IC's, 2007. ISPSD '07. , 27–31 May 2007 . 1-4244-1096-7.
79. *A 0.6 μm CMOS pinned photodiode color imager technology.* **Guidash, R.M.; Lee, T.-H.; Lee, P.P.K.; Sackett, D.H.; Drowley, C.I.; Swenson, M.S.; Arbaugh, L.; Hollstein, R.; Shapiro, F.; Domer, S.;.** Washington, DC , USA : IEEE Electron Devices Meeting, 1997. IEDM '97, 1997.
80. **i-micronews.** Sony and Omnivision to develop BSI architectures for CMOS image sensor markets. [Online] http://www.i-micronews.com/lectureArticle.asp?id=1607.
81. *Progress in CMOS active pixel image sensors.* **Mendis, Sunetra K., et al., et al.** s.l. : Proc. SPIE Vol. 2172, p. 19–29, Charge-Coupled Devices and Solid State Optical Sensors IV, Morley M. Blouke; Ed., 194.
82. *3D Simulation of CMOS Image Sensor.* **Y. G. Xiao, Fred Y. Fu, Simon Li.** Whistler,BC, Canada : CMOS Emerging Technologies, 2010.
83. *Electrically pumped hybrid AlGaInAs-silicon evanescent.* **Alexander W. Fang, Hyundai Park, Oded Cohen, Richard Jones, Mario J.Paniccia, and John E. Bowers.** 20, October 2, 2006, OPTICS EXPRESS, Vol. 14.
84. Semiconductor Industry Association. [Online] http://www.sia-online.org/.
85. **Berkeley, UC.** EECS depart of UC Berkely. [Online] http://www.eecs.berkeley.edu/.
86. Design rule checking. *Wikipedia.* [Online] [Cited: March 25, 2011.] http://en.wikipedia.org/wiki/Design_rule_checking
87. The Spice Page. *UC Berkeley.* [Online] http://bwrc.eecs.berkeley.edu/classes/icbook/spice/.
88. wikipedia lithography. *Wikipedia.* [Online] http://en.wikipedia.org/wiki/Photolithography.
89. *Si–Ge interdiffusion in strained Si/strained SiGe heterostructures and implications for enhanced mobility metal-oxide-semiconductor field-effect transistors.* **Guangrui (Maggie) Xia, Michael Canonico and Judy L. Hoyt.** 4, s.l. : JOURNAL OF APPLIED PHYSICS, 2007, Journal of Applied Physics , Vol. 101. 0021–8979.
90. **Paul R.Gray, Paul J. Hurst, Stephen H. Lewis, Robert G.Mayer.** *Analysis and Design of Analog Integrated Circuits.* s.l. : John Wiley & Sons, Inc., 2004.

Index

MIX
Papier aus verantwortungsvollen Quellen
Paper from responsible sources
FSC® C105338

FSC
www.fsc.org

If you have any concerns about our products,
you can contact us on
ProductSafety@springernature.com

In case Publisher is established outside the EU,
the EU authorized representative is:
Springer Nature Customer Service Center GmbH
Europaplatz 3, 69115 Heidelberg, Germany

Printed by Libri Plureos GmbH
in Hamburg, Germany